Yeast, Mould and Plant Protoplasts

Yeast, Mould and Plant Protoplasts

PROCEEDINGS OF THE THIRD INTERNATIONAL SYMPOSIUM ON YEAST PROTOPLASTS HELD AT SALAMANCA, SPAIN, IN OCTOBER, 1972

Edited by

J. R. VILLANUEVA, ISABEL GARCÍA-ACHA, S. GASCÓN and F. URUBURU

Departamento de Microbiologia, Consejo Superior de Investigaciones Cientificas y Facultad de Ciencias, Universidad de Salamanca, Salamanca, Spain

AP 1973

ACADEMIC PRESS LONDON AND NEW YORK

A Subsidiary of Harcourt Brace Jovanovich, Publishers

ACADEMIC PRESS INC. (LONDON) LTD.
24/28 Oval Road,
London NW1

United States Edition published by
ACADEMIC PRESS INC.
111 Fifth Avenue
New York, New York 10003

Library of Congress Catalog Card Number: 73-7040
ISBN: 0 12 722160 3

Printed in Great Britain by
The Whitefriars Press Ltd., London and Tonbridge

LIST OF PARTICIPANTS

* Indicates contributor to this volume.

E. P. ABRAHAM, Sir William Dunn School of Pathology, University of Oxford, South Parks Road, Oxford OX1 3RE, England.

F. ALTERTHUM, School of Biological Sciences, University of Bath, Claverton Down, Bath BA2 7AY, England.

V. ARIES, Societé d'Assistance Technique pour produits Nestlé, Laboratoire Industrial (Linoz) 1350 Orbe, Switzerland.

A. ARNAUD, Ecole Nationale Superieure Agronomique, 34060 Montpellier Cedex, Montpellier, France.

J. S. D. BACON*, Macaulay Institute for Soil Research, Craigiebuckler, Aberdeen AB9 2QJ, Scotland.

Y. P. S. BAJAJ, School of Botany and Zoology, University Park, Nottingham NG7 2RD, England.

E. BALISH, Department of Medical Microbiology, Medical School, The University of Wisconsin, 1300 University Avenue, Madison, Wisconsin 53706, U.S.A.

S. BARTNICKI-GARCIA*, Department of Plant Pathology, University of California, Riverside, California 92502, U.S.A.

J. M. BASTIDE, Laboratoire d'Inmunologie, Virologie et Parasitologie, Faculté de Pharmacie, 34, Montpellier, France.

M. BASTIDE, Laboratoire d'Inmunologie, Virologie et Parasitologie, Faculté de Pharmacie, 34, Montpellier, France.

C. G. BATALLA, "San Miguel" Fabrica de Cerveza y Malta, S.A. Apartado 67, Lérida, Spain.

H. BAUER, Societé d'Assistance Technique pour Produits Nestlé, S.A. 1814 La Tour-de-Peilz, Switzerland.

C. BIZEAU, I.N.R.A., Ecole Nationale Superieure Agronomique, 34060 Montpellier Cedex, Montpellier, France.

P. BOER, University of Utrecht, Laboratory for Physiological Chemistry, Vondelaan 24a, Utrecht, Holland.

P. BOS*, Laboratory of Microbiology, University of Technology, Julianalaan 6, Delft, Holland.

J. P. BOURGIN*, C.N.R.A. Station de Physiologie Végètale, 78, Versailles, France.

C. BUCKLEY*, Botany Department, University of Nottingham, University Park, Nottingham NG7 2RD, England.

E. CABIB*, National Institute of Arthritis and Metabolic Diseases, National Institute of Health, Bethesda, Maryland 20014, U.S.A.
L. M. CARBONELL*, Centro de Microbiologí ay Biología Celular. Instituto Venezolano de Investigaciones Científicas, IVIC, Apartado 1827, Caracas, Venezuela.
I. CHUPEAU*, Laboratoire de Biochimie Cellulaire et Végètale, CNRA, 78-Versailles, France.
E. C. COCKING*, Department of Botany, University of Nottingham, Nottingham, England.
J. L. COLAS, Laboratoire d'Inmunologie et Virologie, Faculté de Pharmacie, 34, Montpellier, France.
M. CORTAT, Université de Genève, Institut de Microbiologie, Rue de Candolle 3, 1211 Genéve 4, Switzerland.
R. H. A. COUTTS, Department of Botany, University of Nottingham, Nottingham, England.
J. DAUM, Shering A.G., Mullersh. 170-172, 1000 Berlin 65, Germany.
M. R. DAVEY, Department of Botany, University Park, Nottingham NG7 2RD, England.
A. M. P. DOOIJEWAARD-KLOISTERZIEL, Laboratorium voor Microbiologie, Plantage Unidergracht, 14, Amsterdam, Holland.
A. DURAN, Departamento de Microbiología, Facultad de Ciencias, Universidad de Salamanca, Salamanca, Spain.
M. V. ELORZA*, Departamento de Microbiología, Facultad de Ciencias, Universidad de Salamanca, Salamanca, Spain.
P. FAWCETT, Sir William Dunn School of Pathology, South Park Road, Oxford, England.
L. FERENCZY, Institute for Plant Physiology and Microbiology, University of Szeged, Szeged 428, Hungary.
M. GACTO*, Departamento de Microbiolgía, Facultad de Ciencias, Universidad de Salamanca, Salamanca, Spain.
I. GARCIA-ACHA*, Departamento de Microbiología, Instituto de Biología Celular, Facultad de Ciencias, Universidad de Salamanca, Salamanca, Spain.
M. D. GARCIA LOPEZ, Departamento de Microbiología, Facultad de Ciencias, Universidad de Salamanca, Salamanca, Spain.
A. GARCIA-OCHOA*, Departamento de Microbiología, Facultad de Ciencias, Universidad de Salamanca, Salamanca, Spain.
S. GASCON*, Departamento de Microbiología, Instituto de Biología Celular, Facultad de Ciencias, Universidad de Salamanca, Salamanca, Spain,
I. GIL*, Instituto Jaime Ferran de Microbiología, Joaquin Costa 32, Madrid, 6, Spain.

R. K. GIBSON*, University of Nottingham, University Park, Nottingham NG7 2RD, England.

A. HAMDOUME, Laboratoire d'Inmunologie et Virologie, Faculté de Pharmacie, 34, Montpellier, France.

C. HARDISSON, Cátedra de Microbiología, Facultad de Ciencias, Universidad de Oviedo, Oviedo, Spain.

C. de HARO, Departamento de Microbiología, Facultad de Ciencias, Universidad de Salamanca, Salamanca, Spain.

M. P. HATTON, Distillers Co. Ltd., Glenochil Technical Centre, Menstrie, Clacks, Scotland.

P. HERRERO, Departamento de Microbiología, Facultad de Ciencias, Universidad de Salamanca, Salamanca, Spain.

P. H. HOFSCHNEIDER, Max-Planck-Institut für Biochemie, Goethestrasse 31, 8 München 15, Germany.

K. HORIKOSHI*, The Institute of Physical and Chemical Research, Wako-Shi, Saitama 351, Japan

J. HOSSACK*, School of Biological Sciences, University of Bath, Claverton Down, Bath BA2 7AY, England.

D. KERRIDGE, Sub-Department of Chemical Microbiology, Department of Biochemistry, Tennis Court Road, Cambridge, England.

F. KEVEI, Institute of Plant Physiology and Microbiology, S.A. University of Szeged, Szeged 428, Hungary.

T. Y. KOH, Sub-Department of Chemical Microbiology, Department of Biochemistry, Tennis Court Road, Cambridge, England.

G. KRAEPELIN*, Technische Universität, Humboldtstrasse, 33 Braunschweig, Germany.

D. R. KREGER*, Department of Dermatology, State University, Oostersingel 59, Groningen, Holland.

N. J. W. KREGER-VAN RIJ, Department of Dermatology, State University, Oostersingel, 59, Groningen, Holland.

I. S. KULAEV*, Department of Plant Biochemistry, Moscow State University, Moscow, U.S.S.R.

F. LABORDA, Research Station, Long Ashton, Bristol BS18, 9AF, England.

J. O. LAMPEN*, Institute of Microbiology, Rutgers State University, New Brunswick, N.J. 08903, U.S.A.

A. W. LINNANE*, Biochemistry Department, Monash University, Clayton, Victoria 3168, Australia.

E. LIPPMAN, Department of Plant Pathology, University of California, Riverside, California 92502, U.S.A.

A. B. LOSINOV, Institute of Biochemistry and Physiology of Microorganisms, Pustchino-on-the-Oka, Moscow, U.S.S.R.

M. S. MARRIOTT, Department of Biochemistry, Tennis Court Road, Cambridge CB2, 1QW, England.

F. MAYOR, Departamento Interfacultativo de Bioquímica, Facultad de Farmacia, Universidad de Granada, Granada, Spain.

J. MEYER, Institute of General Botany, E.T.H., Universitätstrasse 2, 8006 Zurich, Switzerland.

J. M. MITCHISON*, University of Edinburgh, Department of Zoology, West Mains Road, Edinburgh EH9 3JT, Scotland.

L. MONTES, Department of Dermatology, The University of Alabama in Birmingham, 1919 Seventh Avenue South, Birmingham, Alabama 35233, U.S.A.

P. MOORE, Botany Department, University of Nottingham, University Park, Nottingham NG7 2RD, England.

F. MORENO*, Departamento de Microbiología, Facultad de Ciencias, Universidad de Salamanca, Salamanca, Spain.

G. MOREL*, Station Centrale de Pathologie Végètale, Centre National de Recherches Agronomiques, Etoile de Choisy, Route de Saint-Cyr, 78 Versailles, France.

G. NICOLAS, Cátedra de Fisología Vegetal, Facultad de Farmacia, Universidad de Santiago, Santiago de Compostela, Spain.

T. NURMINEN, Research Laboratory of the State Alcohol Monopoly, Alko, Box 10350, Helsinki, Finland.

W. PACHE, Sandoz Ltd., Pharmaceutical Division, Chemical Research-Microbiology CH-4002 Basle, Switzerland.

I. H. PALLET, Beecham Research Laboratories, Chemotherapeutic Research Centre, Brockham Park, Betchworth, Surrey, England.

J. PASERO, Societé Internationale de Recherche B.P., Laboratories, Boîte Postale 12, Epernon 28, France.

J. F. PEBERDY, Department of Botany, University of Nottingham, University Park, Nottingham NG7 2RD, England.

J. M. PEINADO, Departamento de Microbiología, Fundación Gulbenkian Oeiras, Portugal.

H. J. PHAFF*, Department of Food Science and Technology, University of California, Davis, California 95616, U.S.A.

I. POTRYKUS*, Institut für Botanische, Entwicklungsphysiologie der Universität Hohenheim, 7 Stuttgart 70, Germany.

S. RAMOS, Departamento de Microbiología, Facultad de Ciencias, Universidad de Salamanca, Salamanca, Spain.

F. REYES, Instituto "Jaime Ferrán", Centro de Investigaciones Biológicas, Velazquez, 144, Madrid, Spain.

B. ROCHE, Laboratoire de Chimie Bacterienne CNRS, 32 Chemin Joseph-Aiguier, 13 Marseille (9éme), France.

J. ROMERO, Laboratoire de Pathologie Végètale, Faculté des Sciences Agronomiques, 5800 Gemblooux, Belgium.
L. ROSSINI, CNRS Physiologie Pluricellulaire, 91 Gif-sur-Ivette, France.
N. J. RUSSELL, Sub-Department of Chemical Microbiology, Department of Biochemistry, Tennis Court Road, Cambridge, England.
P. SANCHEZ LAZO, Departamento de Microbiología, Facultad de Ciencias, Universidad de Salamanca, Salamanca, Spain.
F. SCHLENK, Division of Biological and Medical Research, Argonne National Laboratory, Argonne, Illinois 60515, U.S.A.
B. SCHULZ*, Institute of Botany, Technical University, 1 Humboldstrasse, 3300 Braunschweig, Germany.
G. SAN BLAS, Heriot-Watt University, 17 Queen's Crescent, Edinburgh 9, Scotland.
G. SCRYABIN, Institute of Biochemistry and Physiology of Microorganisms, Pustchino-on-the-Oka, Moscow, U.S.S.R.
R. SENTANDREU*, Departamento de Microbiología, Instituto de Biología Celular, Facultad de Ciencias, Universidad de Salamanca, Salamanca, Spain.
R. SERRANO, Departamento de Bioquímica, Facultad de Medicina, Universidad Autónoma, Madrid, Spain.
J. H. SIETSMA*, University of Amsterdam, Laboratory for Microbiology, Plantage Unidergracht 14, Amsterdam, Holland.
J. M. SIERRA*, Departamento de Microbiología, Facultad de Ci encias, Universidad de Salamanca, Salamanca, Spain.
H. SUOMALAINEN*, Research Laboratory of the State Alcohol Monopoly, Alko, Box 10350, Helsinki, Finland.
S. STABEL, Technische Universitát, Humboldstrasse, 33 Braunschweig, Germany.
E. TACORONTE, Departamento de Microbiología, Facultad de Ciencias, Universidad de Salamanca, Salamanca, Spain.
R. R. THEIMER*, Botanisches Institut der Universität, Menzingerstrasse 67, 8 München 19, Germany.
P. TRAVE, Laboratoire d'Inmunologie Virologie et Parasitologie, Faculté de Pharmacie, 34, Montpellier, France.
F. URUBURU, Departamento de Microbiología, Facultad de Ciencias, Universidad de Salamanca, Salamanca, Spain.
H. J. M. VAN RYN, Lab. voor Fys. Chemie, Vondellaan 24[a], Utrecht, Holland.
F. VEZINHET, I.N.R.A., Ecole Nationale Superieure Agronomique, 34060 Montpellier Cedex, Montpellier, France.

J. R. VILLANUEVA*, Departamento de Microbiología, Facultad de Ciencias, Universidad de Salamanca, Salamanca, Spain.

O. M. H. de VRIES*, Biologisch Centrum, Kerklaan 30, Haren (Groningen), Holland.

J. G. M. WESSELS*, Department of Botany, University of Groningen, Kerklaan 30, Haren, Holland.

A. WIEMKEN, Institut für Allgemeine Botanik, Eidg, Technische Hochschule, 8002 Zurich, Switzerland.

R. YU, Max-Planck-Institut für Molekulare Genetik, 1 Berlin 33 (Dahlem) Ihestrasse 63-73, Germany.

J. ZSOLT, Institute of Plant Physiology, University of Szeged, Szeged 428, Hungary.

PREFACE

Until quite recently, studies on fungal and plant protoplasts were mainly directed to their obtention and to microscopical studies of the process of cell wall regeneration. Thanks to this early work, protoplasts are now readily obtained from a rapidly increasing number of species, and botanists, microbiologists, biochemists and geneticists have made use of protoplasts for their studies. Although lytic enzymes and protoplast formation are still a major field of interest, as can be judged from the contents of this book, protoplasts are attracting the attention of people interested in seemingly remote problems. The studies on structure and morphogenesis, biosynthesis and secretion of cell wall material and exocellular enzymes, and of fusion and development of protoplasts, that are well represented in the book, demonstrate the possibilities opened by the availability of protoplasts from varied sources.

This volume contains the Proceedings of the Third International Symposium on Yeast Protoplasts, held in Salamanca (Spain) on October 2 to 5, 1972 organized by the Department of Microbiology of the Faculty of Sciences and under the auspices of the Spanish Research Council and the University of Salamanca.

July, 1973

J. R. VILLANUEVA
ISABEL GARCÍA ACHA
S. GASCÓN
F. URUBURU

CONTENTS

PART III: PHYSIOLOGY AND BIOCHEMISTRY OF YEAST PROTOPLASTS

PART IV: PROTOPLASTS OF MOULDS

PART V: PROTOPLASTS OF PLANTS

OPENING ADDRESS

Julio R. Villanueva
Chairman of the Organizing Committee

Mr. President
Rector Magnificus
Dear Colleagues, Ladies and Gentlemen,

As President of the Organizing Committee of the Third International Symposium on Yeast Protoplasts, it is my privilege to open the present International Conference and my great pleasure, on behalf of the Organizing Committee, to extend a most heartfelt welcome to all participants. We are proud and happy to greet here a range of distinguished microbiologists who have so illuminated their respective fields of investigation and we are grateful to them for visiting Salamanca and its University, a University with more than 760 years of history. At this time I also want to thank the Rector Magnificus of Salamanca University for his continuous help and encouragement.

Since the publication of the first articles on the digestion of yeast cell walls and the liberation of protoplasts by Necas in 1956 and by Eddy and Williamson in 1957, there has been a growing interest in this problem which touches a diversity of biological disciplines. This meeting will be a continuation of previous conferences, held at Jena and Brno, where volumes containing all the contributions were published. In referring to this fact we simply wish to stimulate your interest and to provide a link between topics discussed in those places and the lectures to be delivered here over the next few days.

Conferences of this nature are extremely profitable because they provide a good opportunity to bring together scientists interested in related studies and facilitate international cooperation by stimulating friendship among a large number of scientists from different countries.

During recent years a number of general International Meetings on yeasts have been organized in various countries. Many of you participated in one or the other of those held in Smolenice, Bratislava, Delft and The Hague. The next General Symposium on Yeasts will be held in Vienna, Austria, in July 1974, in response to the needs and wishes of a number of research workers. Apart from these general meetings, a number of specialized symposia were also organized such as that held in Smolenice and another, last March, in Kyoto. A meeting at Otaniemi, Finland, on the subject of "Metabolism and Regulation of Cell Processes" is arranged for June 1973. In this connection we are particularly fortunate to have the Chairman of that meeting, Dr. Heikki Suomalainen, here in Salamanca.

The Council for International Co-operation on Yeast Science was founded on the occasion of the 2nd International Symposium on Yeast held in Bratislava in July 1966, and this Council is now a Commission on Yeast and Yeast-like Microorganisms within the Division of Microbiology of the International Union of Biological Sciences. So, this Council as Professor T. O. Wiken informed at the Opening Address of the Third International Symposium on Yeast held at Delft in 1969, has become an official constituent of the IUBS. On this occasion we must express our gratitude to our Czechoslovak colleagues, practically all those absent from this meeting even though particularly interested in it, for their excellent initiative and all the prolific work that has made the yeast meetings so profitable and fruitful for international cooperation.

In connection with this meeting, we must also mention an interesting Symposium held in the University of Osaka on enzymic degradation of cell wall structures. The Proceedings of this meeting have been published in volume 48 of the *Journal of Fermentation Technology.*

It is of interest to pick out on this occasion the publication of a number of volumes of particular interest in the yeast field. First, there is the appearance of the second edition of "The Yeasts, A Taxonomic Study" by J. Lodder, which clearly reflects the increasing interest in these organisms attained during the twenty years that has elapsed since the publication of the first edition by J. Lodder and N. J. W. Kreger-van Rij. Second, the publication of the first series of volumes on "The Yeasts" by Anthony H. Rose a visitor here today, and J. S. Harrison, which also reflects the progress in a number of fields, namely in the biology of yeasts, in yeast physiology and biochemistry and in the technology of yeasts. Together, the three volumes present an up-to-date account of knowledge on yeasts and their activities. Third it is also of interest to appreciate the efforts made by G. C. Ainsworth and Alfred S. Sussman over the publication of "The Fungi", a series of three volumes that is a great contribution to the work of many scientists in that field. In the same line is also the publication of "Methods in Microbiology" Vol. IV by C. Booth in which a collection of papers on fungi are included.

During recent years a special debt of gratitude is due to Professor Herman J. Phaff, of the University of California and associated editors, for his important effort to stimulate cooperation among research people connected with yeasts by the publication of "Yeast: A News Letter for Persons Interested in Yeast". As many of you will already know, this publication was adopted in the Third International Symposium on Yeasts held at Delft, The Hague, as the official medium of communication between the Council for International Cooperation in Yeast Science and workers in the yeast field. I would like to express, I think, in the name of all scientists present here, our appreciation of the excellent work done by Professor Phaff, who is also here with us, as editor of the Yeast News Letter.

The growing interest in the field of yeast and mould protoplasts has led investigators around the world to concentrate on experimental studies and discuss their results in specialized meetings. The First International Symposium on Yeast Protoplasts was held in 1965 in the Institute of Microbiology at Jena, GDR, and was organized by Dr. R. Müller. The Second International Symposium on the same subject was held in 1969 at Brno, Czechoslovakia, in the Department of Biology, Purkyne University, under the chairmanship of Prof. Oldrin Necas.

Since 1965, numerous review papers have been published on various aspects of the problem of fungal protoplasts. The seven years of fruitful labour after Jena demanded a review and appraisal at a Third International gathering. Moreover, since the range of the subject had expanded vastly, sessions of this Symposium will be held on related branches of the problem. You will notice that there is special emphasis on the morphological, physiological and biochemical aspects of protoplasts.

In the last few years developments in yeast science have taken place at high speed. Yeast cells seem to me extremely useful for several reasons, among them being the bulk of research and knowledge already gathered. Of special importance for general cell biology is the study of the cell surface of yeasts and filamentous fungi; a large number of contributions submitted to this Symposium will be concerned with the structure of the cell wall and cytoplasmic membrane. The interest of studies related to the chemical biosynthesis and morphological organization of the cell surface, as well as those on the biochemical control of morphogenesis, are relevant in this area of research. Of great interest, I am sure, for several participants in this conference, will be the discussions concerning the role of the cell wall and cytoplasmic membrane in the synthesis and secretion of various enzymes.

There is now no doubt that the great effort which has gone into the study of all aspects of *Escherichia coli* is one of the main reasons why molecular biology has progressed so rapidly over the past decades. Similar attention now needs to be directed to some types of eucaryotic cells, preferably concentrating on the molecular biology of one or two of the simplest examples. When one compares yeast cells with other eucaryotic cell types, one finds several reasons that recommend this group of organisms. The relative genetic simplicity of yeasts when compared, for instance, with animal cells, the economy of work with microorganisms, together with the ease with which detailed genetic analysis can be applied to this type of fungi, has made James D. Watson suggest that: "This may be the time for many biologists to work with organisms like yeasts. They are very easy to grow as single cells, exist in either the haploid and diploid state, and are already very well studied genetically."

Another aspect of special interest for meetings like this is the use of yeast and other fungi in which cell differentiation is an important aspect of the life cycle. This type of microorganism can be a useful model system for studying differentiation of higher cells. Luckily both the theoretical background and the experimental techniques are now at hand to permit an attack on the basic problems of structure and function in fungi. This Symposium is an attempt to begin to answer some of these questions at a molecular level.

The programme of this Symposium was planned after ample deliberation and discussion of the Organizing Committee, and it was concluded that the moment was appropriate to arrange a meeting about different aspects of yeast protoplasts, as well as protoplast problems related to filamentous fungi and plants. As has been the case in the two previous Protoplast Symposia, the meeting will provide information about the trends and prospects of both fungal and plant protoplasts so that the links or connections between these different fields may be made clearer.

I want to refer now specifically to some characteristics of this meeting. In the Preliminary Programme a Plenary Lecture given by Professor O. Necas of Purkyne University, Brno, Czechoslovakia, and a Closing Lecture by Professor A. W. Linnane of Monash University, Victoria, Australia were announced. These preliminary plans had unfortunately to be changed last July as soon as Professor Necas let us know that he was unable to attend this Symposium This was a very difficult situation for the Organizing Committee and in order to avoid further complications with the programme it was decided to invite Professor Linnane to deliver the Opening Lecture and also to close the Meeting in a very simple way without any special ceremony. I want to thank Professor Linnane for his acceptance of those changes. There is certainly no need to go out of our way to introduce Professor Linnane since he is so well known already to most of you.

You will also notice that some other small changes have been made in the Preliminary Scientific Programme but they were, we think, of minor importance. We sincerely regret the small alterations in times announced for some of the free communication sessions. These adjustments had necessarily to be made because of the latest news we received about the difficulties of some of the participants.

We are thankful to the invited speakers for having so spontaneously accepted the subjects proposed to them by the Organizing Committee.

In this connexion, the Organizing Committee thanks Professor E. C. Cocking of the University of Nottingham for accepting our invitation to summarize at the Plant Protoplast Session of this Symposium the main topic discussed during the Plant Protoplast Conference held last month in Versailles, in France.

As previously announced, all invited speakers taking part in this Symposium were entitled to submit their contributions for inclusion in a volume with the title "Yeast, Mould and Plant Protoplasts" to be published shortly by Academic Press. This conference comprises 19 invited papers, to be presented in the five sessions of the Symposium, and 46 short papers that will be read as free communications. Due to this high number of submitted papers, and the short time available, it has been necessary to organize and run two long free communication Sessions separately. A small exhibition of scientific pictures related to the sessions of the symposium will be also held in this building.

The Sessions of the Symposium were to be held in the anfiteatro and the lecture rooms of the new Faculty of Science where the Department of Microbiology is located. However, various visitors, among them Professor Anthony H. Rose, suggested that we move everything to these wonderful rooms of the old building of the University of Salamanca, where you may have to face some difficulties but in compensation you will have the opportunity to spend some, I hope memorable hours, in a building several centuries old with a special magnificance and grandeur. It is exactly where the University of Salamanca, one of the three oldest universities of the world, was founded in 1215.

We are conscious that there is not much time free at this conference but we would be most appreciative if in the course of the Symposium, as far as time permits, informal gatherings are arranged by those of you that share our feelings that such activities are important. In this connexion a big room the Aula Francisco de Vitoria labelled "Discussion Room", located in the main hall of this building, will be provided for the participants.

It is also a pleasure to thank the collaboration of various distinguished scientists who accepted our invitation to chair the various Sessions. My thanks also to the Presidents of the Consejo Superior de Investigaciones Científicas, of the Sociedad Española de Microbiología and of the Sociedad Española de Bioquímica who are taking part at this Opening Ceremony. I want also to thank Professor Phaff of California University for accepting our invitation to speak on behalf of the participants.

We are particularly indebted to the Ministry of Education and Science of Spain, to the Consejo Superior de Investigaciones Científicas and to the Sociedad Española de Bioquímica for their generous and significant financial support.

We are also very glad that Salamanca has put at our disposal its hotel accommodation and the University of Salamanca has encouraged us to use the Paraninfo, the Aula de Unamuno and the Aula Salinas in this wonderful centuries-old building to develop this Symposium. In addition, we are thankful to the Ayuntamiento of the City, the Alcalde, the Diputación of Salamanca and the Delegado del Ministerio de Información y Turismo and the sponsoring Industries which enable us to organize this Symposium and to give it a festival framework.

The successful organization of this conference is due largely to the generous cooperation of the invited speakers, all active contributors themselves to the advance of knowledge in this rapidly developing field. It will also reflect the valuable assistance of my collaborators Drs. Isabel García-Acha, Santiago Gascón, Federico de Uruburu,, Jose Manuel Sierra and Rafael Sentandreu as well as that of the students and assistants of the Departamento de Microbiología. Finally, the Symposium would not have emerged in its present form without the continued advice and constructive criticism of our colleagues and friends.

You will notice that we have made a great effort to organize this Symposium in a foreign language. English, in practice the scientific language in biology, was adopted as the official language of the Symposium and no arrangements have been made for simultaneous translation. All the papers presented by the people of our department will be in English and therefore we hope you will understand our natural difficulties and we apologize for them.

It seems to me that it is unnecessary to tell you that the Members of the Organizing Committee in cooperation with the Staff of the Department of Microbiology have worked hard in order to make this Symposium as good and perfect as possible. However, it is possible that despite our best efforts mistakes and omissions may occur. We will try to deal with any difficulties as quickly as possible. We only hope sincerely that the days you spend in Salamanca will be extremely pleasant and profitable from all points of view. Our dearest hope is that we achieve a great success with the discussions and results of this meeting.

Although the number of participants in this meeting is not high I think they are outstanding and very select. Specialists from 24 countries, coming from countries as far as Japan, Australia, Canada, Central and South America, Soviet Union, as well as from most European countries, mostly participants in the previous Symposia, are gathered here. However, we miss some of those present at the earlier meetings. Several of our colleagues were not able to attend this Symposium although they had announced valuable contributions. They have sent letters of apology and wish this meeting the best success. We bear them all in mind.

I still remember the First Yeast Protoplast Symposium in Jena, its high scientific level and the perfect organization due to Dr. R. Müller. We, the organizers of the Third International Symposium, should be glad to obtain the same kind of success as they got.

At this time of my talk and before finishing it, I want to make a special announcement. There is a proposal made by Professor Anthony H. Rose of the University of Bath, Great Britain, to organize the Fourth International Symposium on Yeast Protoplasts at Bath in 1975. It is important to discuss this possibility during these next few days, in order to make a firm decision for the organization of the next Meeting before finishing this Symposium. There is no doubt of the great scientific interest of these gatherings.

We will owe any success of this conference to the scientific contributions made by the participants. My sincere thanks to all the distinguished scientists taking part in this conference for their kind cooperation and helpful suggestions.

I wish for all of you the best time in Salamanca and in Spain, a country well known all over the world for its hospitality, as is clearly attested by the 32 million visitors, a number very close to the population of the country, we will receive this year. And many thanks for your attention.

PART I

Yeast Cell Wall Lytic Enzymes

4

Enzymic Composition of a Lytic Syst
from *Micromonospora chalcea AS*

J. R. VILLANUEVA, M. GACTO and J. M. SIERRA

Departamento de Microbiología, Instituto de Biología Celular (CSIC), Facultad de Ciencias, Universidad de Salamanca, Spain.

I Introduction

The use of enzymes in investigations concerning resolution of the yeast cell wall architecture has notably increased the knowledge we have of structural features of this organelle [1]. The cell wall appears to be composed of a specific arrangement of various polymers in a complex macromolecular organization in such a way that the enzymic attack on selected components can bring about the dissolution of the cell wall and permit the release of the yeast protoplast under appropriate conditions. Much information can be gained by an analysis of the hydrolysing enzymes present in the lytic systems together with an exploration of their contribution to lysis and protoplast formation. Recently, interest has been mainly focussed on the participation of β-glucanases acting on yeast glucan, the wall component that is thought to be the structural support.

A large amount of research has been done using microbial sources for producing lytic systems able to digest the protective coat of filamentous fungi and yeasts. The enzymic composition of these microbial lytic complexes on yeast has frequently been elucidated, for example in the cases of *Cytophaga* and *Bacillus* strains [2, 3]. The studies show that the lytic factors are enzymes which act by cleaving some cell wall component. The lytic capacity is commonly found in actinomycetes, and studies on the distribution of this activity have revealed that several species of the genera *Streptomyces* and *Micromonospora* are among the better producers of these lytic agents. [4]. Preliminary reports on the lytic properties of *Micromonospora chalcea* have already been published [5, 6], but there has been no extensive study on the enzymes present in those preparations. In this paper we summarize the enzymic composition and characterization of a system from this organism active on several yeast species *in vivo*. The study emphasizes that the two major types of activity found in the lytic preparations seem to be involved in the lysis of components that maintain the integrity of the yeast cell wall.

II. Materials and Methods

A. *Production of the Lytic System*

The lytic system was obtained by growing *Micromonospora chalcea* AS (CECT) at 28°C with shaking in a defined liquid medium containing: Laminarin (Koch and Light Lab.) 0.25%; chitin (Sigma) 0.25%; yeast extract (Difco) 0.05%; $NaNO_3$ 0.1%; K_2HPO_4 0.2%; NaCl 0.1% and $MgSO_4 . 7H_2O$ 0.1%. A ratio of one to four of the growth medium to flask capacity was considered adequate for the production of lytic complex. After 6-7 days of incubation the cultures were centrifuged and the cell free supernatants obtained, concentrated to 15% of their original volume and then kept frozen or freeze-dried until required for use. Dialysed samples were analysed for lytic effect and enzymic activity.

B. *Lytic Assays*

Assays for lytic capacity were performed upon several living yeast strains harvested in the logarithmic phase. Methods were essentially similar to those already described [6]. Yeasts were grown in Winge medium (2% glucose, 0.3% yeast extract) for 12-15 h under the conditions of shaking and the temperature described above. The cells were washed three times with distilled water and resuspended in 50 mM phosphate buffer, pH 6.7, to an absorbance of 0.7 measured at 600 nm, in a Bauch-Lomb Spectronic. One ml of this suspension was added to the same volume of enzymic system and incubated at 37° C for various lengths of time. Preparation of protoplasts was carried out in the presence of 1 M $MgSO_4$ as stabilizer.

Similar lytic effects were obtained without added buffer providing the preparations were approximately neutral. The decrease in absorbance of the cell suspensions and observations with a phase-contrast microscope were used to assess lysis.

C. *Substrate Preparations*

β(1-3) glucanase activity was estimated using commercial insoluble laminarin. Pustulan, a β(1-6) glucan, was obtained from the lichen *Umbilicaria pustulata* [7]. β(1-2) glucan was isolated by precipitation and fractionation of culture filtrates of *Agrobacterium tumefaciens* (CECT 89) with ethanol as described by Reese *et al.* [8].

Yeast mannan was extracted from pressed baker's yeast as alkali soluble polysaccharide and purified by precipitation with Fehling's solution; the blue precipitate was dissolved in 3N HCl and reprecipited with methanol to give a copper-free preparation. Complete hydrolysis of this material yielded mannose as the only chromatographycally recognizable sugar. When the above treatment was repeated with *Geotrichum lactis,* a polymer composed of galactose and mannose was obtained.

Commercial chitin was prepared as colloidal chitin by the method of Berger and Reynolds [9] and a relationship established correlating absorbance and chitin concentration at 600 nm. Haemoglobin powder (Schuchardt) and casein (Hammarsten) were used as substrates for estimation of proteolytic activity. Commercial salicin (salicyl-alcohol-β-D glucoside) was used as substrate for β-glucosidase determinations.

Polysaccharides, except laminarin, were thoroughly dialysed first against distilled water and then against the appropriate buffer.

D. *Enzymic Analyses*

Glucanase activities were determined by adding 0.2 ml of an appropriate dilution, if necessary, of the lytic preparation to 0.8 ml of the appropriate glucan (0.3%) previously dissolved in buffer by gentle heating. Assays were run at 37°C for 30 min under pH conditions established as optimal for each activity; similar tests were carried out to detect mannanase activity. The increase in reducing power by enzymic action was expressed in glucose-equivalents, according to the Somogyi-Nelson method [10], using standard curves performed in appropriate buffers. One unit of glucanase activity was defined as the amount of enzyme(s) which release 1 micromole of glucose equivalents under the conditions of incubation. Glucose determinations were carried out using the glucose-oxidase method.

Chitinase activity was similarly estimated by adding 0.2 ml of the system to 0.8 ml of colloidal substrate (1.2% in 50 mM phosphate buffer pH 6.0). After incubation at 37°C for 30 min the experimental tubes were cooled in an ice bath and centrifuged at once. Liberation of *N*-acetylglucosamine was determined in 0.5 ml of supernatant using the dimethylamino-benzaldehyde reagent of Reissig *et al.* [11]. One unit, in this case, is the amount of enzyme that liberates 1 micromole of *N*-acetylglucosamine from colloidal chitin under the stated conditions.

Proteolytic activity was followed by a modification of the Kunitz method [12]. To 1.25 ml of substrate (1% in 100 mM Tris-HCl buffer pH 8.0) 0.25 ml of a diluted sample of the system was added. Incubation was continued for 30 min at 37° C and the reaction stopped with 2.5 ml 5% TCA. After centrifugation, readings were made on the supernatant at 280 nm in a Beckman spectrophotometer. One unit of proteolytic activity liberates 1 micromole of tyrosine equivalent under the above conditions.

In our assay conditions, enzymic activities were proportional to the incubation time and to the amount of enzyme. Estimation of protein was based on the Lowry method [13] using bovine serum albumin (Sigma) as standard. The specific activity of the lytic enzymes is calculated with reference to the protein content of the broth medium after concentration and dialysis.

E. *Paper Chromatography*

The nature of the glucans was determined by chromatography of the products of acid hydrolysis on Whatman paper no. 1, buffered with M/15 phosphate pH 5.5. Butanol-acetone-water (4:5:1 v/v) was employed as solvent and detection of sugar spots was made by spraying with aniline phthalate reagent. Glucose, laminaribiose and laminaritriose (from laminarin), glucose and gentiobiose (from pustulan) and principally glucose (from the crown-gall polysaccharide) were identified in each case. Yeast mannan was free of glucose and galactose.

The pattern of enzyme action on laminarin was also studied by chromatographic analysis of the enzymic hydrolysis products.

F. *Analytical Gel Filtration and Ion Exchange Chromatography*

Preparation of Sephadex gels (Pharmacia) was made following the manufacturers directions. Sephadex G-100 and G-200 columns were equilibrated before using with elution buffers and the void volume was determined using blue-dextran 2000 (Pharmacia). For chromatography in DEAE-Sephadex A-50 columns, the samples to be run were previously equilibrated at a pH and ionic strength similar to the initial buffer and the column developed with an NaCl continuous gradient. Fractions were collected by an LKB automatic collector and the ionic strength checked with a conductivity meter.

For calibration purposes in gel filtration a standard curve of elution volumes with several known molecular weight proteins was prepared; bovine serum albumin, ovalbumin, tripsin and horse heart cytochrome c were used.

III. Results

A. *Lytic Activity and Protoplast Formation*

Lytic activity of the enzymic system from *M. chalcea* AS was assayed with 17 yeast species as mentioned in Materials and Methods. No effect was detected in some cases (Table I), but lysis was evident with *Candida utilis, Geotrichum lactis, Oospora suaveolens, Pichia polymorpha, Saccharomyces cerevisiae, Sacch. elegans, Sacch. fragilis, Sacch. carlsbergensis* and *Schizosaccharomyces pombe.* EDTA (5mM when present) was not an activator in the lytic assays.

TABLE 1

Lytic range of the *Micromonospora chalcea* preparation on living yeast cells in log-phase. + = lysis; – = no lytic effect could be detected even in prolonged incubations.

Candida utilis CECT 1061	+
Candida humicola CECT 1016	–
Cryptococcus diffluens CECT 1073	–
Geotrichum lactis CECT 1102	+
Hansenula anomala CECT 1107	–
Hansenula subpelliculosa CECT 1116	–
Oospora gracilis CECT 1123	–
Oospora suaveolens CECT 1335	+
Pichia polymorpha CECT 1132	+
Rhodotorula glutinis CECT 1137	–
Saccharomyces cerevisiae CECT 1179	+
Saccharomyces carlsbergensis CECT 1166	+
Saccharomyces elegans CECT 1202	+
Saccharomyces fragilis CECT 1207	+
Schizosaccharomyces pombe CECT 1221	+
Sporobolomyces albidus CECT 1268	–
Torulopsis aeris CECT 1280	–

A quick conversion to osmotically sensitive protoplasts could be obtained with *C. utilis, G. lactis, O. suaveolens* and *P. polymorpha,* but several hours were necessary for partial conversion of *Schizossach. pombe* whose protoplasts were comparatively very unstable. Although the system did show lytic effects on cells of *Saccharomyces* species, a low efficiency of protoplast formation was obtained for these strains under our conditions; formation was not enhanced when 0.6 M sorbitol was used as an alternative stabilizer.

Protoplast release in *C. utilis* takes place at the equatorial zone of the cell as has already been described elsewhere [14]. Cell walls without cytoplasmic contents which had been preferentially attacked were frequently seen in the preparations at the earliest stages of the process (Figs 1, 2). Protoplasts of the mycelial form of *G. lactis* are liberated generally from the apical extreme of the hyphae; occasionally they emerge from intercalary zones. In the case of *O. suaveolens* liberation is also terminal. Pictures like those produced by Johnson [15] showing the extrusion of *Schizosacch. pombe* cell contents, supposed to be the result of inhibition of the wall glucan synthesis, are obtained when the lytic system acts upon this fission yeast (Figs 3, 4). The onset of protoplast formation occurs at localized sites on the cell wall which are the zones where active elongation of glucan layer occurs. In other species the point of initial attack was not clearly defined.

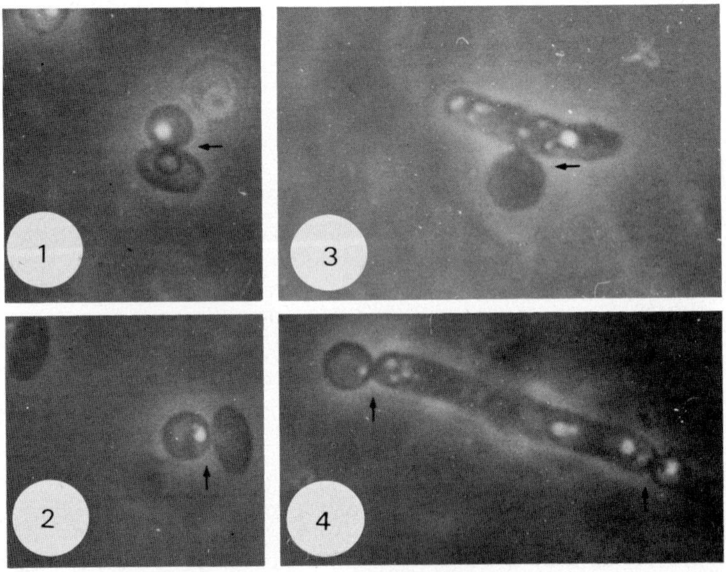

Figs 1, 2, 3, 4. 1 and 2: Release of protoplasts from *Candida utilis.* 3 and 4: Release of protoplasts from *Schizosaccharomyces pombe.*

B. *Enzymic Activities of the System*

Characterization of the complex from the point of view of its enzymic composition was performed in order to discover the enzymes involved in yeast cell wall degradation. To detect the enzyme activity, substrates were chosen which reflected the sort of polymers usually found in the cell wall preparations. Table II

TABLE II
Extracellular lytic enzymes of *Micromonospora chalcea*. One enzymic unit is the amount of enzyme that liberates 1 μmole of glucose equivalent (glucanase activities) or 1 μmole of tyrosine equivalent (proteolytic activity) from the substrate during 30 min at 37°C under standard assay conditions. One unit of chitinase activity releases 1 μmole of *N*-acetylglucosamine under similar conditions.

Activity	Substrate	pH	units/ml	Specific activity
β(1-2) glucanase	Crown-gall polysaccharide	4.5	0	—
β(1-3) glucanase	Laminarin	5.5	22.77	12.94
β(1-6) glucanase	Pustulan	5.5	0.89	0.50
α-mannanase	Yeast mannan	5.5	0	—
Protease	Haemoglobin	8.0	14.14	8.03
	Casein	8.0	24.75	14.06
Chitinase	Colloidal chitin	6.0	0.08	0.04
β glucosidase	Salicin	5.0	0	—

summarizes the enzymic levels found in the lytic culture filtrates. β(1-3) glucanase and protease activities were high as compared with other activities. Only a very weak β(1-6) glucanase activity was present and neither mannanase, β(1-2) glucanase nor β-glucosidase were detected, even on prolonged incubation. As β-glucosidase is absent, action upon laminarin must be due to the β(1-3) glucanase activity; chromatography of the reaction mixture showed both endo- and exo-hydrolytic action.

Galacto-mannan from *G. lactis* was not hydrolysed by the system but nevertheless lysis of this yeast did occur (Table I). Traces of chitinase activity were also present. Although some variation seems to occur in this activity, a low yield appears to be a general rule and chitinous debris is always present in the culture medium at the end of the incubation period.

Some general characteristics of the predominant enzymic activities were determined. β(1-3) glucanase was stable at refrigerator temperature for a long time. Repeated freezing and thawing, however, decreased activity noticeably. No substantial loss was observed when preparations were concentrated by vacuum distillation at 28°C or freeze dried.

Stability of the β(1-3) glucanase activity is greater at slightly alkaline pH values, coinciding with final pH of the growth medium, but activity extends over a wide range. Figure 5 shows the effect of pH on the enzymic activity, maximal action occurring at pH 5.5. No

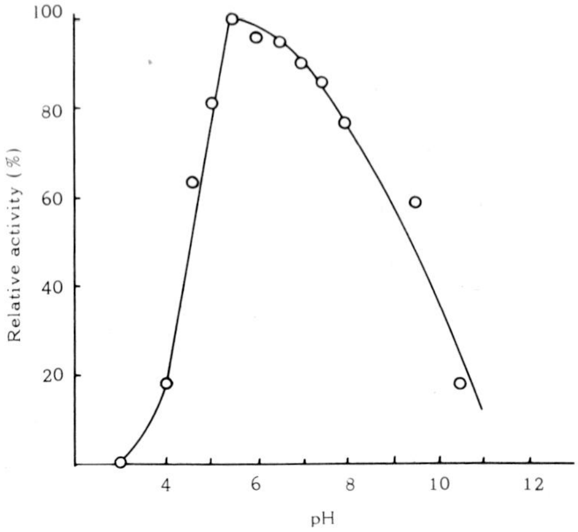

Fig. 5. β(1-3) glucanase pH activity curve.

appreciable differences were found due to the nature of the buffer; 50 mM citrate or acetate buffers pH 5.5 were usually used in the routine assays.

When β(1-3) glucanase was assayed as usual, but the reaction carried out at different temperatures, maximum hydrolysis occurred at 50°C. Thermal stability of the β(1-3) enzyme(s) was also investigated. Enzyme solutions, appropriately diluted, were treated for 10 min at several temperatures, rapidly cooled in an ice-water bath, and then assayed by the standard procedure. Inactivation occurs rapidly at temperatures above 50°C (Fig. 6).

The enzymic activity was substrate dependent and release of reducing groups was constant at concentrations of laminarin above 2 mg/ml. The activity was inhibited by $Cu^{2+}$and particularly by Hg^{2+} ions, the last producing complete inhibition at a final concentration of 1 mM, thus suggesting the need for free —SH groups for enzymic action.

As shown in Table II, the lytic system showed a relatively great proteolytic activity. Casein was more rapidly hydrolysed than haemoglobin. Estimations of hydrolytic action show peaks of activity at more than one pH value and this suggests that the preparation may contain more than one type of enzyme (Fig. 7); greatest activity is reached in the range pH 7-9. The temperature for optimal proteolytic activity was 50°C with both casein and haemoglobin as substrates. The protease preparation, when assayed after 10 min treatment at various temperatures, was stable up to 40°C but

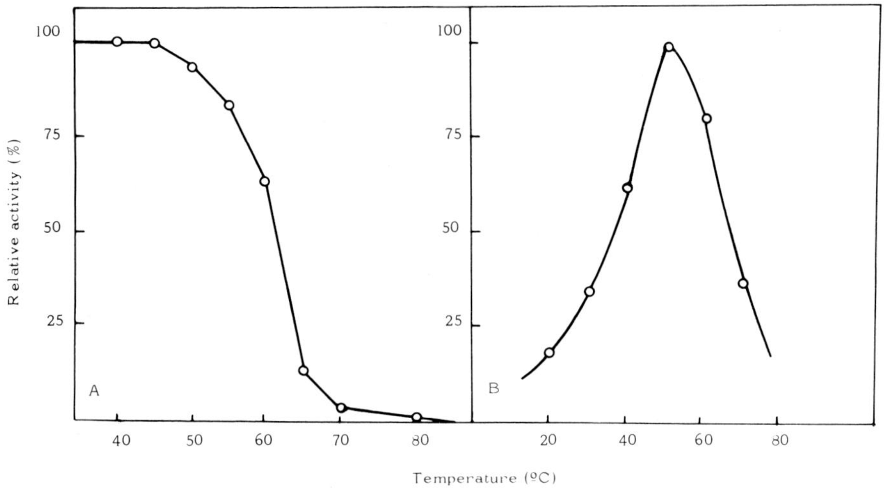

Fig. 6. β(1-3) glucanase activity. (A) Enzymic stability at different temperatures. (B) Effect of temperature on enzymic hydrolysis of laminarin.

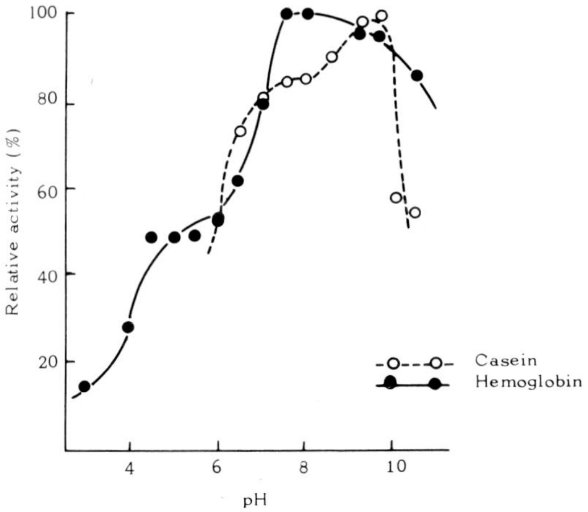

Fig. 7. Effect of pH on proteolytic activity.

at higher temperatures inactivation increased gradually, complete inactivation being obtained at 80° C. The temperature effect was more marked with casein than with haemoglobin (Fig. 8).

C. *Preliminary Fractionation of the Enzymic System*

Initial attempts to fractionate the β(1-3) glucanase and proteolytic activities by ammonium sulphate precipitation were unsatisfactory,

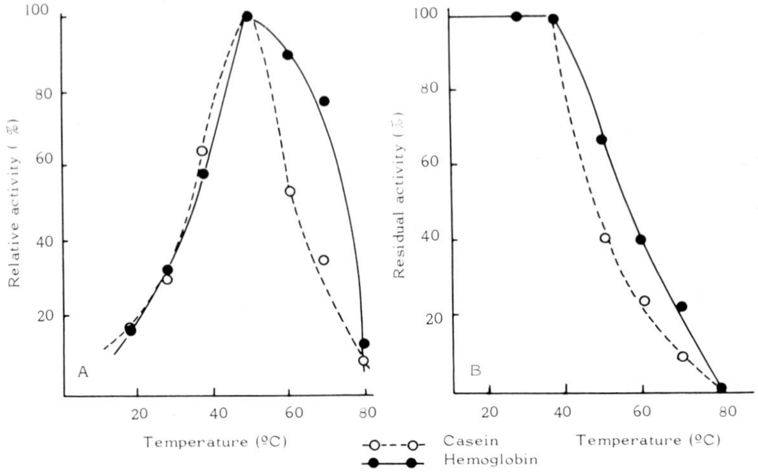

Fig. 8. Proteolytic activity. (A) Influence of temperature on activity. (B) Enzymic stability at different temperatures.

these activities being spread among many fractions. Gel filtration chromatography confirmed the multicomponent nature of the β(1-3) glucanase, agreeing with the analysis of the action pattern. With Sephadex G-200, β(1-3) glucanase activity was found distributed over a large elution range (Fig. 9) but, when gel filtration was on Sephadex G-100, two peaks were clearly distinguishable. The main

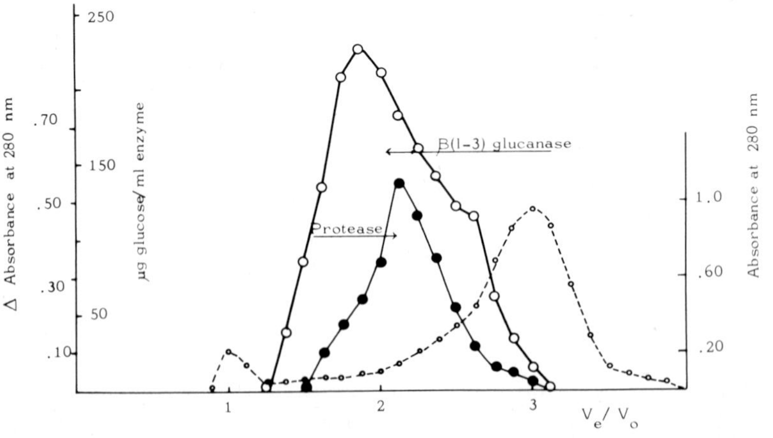

Fig. 9. Elution patterns of β(1-3) glucanase and protease activities after filtration of dialysed freeze-dried culture filtrates through a Sephadex G-200 column. Alkaline protease was determined on casein.

component exhibited an endohydrolytic action, oligosaccharides of the laminarin series being the products of enzymic activity. A second β(1-3) glucanase peak was eluted later that showed both types of enzymic activity, though endo-action was predominant (Fig. 10). On DEAE Sephadex there were also two β(1-3) glucanase components;

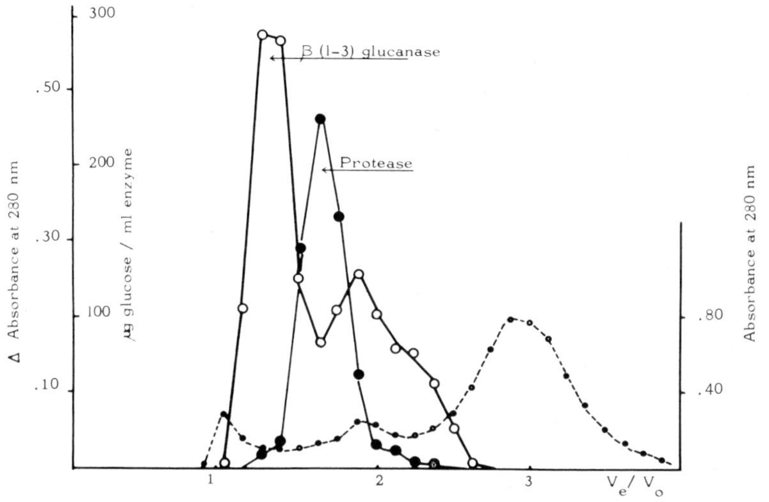

Fig. 10. Elution patterns of β(1-3) glucanase and protease activities through a Sephadex G-100 column. Alkaline protease was determined on casein.

the first one, having no affinity for the ionic exchanger, showed both hydrolytic patterns; the second was liberated at low ionic strength and showed endohydrolytic action (glucose was feebly detected but only after long incubation) (Fig. 11).

The absence of β-glucosidase activity allow one to state that at least three β(1-3) glucanase activities can be recognized as present in the lytic system from the combined evidence of their substrate profile, gel filtration and ion-exchange chromatography. The sharp peak, eluting from Sephadex G-100 at a V_e/V_o ratio close to 1:3, does not seem to correspond entirely to the peak lacking affinity toward the anion exchanger, since that peak had no exohydrolytic activity. Exoglucanase was associated in gel filtration with the minor endohydrolytic component and the opposite occurred in DEAE chromatography.

Proteolytic activity appeared on Sephadex G-100 as a single component overlapping the glucanase activities, but the complex

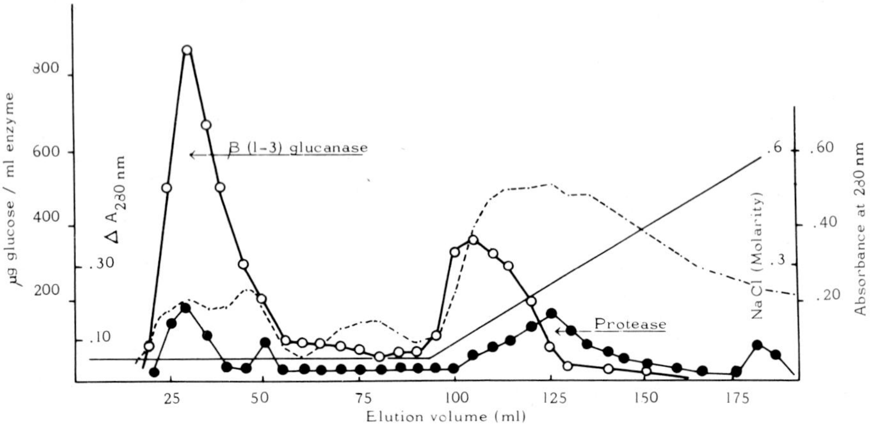

Fig. 11. Fractionation of dialysed freeze-dried culture filtrates by chromatography on DEAE- Sephadex A-50. The sample was dissolved in initial elution buffer (50 mM Tris-HCl pH 7, 5) and subjected to a linear gradient elution with NaCl. Alkaline protease was determined on haemoglobin. Protein distribution is shown as A_{280}nm.

nature of this activity was further demonstrated by resolution on DEAE columns. As indicated above, protease activity detected at alkaline pH is always higher than the corresponding neutral or acid proteolytic activities.

Although complete resolution was not obtained by isolated procedures, preliminary information about the two major activities in the lytic system was gained and experiments showed that fractions with β(1-3) glucanase activity, of the endohydrolytic type, and even low proteolytic activity produced protoplasts from *C. utilis* and *G. lactis* in the absence of any other detectable enzymic activity. This suggests that these enzymes must be involved early in lytic action against the cell wall.

A similar tentative conclusion could be reached from experiments made with the growth media of *Micromonospora* to study any correlation between β(1-3) glucanase content and lytic action. Lysis of young *C. utilis* cells (measured as percentage decrease in absorbance, during a three hours assay) and β(1-3) glucanase level are quite well correlated (Fig. 12). The same results can be established for proteolytic activity, both activities following a parallel course in the culture filtrates (Fig. 13).

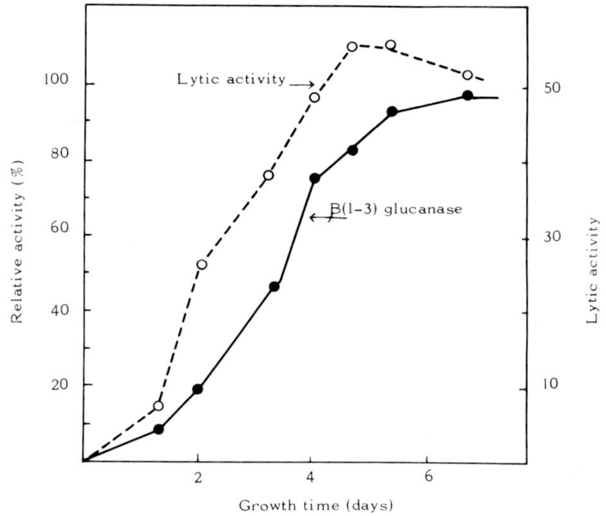

Fig. 12. A typical time-course production of the lytic system as assayed on *Candida utilis* cells.

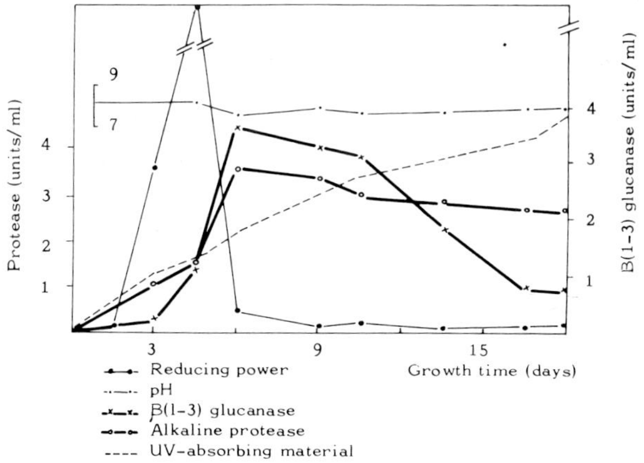

Fig. 13. Appearance of β(1-3) glucanase and protease in the grown medium of *Micromonospora chalcea*. Values for reducing power and absorbance at 280 nm are given in relative arbitrary units.

D. *Isolation of β(1-3) Glucanase and its Action on Yeast Cells*

In order to clarify the specific role of glucanase and protease activities in the lysis of living yeast cells it was decided to isolate the major endoglucanase component in the system. The isolation and purification involved concentration of culture filtrates, ammonium sulphate precipitation, gel filtration and column chromatography on DEAE-Sephadex. Some salts from the growth media were discarded after vacuum distillation and centrifugation (12,000 xg for 20 min) at 4° C. The supernatants were subjected to ammonium sulphate fractionation and the precipitates obtained between 20 and 70% saturation collected, after allowing to settle overnight, by spinning at 23,000 xg for 20 min. A great amount of interfering pigment diffuses from the resuspended precipitates during dialysis.

These preparations, when eluted through gel filtration, show almost the same elution profile as the original complete system, indicating that the initial glucanase activities are still present (Fig. 14). Ultraviolet-absorbing pigment is completely removed from the enzymic fractions under study at this step.

The major β(1-3) glucanase component, endohydrolytic in nature, was collected in the fractions indicated by the arrows, care being taken to obtain as little proteolytic contamination as possible. Further chromatography on DEAE- Sephadex with 25 mM Tris-HCl buffer pH 7.5 yielded endoglucanase activity with a degree of purification above 30-fold and lacking other lytic activities when

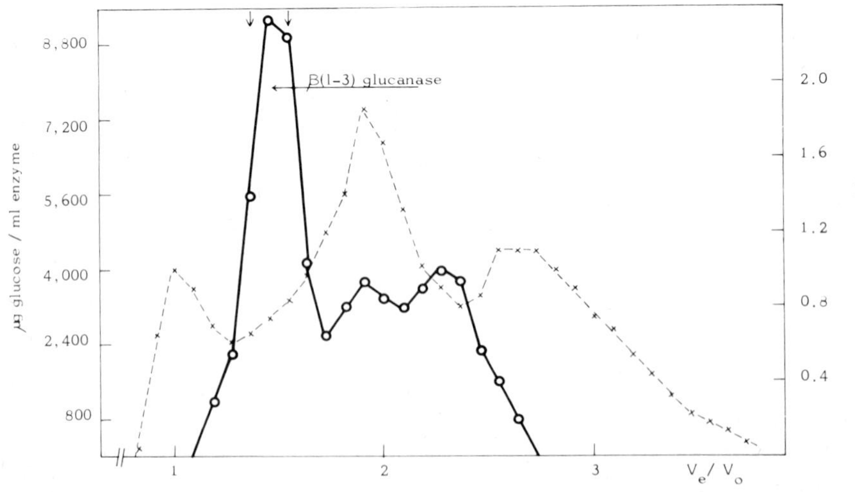

Fig. 14. Gel filtration of the ammonium sulphate precipitate (20-70% sat.) through Sephadex G-100.

tested on the respective substrates over long periods of incubation (Fig. 15). Results from such a purification procedure are given in Table III. Pigment interference can account, at least in part, for the disparity in values of protein as estimated by both the Lowry and the spectrophotometric methods. The final recovery was about 6%, this

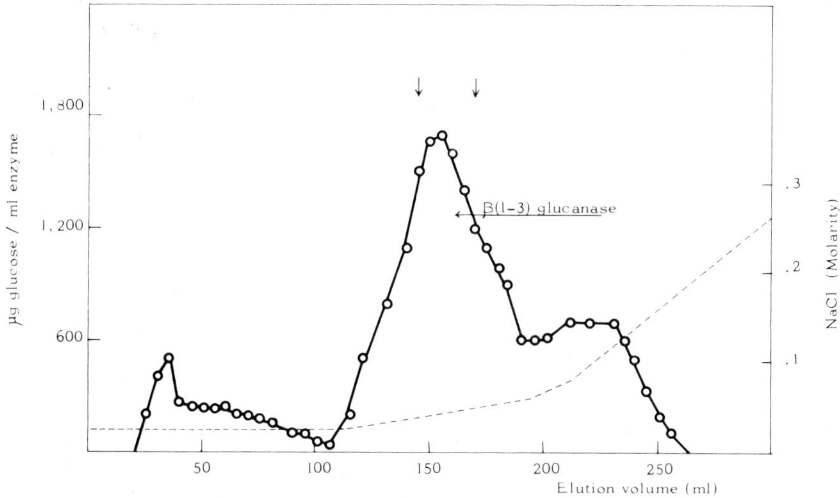

Fig. 15. Chromatography on DEAE- Sephadex A- 50 of the fractions collected by gel filtration. The elution buffer was 25 mM Tris-HCl pH 7,5 with a linear gradient of NaCl. Bed dimensions, 23 x 1.5 cm.

TABLE III

Purification of a β(1-3) glucanase from culture filtrates of *Micromonospora chalcea.* (1) Protein measurement by the method of Lowry *et al.* (1951). (2) Estimation of protein according to the spectrophotometric procedure (Warburg and Christian, 1941).

	Total enzyme units	Protein (mg) (1)	Protein (mg) (2)	Specific activity (1)	Specific activity (2)	Purification (1)	Purification (2)	% Recovery
Culture filtrate	3750	350	960	10.7	3.9	—	—	100
Concentrated centrifuged fluid	3645	288	630	12.6	5.7	1.1	1.5	97
Dialysed ammonium sulphate precipitate	2437	47.5	98.8	51.3	24.6	4.8	6.3	65
Gel filtration in Sephadex G-100	765	7.4	9.5	103.3	80.5	9.6	20.6	20
DEAE-Sephadex chromatography	198	0.6	1.2	330.0	165.0	30.8	42.3	6

value seeming reasonable in view of the multicomponent nature of the system and the restricted number of fractions collected. It seems surprising, however, that β(1-3) glucanase activity represents such a very low proportion of the protein in broth cultures.

The molecular weight of the isolated glucanase, as deduced by interpolation on a plot of elution volumes of known molecular weight proteins on a calibrated Sephadex G-100 column, was about 73,000 daltons.

As β(1-3) glucanase preparations hydrolysed glucan suspensions obtained by treatment of whole cells of baker's yeast [16] and since they rapidly digested alkali-treated cells, the lytic nature of this enzyme on the wall glucan was assumed. However, when lytic ability of the purified β(1-3) glucanase was assayed on living cells of *C. utilis, Schizosacch. pombe, O. suaveolens, Sacch. cerevisiae* and *G. lactis,* no effects in terms of cellular lysis were detected, in contrast to the lytic action achieved by the whole enzymic system acting on these yeasts. In spite of this fact, cells of *C. utilis* treated by aqueous alkali (3%) for 30 min were attacked and almost complete loss in turbidity could be observed in parallel assays. In this case, structures very similar to those described as bud scars by several workers in *Sacch. cerevisiae* were the only residue. These structures probably represent typical material from this region of the cell and which have been described elsewhere. Their resistance to glucanase action points to a chitinous composition (Fig. 16).

From these observations it can be deduced, as Bacon *et al.* do for *Sacch. cerevisiae* cells [2], that in their natural state the walls of these yeasts have no structural glucan on the outer surface directly available for the lytic action of the enzyme. The results show clearly that some external component in whole living yeast cells prevents the enzymic dissolution of the structural polymer at the cell wall and that factor(s) other than β(1-3) glucanase must be related with the lytic process.

As previous experiments suggested, there is a possibility that the attack on the inner glucan layer by β(1-3) glucanase might follow the action of the proteolytic components in the lytic system on a proteinaceous surface layer. Although some attempts have been made, until now we have not succeeded in isolating the proteolytic activity of the system free of glucanase activity in order to make a detailed study of its role. Preliminary results from this sort of study seem to show that ultraviolet absorbing material is released from the walls of *C. utilis* cells by commercial papain (Schuchardt) and no other activities than protease were detected in the assayed preparations. When mild proteolytic treatments were used, cells kept their

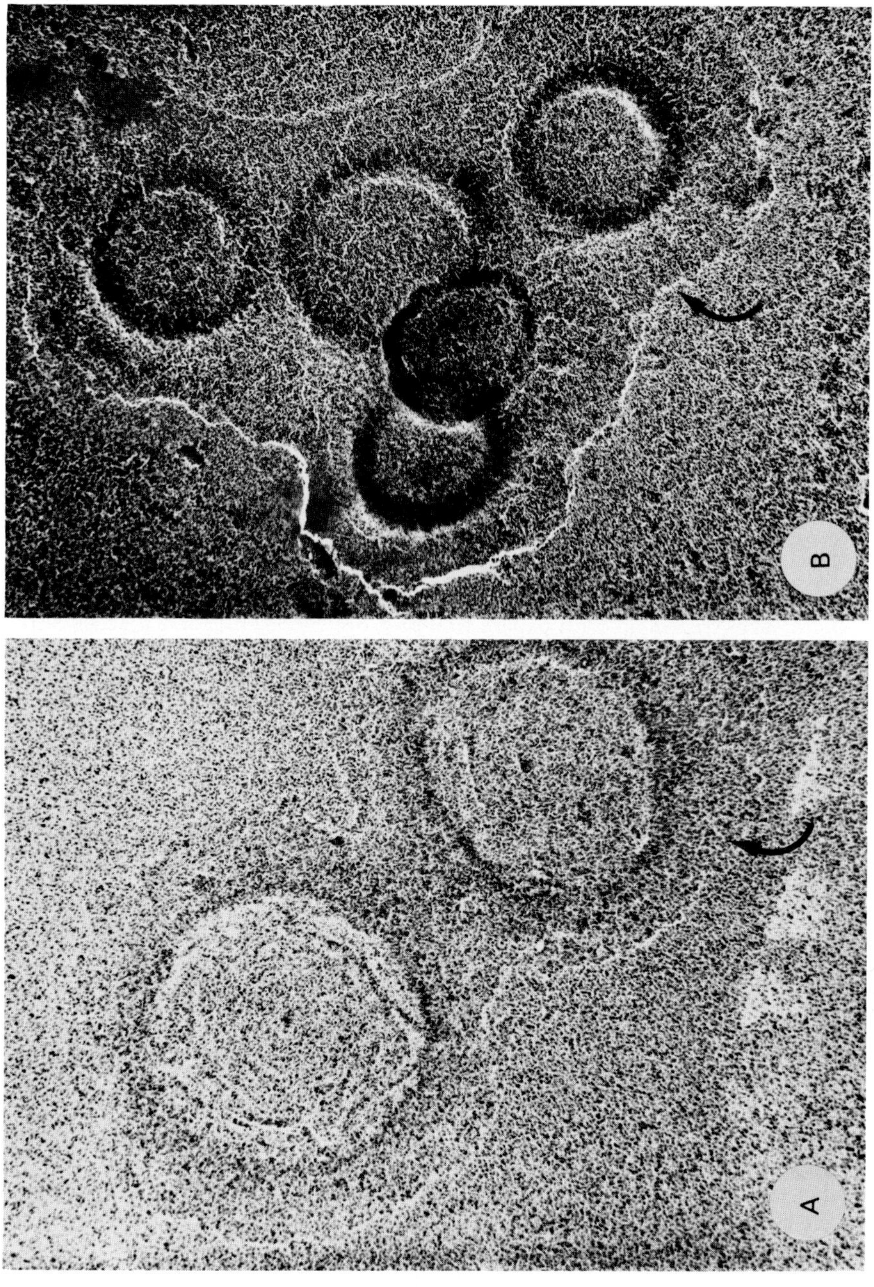

Fig. 16. (A) Bud-scar residues from alkali-extracted cells of *Candida utilis* after enzymic treatment by β(1-3) glucanase; x60.000. (B) Bud-scars from *Saccharomyces cerevisiae* after chemical solubilization of the structural cell wall glucan (Bacon *et al.*, 1966); x45.000.

initial shape and the subsequent action of added glucanase produced a decrease in the initial absorbance of pre-treated cell suspensions when compared with preparations that had undergone the pre-treatment alone (Fig. 17). The structural disposition of protein complexes in the cell wall might therefore be the factor limiting the action of glucanase on whole cells.

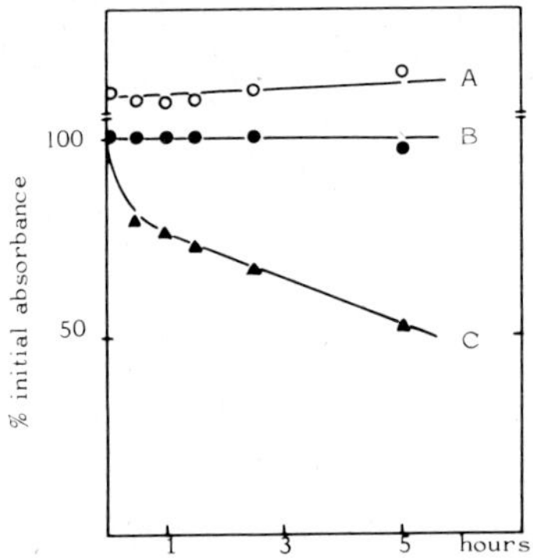

Fig. 17. *C. utilis* cells were suspended in 0.5 mM EDTA, 1 mM 2-mercaptoethanol, 50 mM Tris-HCl buffer pH 8.0 to give an initial absorbance of 0.50 at 600 nm; after treatment with papain (0.9 units on casein/ml) for 2 h at 37° C the cells were washed and resuspended in the same volume with 50 mM acetate buffer pH 5.5. (A) Untreated cells without or with β(1-3) glucanase. (B) Papain-treated cells. (C) Papain treated cells with added β(1-3) glucanase; no lysis prior to the addition of glucanase was detected.

IV. Discussion

The behaviour of the lytic system examined here is consistent with the results obtained when pressed baker's yeast was used as carbon source for the growth of the same *Micromonospora* species [6]. With regard to the predominant enzymic activities, the preparation appears to be similar to that recently shown in *Micropolyspora,* a related actinomycete [17]. The culture system we described here is probably simpler in composition; nevertheless, it exhibits action on yeast cells and allows protoplast formation in suitable conditions of osmotic pressure.

The different response of the cell walls to the action of the lytic system is a reflection of the different nature or specific arrangement of its components *in vivo*. The low chitinase activity may account for the absence of lytic action on cell walls of *Rhodotorula* and *Sporobolomyces,* the apparent resistance being due to the high chitin content of their walls [18]. Application in our laboratory of the *Streptomyces venezuelae RA* system shows a larger lytic effect on fungal walls having higher chitin content, and enzymic analysis revealed a stronger chitinase activity than the *Micromonospora* system.

Whenever primary attack on yeast walls is localized it is possible that the preparation acts where a susceptible structural polymer is more available from the outside. Very thin and insoluble residues from yeast cells having the original shape remain after prolonged incubation in some cases, suggesting that if glucan contributes to the cellular form it is not completely digested by β-glucanases. The existence of α-linked glucans has also been demonstrated in yeasts [1].

Although specific variations probably occur in the yeast cell wall architecture, the rigidity of the structure is generally ascribed to a predominant β(1-3) glucan closely associated with other components. Mannan-protein complexes seem to form the external layers of the cell wall and existence of specific enzymic agents for its dissolution has been claimed. McLellan *et al.* [19, 20] described the action of a phosphomannanase capable of removing cell wall mannan and a very small amount of protein without affecting yeast glucan, to be a preliminary requirement for action on the glucan. Release of mannan-protein from cell walls has also been achieved by β(1-6) glucanase from *Cytophaga* cultures [2] (suggesting some structural relation of that fraction with a glucan component) and by action of trypsin and papain [21].

The primary importance in the study of the lytic mechanism by our system has been conceded to the β(1-3) glucanase and protease preparations, the two main hydrolytic components. If the proposed model can be applied to some extent to other yeasts, and specially to the *C. utilis* cell wall, action of β(1-3) glucanase on the inner glucan might be facilitated by proteolytic action on the surface components enclosing the glucan, its absence preventing cell wall disruption. This suggestion seems possible, since our preparations had no apparent effect on mannan and isolated glucanase does not seem to have access in the assayed yeasts to the cell wall supporting polymer. No lytic factor, in terms of cellular lysis, is found either in the purified β(1-3) glucanase or in commercial papain in our experiments and the use of any of these enzymes alone has no action. Lytic effect appears only when the conjoint action of both activities takes place on cell wall components.

Our system is almost devoid of β(1-6) glucanase and chitinase, two activities concerned with cell wall components. The participation of these enzymes in steps leading to further disintegration cannot be excluded.

Protease does not seem to play an essential role in cell wall degradation by other lytic systems [20, 22]. However, it must be expected that if, in fact, part of the mannan and protein lies outside the glucan layer, the action on the protein could alter the external structure of the cell wall and facilitate the subsequent action of other enzymes. It is now well known that a similar situation is obtained by treatments of whole cells under defined alkaline conditions in which mannan goes into solution [16]. Thiol treatment to increase lytic efficiency can be interpreted in some systems as also acting at this level by breaking somewhere in the mannan-peptide fractions [2] and we thought that a similar effect could be exerted by some proteases.

Extensive work has been done on enzymes involved in lysis, but proteases have received comparatively little attention. Nevertheless, the work of Furuya and Ikeda [23] provided proof that this activity was critical in a *Streptomyces* system acting on some kinds of isolated wall preparations, and the same conclusion is reached from the study of the mycolytic enzymes from the basiomycete *Coprinus* acting on yeast cells [24].

A conflict concerning protoplast formation is apparent between the results of Millbank and Macrae [22] on the one hand and those reported by Monreal *et al.* [25] on the other. The first authors suggested that attack on the glucan alone does not of itself lead to protoplast appearance. The second that results can be explained by implicating protease function, the report showing that their β(1-3) glucanase enzyme preparation was not completely free of proteolytic activity.

The suggestion that the inner layer of the cell wall is protected from glucanase action has been widely expressed, and our results are in agreement with such a picture. However, the action on living yeast cells of an isolated enzyme from *Rhizopus* [26], tentatively considered as glucanase of endohydrolytic type, seems to indicate a new situation for glucan location.

In view of the complex organization of the yeast cell wall and the partial knowledge we have about the inter-relationship of the various components, a correct interpretation of the lytic process is not easy. Exogenous lysis can be the result of various enzymic processes; some of the better known lytic enzymes (glucanases, mannanases or chitinases among others) often seem to be involved, but we cannot ignore the possibility that lysis may arise also from a very low concentration of a specific enzyme, difficult to detect, acting on some restricted wall component. The chance that enzymic factors other than those usually considered might be present in enzyme preparations must be emphasized as the critical point in the elucidation of wall structure by enzymic techniques and, in fact, seems to limit extensive generalizations.

The models proposed hitherto do explain some features of the structure of the cell wall but in a number of aspects they are not consistent and leave unresolved questions. From the accumulated information to date, two regions of obscurity seem to emerge from the study of the cell wall structure by lytic enzymes and in the interpretation of the lytic processes; one concerns the unresolved relationship between the cell wall components, and other the specificity and order of enzymic action against the intact wall. A greater knowledge of both aspects will provide a better understanding of the yeast cell wall problem.

Acknowledgements

This work was done with the support of the Ministerio de Educación y Ciencia of Spain (through a fellowship to M. Gacto and J. M. Sierra) and Fundacion Juan March.

References

1. Phaff, H. J. (1971). In "The Yeast" (A. H. Rose and J. S. Harrison eds.), vol. II, p. 135. Academic Press, London.
2. Bacon, J. S. D., Gordon, A. H., Jones, D., Taylor, I. F. and Webley, D. M. (1970). *Biochem. J.* **120**, 67.
3. Tanaka, H. and Phaff, H. J. (1965). *J. Bacteriol.*, **89**, 1570.
4. Gascón, S. and J. R. Villanueva (1963). *Can. J. Microbiol.*, **9**, 651.
5. Gascón, S., Ochoa, A. G. and Villanueva, J. R. (1965). *Can. J. Microbiol.*, **11**, 573.
6. Gascón, S., Ochoa, A. G. and Villanueva, J. R. (1965). *Arch. Mikrobiol.*, **51**, 156.
7. Reese, E. T., Parrish, F. W. and Mandels, M. (1962). *Can. J. Microbiol.*, **8**, 327.
8. Reese, E. T., Parrish, F. W. and Mandels, M. (1961). *Can. J. Microbiol.*, **7**, 309.
9. Berger, L. R. and Reynolds, D. M. (1958). *Biochim. biophys. Acta*, **29**, 522.
10. Somogyi, M. (1952). *J. biol. Chem.*, **195**, 19.
    Nelson, N. (1944). *J. biol. Chem.*, **153**, 375.

11. Reissig, J. L., Strominger, J. L. and Leloir, L. F. (1955). *J. biol. Chem.*, **217**, 959.
12. Kunitz, M. (1947). *J. gen. Physiol.*, **30**, 291.
13. Lowry, O. H., Rosebrough, N. J., Farr, A. L. and Randall, R. J. (1951). *J. biol. Chem.*, **193**, 265.
14. Villanueva, J. R., Elorza, M. V., Monreal, J. and Uruburu F. (1966). *Proc. Int. Symp. Yeasts. Znd. Bratislava.*
15. Johnson, B. F. (1968). *J. Bacteriol.*, **95**, 1169.
16. Bacon, J. S. D., Farmer, V. C., Jones, D. and Taylor, I. F. (1969). *Biochem. J.*, **114**, 557.
17. Okazaki, H. and Iizuka, H. (1970). *J. gen. app. Microbiol. Tokyo*, **16**, 537.
18. Crook, E. M. and Johnston, I. R. (1962). *Biochem. J.*, **83**, 325.
19. McLellan, W. L. and Lampen, J. O. (1968). *J. Bacteriol.*, **95**, 967.
20. McLellan, W. L., McDaniel, Ll. E. and Lampen, J. O. (1970). *J. Bacteriol.*, **102**, 261.
21. Eddy, A. A. (1958). *Proc. R. Soc. B.*, **149**, 425.
22. Millbank, J. W. and Macrae, R. M. (1964). *Nature, Lond.*, **201**, 1347.
23. Furuya, A. and Ikeda, Y. (1960). *J. gen. appl. Microbiol. Tokyo*, **6**,49.
24. Kawai, M. (1970). *J. Ferment Technol.*, **48**, 397.
25. Monreal. J., Uruburu, F. and Villanueva, J. R. (1967). *J. Bacteriol.*, **94**, 241.
26. Yamamoto, S., Shiraishi, T. and Nagasaki, S. (1972). *Biochem. biophys. Res. Commun.*, **46**, 1802.

Comparative Studies on β-1,3-Glucanases of Microorganisms

K. HORIKOSHI

The Institute of Physical and Chemical Research,
Wako-shi, Saitama, Japan

In a previous paper of this series [1], it was reported that an enzyme produced by *Bacillus circulans* IAM 1165 had strong lytic activity on the cell walls of *Aspergillus oryzae,* releasing various kinds of sugars from them. Horikoshi *et al.* [2] found that the enzyme from *B. circulans* had β-1,3-glucanase activity which hydrolysed laminarin and yielded glucose, laminaribiose, laminaritriose etc. Using this enzyme, fine structure of the cell walls of *Asp. oryzae* was studied [3, 4, 5], chitin fibre was coated with polysaccharide which was hydrolysed by the enzyme. The cell walls treated with this glucanase were lysed with chitinase of *Streptomyces griseus.*

Recently we have isolated alkalophilic microorganisms which can grow in high alkaline media containing 1% Na_2CO_3, and these microorganisms produce many enzymes which have optimum pH in alkaline regions, e.g. alkaline protease, pH 11.5; alkaline amylase, pH 10.5; alkaline pectinase, pH 10.0 and alkaline catalase, pH 10 [6-9]. By analogous method, we isolated two strains of bacteria, No. 221 and No. K-12-5 which produce β-1,3-glucanases.

The present paper deals with purification and some properties of β-1,3-glucanases of these bacteria. Comparative studies of these enzymes are also discussed.

I. Materials and Methods

A. *Preparation of Crude Enzyme Solutions*

Bacillus circulans IAM 1165 was grown aerobically at 37°C in the medium containing pachyman, 5 g; polypeptone, 10 g; yeast extract, 5 g; K_2HPO_4, 1 g; $MgSO_4$ $7H_2O$, 0.2 g and 1 litre of 0.05 M Tris-HCl buffer (pH 8.2). Alkalophilic bacteria No. 221 and No. K-12-5 were grown aerobically at 37° C in the medium containing pachyman, 20 g; polypeptone, 5 g; yeast extract, 5 g; K_2-HPO_4, 1 g; $MgSO_4 7H_2O$, 0.05 g; Na_2CO_3, 10 g and 1 litre of water. Sodium carbonate was sterilized separately and added to the medium. Pachyman powder was extracted with ether for 20 h in a Soxhlet apparatus. After three days of cultivation with continuous shaking, cells were removed by centrifugation at 6000 xg for 10 min and the supernatant fluid was examined as a crude enzyme solution.

B. *Assay of β-1,3-glucanase Activity*

The enzyme solution (0.01 ml) containing up to 20 mU of the enzyme in 0.01 M Tris-HCl buffer (pH 7) was mixed with 0.2 ml of 1% laminarin solution (made up with 0.05 M Tris-HCl buffer, pH 8.0, for No. 221 and No. K-12-5 enzymes, and McIlvain buffer, pH 5.6 for *B. circulans* enzyme). After 10 min incubation at 40° C, 1 ml of dinitrosalycilic acid solution [10] was added, and the mixture was heated in a boiling water bath for 5 min, and then 4 ml of water were added. An absorbance at 510 nm was measured. One unit of the enzyme activity is defined as the amount of enzyme which produces 1 mg of reducing sugar calculated as glucose per minute under the condition described above.

C. *Preparation of Fungal Cell Walls*

Preparation of the cell walls were reported previously [4].

D. *Other Assay Methods*

Protein concentration was estimated by the method of Warburg and Christian [11]. Ampholine electrofocusing was done by the method of Vesterberg [12]. Molecular weight was estimated by the sodium dodecyl-sulphate (SDS)-polyacrylamide gel electrophoresis method [13].

E. *Identification of Enzymatic Digests*

The hydrolysates were chromatographed on Whatman No. 1 paper with the solvent system of pyridine-ethylacetate-water (5 : 12 : 4 v/v) at room temperature. Ammoniacal silver nitrate was used as a spraying reagent.

F. *Reagents*

Laminarin was obtained from K and K Laboratory, U.S.A. Ampholine carrier ampholyte (pH 3-10) was from LKB-Producter AB, Sweden.

II. Results

A. *Purification of the Enzymes*

Purification of No. 221 and No. K-12-5 enzymes. The crude enzyme (about 3-5 U/ml) was passed through a column of DEAE-cellulose equilibrated with 0.1 M Na_2CO_3 (pH was adjusted to 9.0 with HCl). The column was successively washed with 0.1 M phosphate buffer (pH 8.0) containing 0.3 M NaCl. The enzyme was eluted with 0.2 M phosphate buffer (pH 5.6) containing 0.5 M Nacl. The enzyme was precipitated with 70% saturation of ammonium sulphate and dissolved in 0.01 M phosphate buffer (pH 7.0). The enzyme solution was passed through a Sephadex G-100 column equilibrated with 0.01 M phosphate buffer (pH 7.0). The enzyme fraction was loaded on to a hydroxylapatite column which had been equilibrated with 0.01 M phosphate buffer (pH 7.0). Elution was carried out increasing phosphate concentration of the washing solution stepwise. The enzyme was detected in a fraction which was eluted with 0.05 M phosphate buffer (pH 7.0). Finally this fraction was passed through a Sephadex G-100 column. No. 221 enzyme was crystallized in 30% saturation of ammonium sulphate.

Purification of *B. circulans* enzyme. The enzyme was purified by the method described in the paper by Horikoshi *et al.* [2] and further purified by passing through a Sephadex G-50 column equilibrated with Tris-HCl buffer (pH 7.0) containing 0.1 M NaCl. The enzyme fraction was collected and brought to 70% saturation with ammonium sulphate. The precipitate was collected, dissolved in water and dialysed against 0.02 M Tris-maleate buffer (pH 6.0) containing 10 mM $CaCl_2$ The following experiments were carried out by using these purified preparations.

B. *Physical Properties of the Enzymes*

The purified enzymes showed only one peak throughout the duration of a run of 90 min at 55,430 r.p.m. The sedimentation constant was measured in 0.05 M Tris-HCl buffer (pH 7.2). Ultraviolet absorption spectrum of the enzymes showed a typical UV absorption of protein. The ratio of absorption at 280 nm to 260 nm was about 1.8 which indicated the absence of nucleic acid. An estimation of the molecular weights of the enzymes were made using SDS-polyacrylamide gel electrophoresis method. Only one band was observed in the gel. The Ampholine electrofocusing showed that a single peak of the activity was detected. These results are summarized in Table I.

TABLE I
Properties of β-1,3-glucanases

	B. circulans	No. 221	No. K-12-5
Molecular weight	28,000	36,000	40,000
S_{20}	3.0	3.2	3.6
Isoelectric point	5.4	4.1	3.5
Optimum pH	5.6	8.0	6-7.0
Specific activity ($U/A_{280\,nm}$)	17	11	15

C. *Effect of pH on Activity*

Figure 1 shows the effect of pH on the enzyme activity. The pH was adjusted with 0.05 M acetate buffer (pH 3-5.5); 0.05 M Tris-maleate buffer (pH 6-8.5); Tris-HCl buffer (pH 7.5-9) and 0.05 M glycine-MaOH-NaCl buffer (pH 9-11). Other conditions were the same as those of the standard assay method. *B. circulans* is most active at pH 5.6; No. 221 at pH 8 and K-12-5 at pH 6-8.

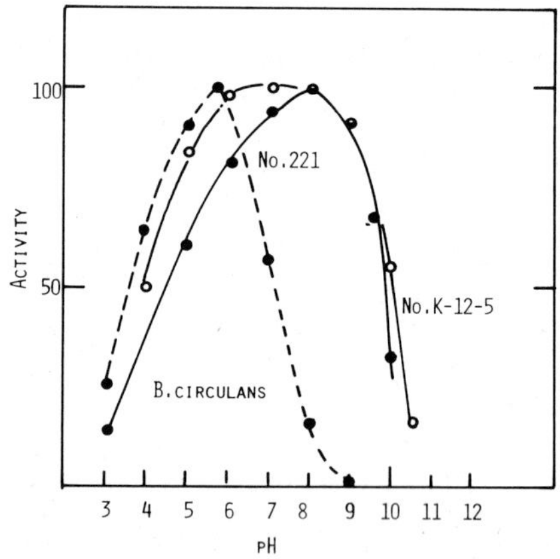

Fig. 1. Effect of pH on enzyme activity.

D. *Effect of pH on Stability of the Enzyme*

Stability of the enzyme was investigated in buffer solutions of various pH values. The mixture was incubated at 65° C for 10 min and the residual activity was determined. The following buffer systems were used: 0.05 M Tris-maleate buffer (pH 5-8.5) and 0.05 M Tris-HCl buffer (pH 7.5-9.5). *B. circulans* enzyme was incubated in the presence of 10 mM $CaCl_2$. No. 221 and No. K-12-5 enzymes are most stable at pH 7 and *B. circulans* enzyme, pH 6.

E. *Thermal Stability of the Enzymes*

The enzyme was dissolved in 0.05 M Tris-HCl buffer (pH 7.0) (for *B. circulans* enzyme, Tris-maleate buffer, pH 6.0, 0.05 M), heated at the indicated temperatures for 10 min and the residual activity measured. Measurement was done in the presence or absence of Ca^{2+} (10 mM). The results are shown in Fig. 2.

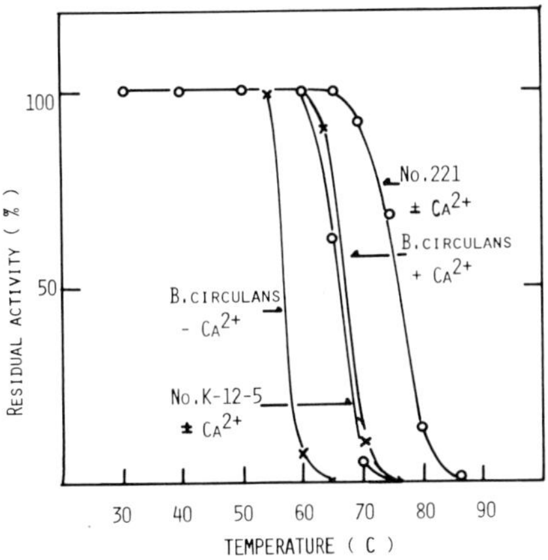

Fig. 2. Effect of temperature on the stability of the enzymes.

F. *Mode of Action of the Enzymes*

Hydrolysis of laminaribiose. About 1 mg of laminaribiose was dissolved in about 0.05 ml of the enzyme solution (1 U in 10 mM Tris-maleate buffer, pH 5.6 or pH 8.0) and incubated for 10 h at 40° C. Hydrolysis of laminaribiose was measured by glucose oxidase method. No remarkable liberation of glucose was observed.

Hydrolysis of laminaritriose. Laminaritriose solution (0.15 ml, 150 μg) was mixed with 0.05 ml of the enzyme solution described above and incubated for 0, 5, 30, 60 and 120 min at 40° C. Glucose released was measured by glucose oxidase method. As shown in Fig. 3, *B. circulans* enzyme hydrolysed laminaritriose and yielded glucose and laminaribiose more rapidly than others. No. 221 enzyme seemed to be inactive to laminaritriose under the tested condition.

Hydrolysis of laminaritetraose. About 1 mg of the sample was dissolved in about 0.05 ml of the enzyme solution described above and incubated for 2 h at 40°C. The reaction mixtures were chromatographed on Whatman No. 1 paper. *B. circulans* enzyme hydrolysed the sample and yielded glucose, laminaribiose and laminaritriose. No. 221 and No. K-12-5 enzymes also hydrolysed

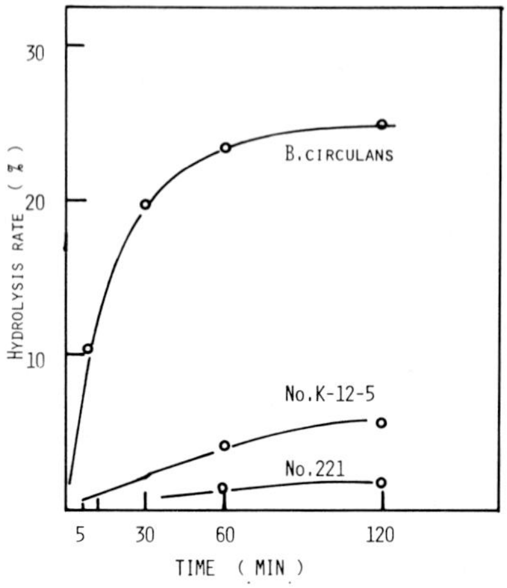

Fig. 3. Hydrolysis of laminaritriose by β-1,3-glucanases.

laminaritetraose and yielded glucose and laminaritriose, not laminaribiose.

Hydrolysis of laminarin. About 10 mg of laminarin was dissolved in 1 ml of water and its pH was adjusted to 5.6 and 8.0. To the laminarin solution was added 0.1 ml of the enzyme (11 U) and the mixture was incubated at 40°C. Aliquots of the reaction mixture were periodically withdrawn and chromatographed. If the enzymes tested were single entity, they were endo-β-1,3-glucanases. Maximum hydrolysis rates measured by reducing power are as follows: *B. circulans* enzyme, 80%; No. 221, 63%; No. K-12-5 enzyme, 68%.

G. *Hydrolysis of Fungal Cell Walls*

Fungal cell walls were mixed with 0.5 ml of buffer (pH 8.0 Tris-HCl buffer, 0.05 M, for No. 221 and No. K-12-5 enzymes; pH 5.6 acetate buffer, 0.05 M for *B. circulans* enzyme; pH 5.0 acetate buffer, 0.05 M, for *Basidiomycetes* QM 806 [14]) containing 1.5 U of the enzyme and incubated for 30 min at 40°C. Hexoses liberated from the cell walls were determined by anthrone method. As shown in Table II, *Mucor* sp. is not hydrolysed by the enzymes tested. No. 221 enzyme did not hydrolyse the cell walls after long incubation (3 h).

III. Discussion

Three strains of *Bacilli* were used for comparative studies of β-1,3-glucanases: *Bacillus circulans* IAM 1165, from which PR factor

TABLE II
Hydrolysis of cell walls by β-1,3-glucanases

Enzymes	Cell walls *Mucor* sp.	*Fusarium* sp.	*Asp. oryzae*
B. circulans	1.2%	8.1%	7.2%
No. 221	0	0	0
No. K-12-5	0	0.5	2.5
QM 806	—	9.5	8.2

was isolated [15] ; No. 221 and No. K-12-5 which belong to alkalophilic *Bacilli*. The β1,3-glucanases of these bacteria were extensively purified and their properties were compared. As shown in Table I, physical properties are different between *B. circulans* enzyme and alkalophilic *Bacilli* enzymes. Most remarkable point is effect of Ca^{2+}; *B. circulans* enzyme is stabilized by calcium ion but others are not. Actions of these enzymes to laminaritriose are completely different. *B. circulans* enzyme can hydrolyse laminaritriose, but the others either do not do so or only weakly hydrolyse it. Enzymes No. 221 and No. K-12-5 can also hydrolyse laminaritetraose and yield glucose and laminaritriose, not laminaribiose. Therefore, maximum hydrolysis rates are different, with *B. circulans* enzyme being the highest. Enzymatic hydrolysis of cell walls of fungi were also investigated. *B. circulans* enzyme hydrolysed the cell walls of *Asp. oryzae* and *Fusarium* sp. extensively. However other enzymes could not hydrolyse the cell walls tested. The purified β-1,3-glucanase from *Basidiomycete* QM 806 [14], kindly donated by Dr. Kikumoto, Taito Company, Tokyo, also had strong lytic activity towards *Fusarium* sp. and *Asp. oryzae* cell walls and released only glucose. These results suggest that β-1,3-glucanase which can hydrolyse laminaritriose may have lytic activity to fungal cell walls. Further works on this point are in progress.

References

1. Horikoshi, K. and Iida, S. (1958). *Nature, Lond.*, **181**, 917.
2. Horikoshi, K., Koffler, H. and Arima, K. (1963). *Biochim. biophys. Acta*, **73**, 268.
3. Horikoshi, K. and Iida, S. (1959). *Nature, Lond.*, **183**, 186.
4. Horikoshi, K. and Iida, s (1964). *Biochim. biophys. Acta*, **83**, 197.
5. Horikoshi, K. (1967). *J. Agr. Chem. Soc. Japan*, **41**, R21.
6. Horikoshi, K. (1971). *J. Agr. Chem.*, **35**, 1407.

7. Horikoshi, K. (1971). *J. Agr. Chem.*, **35**, 1783.
8. Horikoshi, K. (1972). *J. Agr. Chem.*, **36**, 285.
9. Kurono, Y. and Horikoshi, K., in preparation.
10. Sumner, J. R. and Somers, G. F. (1944). *Laboratory Experiments in Biological Chemistry.* Academic Press, New York and London.
11. Warburg, O. and Christian, W. (1942). *Biochem. Z.*, **310**, 384.
12. Vesterberg, O. and Svesson, H. (1966). *Acta Chem. Scand.*, **20**, 820.
13. Davis, B. J. (1964). *Ann. N.Y. Acad. Sci.*, **121**, 404.
14. Huotari, F. I., Nelson, T. E., Smith, F. and Kirkwood, S. (1968). *J. biol. Chem.*, **243**, 952.
15. Nagasaki, S., Neumann, N. P., Arnow, P., Schnable, L. D. and Lampen, J. O. (1966). *Biochem. biophys. Res. Commun.*, **25**, 158.

Effect of Glucanases of Yeast and Bacterial Origin on Cell Walls of *Schizosaccharomyces* Species

GRAHAM H. FLEET and HERMAN J. PHAFF

Department of Food Science and Technology, University of California, Davis, California, U.S.A.

The cell wall composition of species of *Schizosaccharomyces* differs in several ways from that of *Saccharomyces* [1]. The fission yeasts appear to contain approximately 30-35% of an alkali-soluble, α-(1→3)-linked glucan, which precipitates upon neutralization of the alkali extract. Such a glucan has not been demonstrated in *Saccharomyces* species. Another component not found in the last mentioned genus is a galactomannan. Moreover *Schizosaccharomyces* species appear to contain a reduced amount (about 10%) of β-linked glucan of the type found in *Saccharomyces* spp. and they lack chitin altogether.

In view of the rapid lysis of the ascus walls of the species of *Schizosaccharomyces* it seemed likely that these organisms would synthesize active glucanase(s) which may be expected to be involved in this process as well as in conjugation between cells which precedes sporulation. High concentrations of exo-β-glucanases were shown to be present both intracellularly and extracellularly in *Kluyveromyces* (*Fabospora*) *fragilis* and in *Hansenula anomala* [2], two budding yeasts whose asci lyse rapidly. In *Sacch. cerevisiae,* whose asci do not lyse, the exo-β-glucanase level was low inside the cells and almost undetectable in the culture medium, suggesting a causal relationship between exo-β-glucanase level and lysis of the asci. The enzymes of the various species, which were extensively purified, hydrolysed both laminarin and pustulan (β-1,3 and β-1,6-linked glucans, respectively) in an endwise fashion. A single protein appears to be responsible for both activities. Abd-El-Al and Phaff [2] did not obtain any evidence for the presence of an endo-β-glucanase in these species. However, in *Hanseniaspora valbyensis* and *H. uvarum* they demonstrated a β-1,3 endoglucanase in cell-free extracts and in the culture medium, whereas an exo-β-glucanase appeared to be absent [3].

Evidence for the presence of an endoglucanase in multilaterally budding yeasts is fragmentary. Bacon *et al.* [4, 5] briefly mentioned a laminarinase associated with the cell wall of baker's yeast. Maddox and Hough [6] showed that a crude extract obtained by chloroform-induced autolysis of brewer's yeast at room temperature, possessed an enzyme(s) which hydrolysed laminarin with the intermediate formation of oligosaccharides. The crude enzyme solution, however, showed no hydrolytic activity on yeast cell wall glucan. Arnold [7] partially purified two glucanase fractions from a cell-free extract (obtained by passage through a French pressure cell) of baker's yeast. One appeared to have primarily endo-β-1,3-glucanase and the other endo-β-1,6-glucanase activity. Recent work by Matile and coworkers [8, 9] has revealed the existence of glucanase-containing vesicles which may be involved in cell wall modification during budding. These vesicles were isolated and shown to contain exo- and endo-β-1,3 glucanases as well as mannan and protein. The level of these enzymes was highest in the very early budding stage of the cell cycle [6, 9] and nearly absent from stationary phase cells.

Recently a report appeared [10] describing the presence of an endo-β-1,3-glucanase in the cell wall of *Schizosaccharomyces pombe*. No activity was detected in the culture medium and in a cell homogenate the activity was confined to the sedimentable fraction. Isolation and purification of the endoglucanase after autohydrolysis of wall preparations were unsuccessful due to the rapid inactivation of the solubilized enzyme.

Studies on the hydrolysis of cell walls of *Schizosaccharomyces* by bacterial enzymes [11] have shown that *Bacillus circulans* causes extensive clearing of *Schizosaccharomyces* cell walls incorporated in agar plates provided this bacterium is grown on these walls. However, if *B. circulans* is grown on baker's yeast cell walls the induced enzymes can clear agar plates containing *Saccharomyces* walls but plates containing *Schizosaccharomyces* walls show very incomplete clearing.

The work presented below was undertaken to characterize and purify the glucanases associated with *Schizosaccharomyces* cells. Some preliminary results are given on the action of a purified exo-β-glucanase and an endo-β-1,3-glucanase from *Sch. japonicus* var. *versatilis* on isolated cell walls. Glucanases produced by *Bacillus circulans* were used for comparison.

I. Methods and Materials

A. *Microorganisms and Culture Conditions*

The following cultures were obtained from the culture collection of the Department of Food Science and Technology, University of

California, Davis, under the numbers indicated. *Schizosaccharomyces pombe* C-277; *Schizosaccharomyces japonicus* var. *versatilis* 60-255 (for convenience referred to in the text as *S. versatilis*); *Saccharomyces cerevisiae* C-299; *Saccharomyces rosei* C-450; *Kluyveromyces fragilis* C-293; *Bacillus circulans* WL-12. Compressed baker's yeast (Red Star Yeast Co., Oakland, California) was purchased in 1 lb blocks at a local market. Yeast cultures were maintained on yeast autolysate (Albimi, Pfizer and Co., New York) 0.5%, glucose 5%, agar slants. For enzyme production and cell wall preparations the yeasts were grown in liquid medium of the above composition dispensed in 1-litre volumes in Fernbach flasks. These were incubated on a rotary shaker (about 200 r.p.m.) at room temperature. The cells were harvested in a refrigerated RC-2B Sorvall centrifuge. For large scale enzyme and cell wall preparations the yeasts were grown in 50-litre batches in an 80-litre fermentor (Stainless Steel Products, St. Paul, Minnesota, U.S.A.). A 10% inoculum of a 24 h culture was used and the medium was aerated at approximately 20 litres of air/min with stirring at 200 r.p.m The temperature of growth was maintained at 26° C and cells were harvested after 20 h by centrifugation in a large Sharples centrifuge. *Bacillus circulans* WL-12 was maintained on yeast autolysate-cell wall medium [12].

B. *Preparation of Cell Walls*

Walls of *S. pombe* for analytical studies were prepared in the following manner. The yield (1600 g wet weight) from 2 × 50 litre fermentor cultures was washed three times with 3-litre quantities of 0.1 M Tris buffer pH 8.5. The cells were finally suspended as a very thick slurry in the same buffer and broken by mechanical shaking with an equal volume of glass beads (0.45-0.50 mm in diameter) in a Braun homogenizer (Bronwill Scientific) for 2 minutes at 4000 oscillations per minute. Sufficient liquid CO_2 was delivered to the shaking chamber to prevent heating. Phase contrast microscopy generally indicated approximately 95% cell breakage. After disruption the homogenate was diluted with Tris buffer pH 8.5 so that the glass beads readily settled. The homogenate was poured off and the cell walls were sedimented by centrifugation in a Sorvall centrifuge at 1000 × g for 20 min. The walls were then washed three times in buffer and finally resuspended as a thick slurry for one additional treatment in the Braun homogenizer for 1 minute. This step resulted in the disruption of the remaining unbroken cells. The walls were recovered as described above and washed approximately 20 times with buffer and finally three times with distilled water.

Perchloric acid extracts [13] of the walls at this stage did not reveal any nucleic acids indicating that cytoplasmic contamination had been eliminated. The walls were then lyophilized and stored under vacuum at 1° C. The yield of purified walls was approximately 80 g or 18% based on the dry weight of harvested yeast. All steps in this procedure were carried out at 4° C or lower and, with the exception of the final washings, in the presence of 0.1 M Tris buffer pH 8.5. At this pH value and low temperature endogenous glucanase action on the cell walls was prevented.

C. *Preparation of Extracts and Walls for Enzymic Assays*

These materials were routinely prepared as above except that only one disruption in the homogenizer was performed (2-3 min) and 0.1 M sodium succinate buffer pH 5.5 was substituted for the Tris buffer. In addition the walls, after the first centrifugation, were washed only six times with buffer. They were then resuspended in buffer for enzyme assays. The supernatant fluid of the first wall centrifugation at 1,000 x g was recentrifuged at 20,000 x g for 30 min. This supernatant was then dialysed against 0.01 M sodium succinate buffer pH 5.5 and used as the intracellular extract.

D. *Substrates*

Laminarin was purchased from Nutritional Biochemicals Co. (Cleveland, Ohio, U.S.A.). It was poorly soluble in cold water but readily dissolved in hot water. Upon cooling, it remained in solution sufficiently long to allow enzyme assays without interference by precipitation. Laminaribiose, laminaritriose, laminaritetraose, pustulan, oat glucan, cellulose dextrins, and yeast mannan were obtained from the laboratory collection [2, 3]. A laminarin preparation modified at both ends of the molecule was prepared by periodate oxidation [14]. Pachyman, a primarily β-1,3 linked polysaccharide of high molecular weight, was purified from ground *Poria cocos* fruiting bodies [15, 16]. Pseudonigeran (α-1,3-glucan) was prepared by a modification of the procedures of Johnston [17] and Hasegawa *et al.* [18] from mycelium of *Aspergillus niger* NRRL 326. The product was a white powder $[\alpha]_D^{25}$ + 237.6; it gave the same infrared spectrum as published elsewhere [19]. No increase in reducing sugars could be detected when samples of the product were incubated in the presence of purified preparations of endo-β-1,3-glucanase, endo-β1,6-glucanase, exo-β-1,3-glucanase, or α-amylase. Pseudonigeran is insoluble in water. For enzyme assays this substrate was pretreated as follows. A known weight of the polysaccharide was dissolved in 1 N NaOH after which the solution was very slowly neutralized with 1 N acetic acid while stirring. At the point when the solution gels the glucan can be separated by gentle centrifugation, washed three times with distilled water and then suspended in buffer to the desired concentration. Aliquots of this suspension were used as substrate. If the α-1,3-linked glucan is prepared in this way it does not settle out during the enzyme assay and remains suspended, thus promoting maximum surface contact between enzyme and substrate.

E. *Enzyme Assays*

Routine assays were done as follows: 0.5 ml of glucan (10 mg/ml) in 0.1 M sodium succinate buffer at pH 5.5 or 5.0 was equilibrated at 30°C in a water bath. Enzyme solution (0.5 ml) separately equilibrated was then added. Reactions were stopped by the addition of 0.35 ml of 1.0 M Na_2CO_3 followed by 1.0 ml of distilled water. One-ml samples of the inactivated reaction mixtures were then used for the determination of sugar reducing equivalents [20]. The reaction rates were linear over the first 30 min, during which at least three samples were analysed. For enzyme preparations of very low activity longer reaction periods were used. One unit of activity represents the amount of enzyme that releases 1 μmole of reducing equivalents (expressed as glucose) per minute under the above conditions. Proteins for specific activity measurements were determined by the method of Lowry *et al.* [21]. When glucanase was present in cell wall preparations the mixture of walls and substrate was gently agitated during the assay. Cell walls were removed by centrifugation *after* reaction with the alkaline copper reagent [20].

II. Results

A. *Survey of Glucanases in Fission Yeasts*

Prior to the appearance of the publication by Barras [10] we had also noted the very low stability of β-1,3-glucanase activity in extracts of *S. pombe,* a phenomenon also known for extracts of *Hansenula anomala* but not for *Kluyveromyces fragilis* or *Sacch. cerevisiae* [2]. We therefore compared the glucanases produced by *S. pombe* with those from *S. versatilis,* another species of the fission yeasts whose abundantly formed asci lyse very rapidly after their formation, which in turn is preceded by conjugation.

Intracellular extracts, cell wall preparations and culture filtrates of

S. pombe and *S. versatilis* were therefore assayed for activities of β-1,3-glucanase (hydrolysis of laminarin), β-1,6-glucanase (hydrolysis of pustulan) and α-1,3-glucanase (hydrolysis of pseudonigeran). As seen in Table I the extract, walls and culture filtrate of *S. versatilis* showed substantial β-1,3-glucanase activity. Low levels of β-1,6-glucanase activity were observed in the cell extract and culture filtrate from this yeast but none was detected in the wall fraction. Extracts and walls of *S. pombe* exhibited much lower levels of β-1,3-glucanase activity than in *S. versatilis.* A very low level of β-1,6-glucanase activity was apparent in the extract of *S. pombe* but none could be detected in the cell wall preparation. No glucanase activity could be detected in the culture filtrate of *S. pombe.* Attempts to demonstrate the presence of an α-1,3-glucanase in any of the preparations were unsuccessful.

Two other species of *Schizosaccharomyces* were briefly surveyed. *S. malidevorans* showed somewhat lower enzyme levels than *S. versatilis,* while *S. octosporus* produced glucanases in concentrations of the order of *S. pombe.*

Apart from the low level of enzymes, extracts of *S. pombe* presented further complications. The undialysed intracellular extract exhibited progressive increases in sugar reducing equivalents on standing, thus complicating glucanase assays. This problem (apparently due to an endogenous trehalose-trehalase interaction) could be eliminated by dialysis of the extract. However, this resulted in an appreciable loss of the glucanase activity. Storage of the undialysed extract also resulted in rapid activity losses (approximately 50% overnight at 4° C). Extracts of *S. versatilis* did not suffer from these problems, the glucanase activities being quite stable to storage and dialysis. Surprisingly, mixing of *S. pombe* extract with *S. versatilis* extract followed by storage did not result in such losses of the glucanase activity. *S. versatilis* was chosen for a more detailed study of the glucanases.

When grown in yeast autolysate-glucose medium *S. versatilis* passed through a series of morphogenic stages. Initially, normal vegetative growth occurs where the cells divide by fission. This is followed by a stage where the cells in the culture clump together. Microscopic observation of the culture at this point reveals that the cells are undergoing extensive conjugation and zygote production. Ascus development and sporulation then occur and the spores are eventually released into the medium by lysis of the asci. Preliminary analysis of intracellular extracts of cultures at the different stages of growth showed that glucanase activity on the basis of dry weight of

TABLE I

A comparison of glucanase activities in various fractions of *S. versatilis* and of *S. pombe*. The organisms were grown under similar conditions as detailed in Section I. All protein values were determined after dialysis

Enzyme source	*S. versatilis*				*S. pombe*			
	β-1,3-glucanase	β-1,6-glucanase	α-1,3-glucanase	Trehalase	β-1,3-glucanase	β-1,6-glucanase	α-1,3-glucanase	Trehalase
Intracellular extract	0.630[a]	0.008[a]	0	0.0016[a]	0.0019[b]	0.0021[b]	0	0.0088[a]
Culture filtrate	0.15[c]	0[d]	0		0	0	0	
Wall preparation[e]	4.3	0	0	0	0.33	0	0	0

[a] Units/mg of protein after dialysis. [b] Activity determined before dialysis, protein measured after dialysis. [c] Activity in units/ml. [d] A low activity could be measured after concentration. [e] Units/g of wall (dry weight basis).

cells was about three times as high during the stage of conjugation and zygote development as compared to cells in the exponential phase of growth. Brock [22] observed a similar increase in conjugating *Hansenula wingei* cells.

B. *Nature of the Intracellular β-1,3-glucanase*

The dialysed intracellular extract was incubated with laminarin and the products of the reaction examined chromatographically. Glucose was the only product observed suggesting that the enzyme acted by terminal release of glucose units from the substrate, i.e. an exo-mechanism. Further evidence for this mode of attack can be seen from Table II. Laminarin, in which the non-reducing as well as the

TABLE II
Evidence for the exo-mechanism of action of the β-1,3-glucanase in the dialysed intracellular extract of *S. versatilis*

Substrate	Reducing group release μg of glucose equiv./ml of enzyme		Glucose release μg of glucose/ml	
5 mg/ml	15 min	30 min	15 min	30 min
Laminarin[a]	416	824	424	850
Periodate-oxidized laminarin[a]	8.0		12.0	
	60 min	180 min		
Oat glucan[b]	0	0		

[a] Assays performed under identical conditions. [b] Reaction mixture contained 10 x the enzyme concentration as compared to (*a*).

reducing ends of the molecule had been modified by periodate oxidation, was not hydrolysed. The very small increase in reducing value found occurred during the initial stage of the reaction only and may represent incompletely oxidized laminarin. Oat glucan which is known to be cleaved by certain endo-β-1,3-glucanases [12] showed no increase in reducing value when incubated with 1 unit/ml of the enzyme for 3 h. Calculation of the glucanase activity in the crude extract of *S. versatilis* by the separate methods of reducing group formation and specific glucose release (glucose oxidase reagent) gave essentially the same values (Table II).

It thus appears that the cell extract prepared with a Braun cell homogenizer contains only exo-glucanase activity.

C. *Purification of the Intracellular and Extracellular exo-β-glucanases of* S. versatilis

The intracellular extract was obtained from the growth in a 50-litre culture of the yeast harvested during the flocculation and conjugation stage. All of the subsequent steps are summarized in Table III.

The centrifuged extract was dialysed against 5 mM sodium succinate buffer pH 5.5 and lyophilized. No loss of activity occurred during this step. All of the dry powder was reconstituted in distilled water giving 91 ml of solution after centrifugation at 10,000 × g for 20 min. Ten-ml samples of this supernatant liquid were applied to a Sephadex G-100 column (2 × 100 cm) equilibrated with 0.05 M sodium succinate buffer pH 5.5. Elution was carried out with the same buffer. β-1,3-glucanase was eluted in a single peak. The active fractions were combined and dialysed against 0.01 M succinate buffer pH 5.5. The enzyme was then loaded onto a DEAE-cellulose column (1.5 × 25 cm) equilibrated with 0.01 M sodium succinate buffer pH 5.5. Under these conditions of application much inactive protein did not adsorb to the column. Elution was performed in two steps: (i) a linear gradient from 0.01 M sodium succinate buffer pH 5.5 to 0.02 M sodium succinate buffer at pH 5.5 containing 0.2 M NaCl in a total volume of 400 ml and (ii) a linear gradient from 0.02 M succinate–0.2 M NaCl buffer pH 5.5 to 0.02 M succinate–0.75 M NaCl buffer pH 5.5 in a total volume of 900 ml. The first gradient served to remove substantial contaminating material while the β-1,3-glucanase activity was eluted during the second gradient at 0.33 M NaCl. The data in Table III show an approximately 420-fold purification of the enzyme.

The extracellular β-1,3-glucanase was purified by a variation of the above procedure. Because of the large volume of relatively dilute enzyme solution, the culture filtrate (20 litres) was concentrated by ultrafiltration rather than lyophilization. This was done in an Amicon ultrafilter cell model 2000 (Amicon Corporation, Lexington, Mass.) with an associated reservoir and a PM- 10 membrane. The filtrate was concentrated to 600 ml over a three day period at 1°C without loss in enzyme activity. This procedure avoided the problem of dialysing large volumes of the culture filtrate. The concentrated enzyme solution was then dialysed against 0.01 M sodium succinate buffer at pH 5.5, which resulted in some volume increase. It was found that a better purification was achieved if the DEAE- cellulose chromatography preceded the Sephadex gel filtration step. The

TABLE III

Summary of various steps applied in the purification of intracellular exo-β-glucanase from *Sch. japonicus* var. *versatilis*. Activity units were determined with 0.4% laminarin at pH 5.5 as the substrate. The total eluate from the Sephadex column was the result of nine separate applications.

Step	Volume (ml)	Concentration of protein (mg/ml)	Exo-β-glucanase activity (units/ml)	Specific activity (units/mg of protein)	Purification	Yield (%)
Dialysed intracellular preparation	3000	1.71	0.73	0.43		100
Lyophilization reconstitution and centrifugation	91	47.6	24.74	0.52	1.21	103
Sephadex G-100	810	0.17	2.03	11.95	27.8	75
DEAE-cellulose[a]	(I) 700	0.006	1.08	180.0	418.6	34.4
	(II) 100	—	0.088	—	—	0.4

[a] Two peaks (I and II) representing enzyme activity were observed. The minor peak II, representing only traces of activity, was not studied further.

reason for this change was that the spent yeast autolysate glucose medium contained a dark brown pigment which was largely retained by the ultrafilter and could not be separated satisfactorily from the enzyme by chromatography over Sephadex G-100. On DEAE-cellulose, however, an almost complete separation was achieved and a more satisfactory resolution then took place on Sephadex. Elution properties of the enzyme were identical to those of the intracellular enzyme. Table IV summarizes the results of the purification steps.

D. *Properties of the Purified exo-β-1,3-glucanases*

Substrate specificity. A number of substrates were incubated in the presence of the purified enzymes in 0.1 M sodium succinate buffer at pH 5.5 and release of reducing groups was determined. As seen in Table V, apart from hydrolysing laminarin and its oligosaccharide derivatives (laminaribiose and laminaritriose), the enzymes also hydrolysed the β-1,6-glucan, pustulan, and the corresponding disaccharide gentiobiose (the latter very weakly). In all cases the only reducing sugar produced was glucose as determined by paper chromatographic evidence. None of the other substrates was hydrolysed under the conditions employed. In addition, proteolytic and phosphatase activities were not present in measurable concentration in the enzyme preparations. The last two properties were measured since knowledge about their presence or absence is important if such enzymes are tested with yeast cell walls as the substrate (see next section).

The ability of the purified enzymes to hydrolyse both β-1,3 and β-1,6 glucans is consistent with observations on the exo-β-glucanases of other yeasts [2, 24]. The exo-β-glucanases used by the authors cited also hydrolysed p-nitrophenyl-β-D-glucoside as did the enzyme of *S. versatilis*. Evidence was also presented [2, 24] that a single protein possessed both β-1,3 and β-1,6 glucanase activities.

It appears from our work that the exo-β-glucanases of *S. versatilis* follow this same pattern. The ratio of the β-1,3-glucanase activity to that of the β-1,6-glucanase remained constant (approx. 100:1) throughout the various purification steps. This ratio applies to a laminarin concentration of 0.5% and pustulan concentration of 0.4% in the reaction mixtures.

We also studied the inhibition of the two activities by glucono-δ-lactone as described in detail by Abd-El-Al and Phaff [2]. Both activities were similarly inhibited with K_i values for each of 9.0×10^{-4} M glucono-δ-lactone.

TABLE IV

Summary of the various steps applied in the purification of extracellular exo-β-glucanase from *Sch. japonicus* var. *versatilis*. Activity units were determined with 0.5% laminarin at pH 5.5 as the substrate.

Step	Volume (ml)	Concentration of protein (mg/ml)	Exo-β-glucanase activity (units/ml)	Specific activity (units/mg of protein)	Purification	Yield (%)
Concentrated dialysed culture filtrate (see text)	715	1.2	2.82	2.35		100
DEAE cellulose	900	0.027	1.94	97.2	41.4	86.7
Concentration by ultrafiltration	18	0.85	81.4	95.8	40.8	72.6
Sephadex G-100[a]	630	0.008	1.75	218.7	93.1	54.6

[a] Eluate of several columns.

TABLE V
Substrate specificity of the exo-β-glucanases of *S. versatilis*

Substrate	Main linkage	Activity observed[a]
Laminarin	β-1,3	+
Laminaribiose	β-1,3	+
Laminaritriose	β-1,3	+
Pustulan	β-1,6	+
Gentiobiose	β-1,6	Very Weak
Cellulose dextrins	β-1,4	—
Oat glucan	β-1,3; β-1,4	—
Pseudonigeran	α-1,3	—
Starch	α-1,4	—
Yeast mannan	α-1,6; α-1,3; α-1,2	—
Trehalose	α-1,1	—
Methyl β-D-glucoside		—
phenyl β-D-glucoside		—
p-nitrophenyl-β-D-glucoside		+
Denatured hemoglobin[b]		—
Sodium pyrophosphate[c]		—
β-glycerol phosphate[c]		—

[a] Purified preparations of both the intracellular and extracellular enzymes were tested separately. Substrates (5 mg/ml) were incubated in the presence of 1 unit/ml of enzyme for 3 h. [b] Released amino acids measured by ninhydrin. [c] Phosphate release measured according to (23).

Heat denaturation curves for both activities were also similar. The enzyme was incubated in 0.05 M sodium succinate buffer at pH 5.0 for 30 min at 50, 53, 55 and 60°C. At 50°C there was virtually no loss of activity, whereas at the higher temperatures the two activities were inactivated at the same rate at each temperature. At 60° C over 90% of the activity was destroyed in 5 min. Electrophoretic evidence is currently under study to confirm the unienzymatic nature of the two activities.

Kinetic properties. Values for K_m and V_{max} were determined for both laminarin and pustulan as the substrates. Both the intracellular and extracellular enzyme preparations were found to have identical parameters. The data are presented in Table VI. For comparison the kinetic parameters of the exo-β-glucanases in other yeasts [2] are also included. It is noteworthy that the exo-β-glucanase of *S. versatilis* has comparatively low affinity for pustulan.

The pH optimum of activity on laminarin was 5.0. The activities were measured with sodium citrate buffer from pH 3-5.5, with sodium succinate buffer from pH 4.5-6.0 and with sodium phosphate

TABLE VI
Properties of the exo-β-glucanases from several yeast species

Source of purified enzyme	V_{max} (μmoles of glucose /min/mg protein)		K_m (mg/ml)	
	Laminarin	Pustulan	Laminarin	Pustulan
S. versatilis	344.8	52.6	6.25	166.6
S. cerevisiae[a]	9.5	3.8	14.0	3.6
K. fragilis[a]	82.6	7.4	1.2	1.8
H. anomala[a]	54.0	7.1	5.0	5.9

[a] From Abd-El-Al and Phaff (*2*)

from pH 5.5-8.0. There was no noticeable effect of the different anions on the reactions.

E. *Action of the exo-β-glucanase on Cell Walls of* S. pombe

Purified exo-β-glucanase was studied for its ability to digest cell walls of *S. pombe.* Initial studies were done with agar plates into which a lawn of cell walls had been incorporated [12]. Enzyme solution (1 unit/ml) placed into small cups in such plates did not cause any visible clearing or lysis of the walls at pH 5.0.

The action of exo-β-glucanase on wall suspensions of *S. pombe* was studied next. Walls were suspended in distilled water, heated for 2-3 min at 100° C to inactivate endogenous glucanases and then uniformly dispersed by ultrasonic vibration for 1-2 min. Enzyme in 0.1 M sodium succinate buffer pH 5.0 was then added to give a final enzyme concentration of 1 unit/ml and a cell wall concentration of 5 mg/ml. Sodium azide (0.01%) was added to prevent microbial contamination. This inhibitor did not have any effect on the activity of the exo-glucanase. The suspensions were gently agitated on a Rollordrum (New Brunswick Scientific Co.). Samples were removed as a function of incubation time for determination of reducing sugar equivalents. The enzyme remained active throughout the incubation. A parallel series of experiments were performed where the walls had been pretreated with 0.02 M dithiothreitol (DTT) for 2 h at 30° C and then washed free of this sulfhydryl reagent. As seen from Table VII there was very little release of reducing equivalents upon incubation, suggesting a limited ability of the enzyme to attack the glucan in the walls. It would appear that the non-reducing terminals of the wall glucan are inaccessible to the enzyme. The small initial

TABLE VII

Action of purified exo-β-glucanase (1 unit/ml except as noted) from *S. versatilis* on cell walls of *S. pombe* and on β-glucans. All substrates were used at concentrations of 5 mg/ml. Temperature 30°C. The values represent μmoles of glucose produced per ml.

Reaction mixture	Incubation time (h)						
	0	2	6	10	22	26[a]	22
Walls, control no enzyme	0.050	0.050	0.050	0.050	0.050	0.050	0.050
Walls + enzyme	0.050	0.220	0.254	0.250	0.280	0.254	0.060
Walls pretreated with DTT + enzyme	0.050	0.170	0.220	0.230	0.250	0.250	
Laminarin + enzyme[b]	0.340	10.72					
Pachyman + enzyme[c]	0.117	0.352	0.810				

[a] At this time the walls were retrieved by centrifugation, washed, and resuspended in fresh enzyme and incubated for a further 22 h. [b] Enzyme concentration approx. 0.066 unit/ml. [c] Enzyme concentration approx. 0.33 unit/ml.

increase in reducing equivalents is probably due to digestion of some extensively degraded fragments of walls present in the suspension. Pretreatment of the walls with DTT did not provide improved access of the enzyme to the walls. For comparison, Table VII also includes the release of reducing equivalents by the exo-β-glucanase from laminarin and from the insoluble β-1,3-glucan, pachyman. It is evident that with the concentrations of enzyme given, the soluble laminarin releases glucose at a much more rapid rate than does the insoluble pachyman.

F. *Glucanase Associated with the Cell Walls of* S. versatilis

When freshly prepared washed cell walls of *S. versatilis* or *S. pombe* were incubated at 30°C in the presence of 0.05 M succinate buffer pH 5.5 a substantial release of reducing groups was observed. If these wall preparations were incubated in the presence of laminarin an additional release of reducing sugars occurred. In both cases, walls of *S. versatilis* showed much greater reducing group release than walls of *S. pombe.* Preliminary paper chromatographic analyses of the products of such reactions showed release of glucose and oligosaccharides of the laminarin series. These results suggested the association of an endo-β-1,3-glucanase with the cell wall preparations. Because of the much higher enzyme level associated with the walls of *S. versatilis* (Table I) this yeast was studied in more detail.

Autohydrolysis: Cell walls were freshly prepared from *S. versatilis* (see Section I) and washed six times in 0.05 M succinate buffer pH 5.0. A portion of the walls was then suspended in the same buffer and incubated at 30° C with occasional shaking. Five-ml volumes were removed as a function of time and the cell walls sedimented by centrifugation. The supernatant was carefully removed and stored. The wall pellet was washed with 2 ml of buffer and recentrifuged. This supernatant was added to the previous one. The walls were resuspended in buffer to 5 ml and assayed for glucanase activity with laminarin as the substrate. Samples of the supernatant solution were used for determination of reducing equivalents and for paper chromatographic analysis, while the remainder was dialysed against 0.01 M succinate buffer pH 5.0. After dialysis the solutions were assayed for β-1,3-glucanase activity and protein. The results of a typical experiment are shown in Table VIII. As seen from the data presented glucanase associated with the wall is gradually released as incubation continues and an approximately corresponding activity is recovered in the supernatant. This quantitative transition is

TABLE VIII
Autohydrolysis of washed cell walls of *S. versatilis*. Cell wall concentration was 21.5 mg/ml (dry weight basis). Temperature 30°C.

Incubation time (h)	Total glucanase units in wall fraction per 5 ml	Total glucanase units in dialysed supernatant per 5 ml	Protein release mg/ml	Reducing sugar release μmoles of glucose equivalents/ml
0	0.500	0.070	0.045	0.12
3	0.460	0.094	0.082	4.10
6	0.410	0.124	0.110	7.70
12	0.330	0.186	0.120	12.80
24	0.250	0.270	0.130	13.05
36	0.210	0.240	0.135	

accompanied by a release of reducing sugars and protein into the supernatant (last two columns). Release of the enzyme from the wall never went to completion, even upon extended incubation. However, some further release could be achieved if the walls were centrifuged and resuspended into fresh buffer. One possibility is that the accumulation of hydrolysis products progressively inhibit the reaction. We were unable under any of the conditions tried to extract more than 80% to 90% of the enzyme.

The supernatant solutions after several hours of wall hydrolysis also contained some non-dialysable alcohol precipitable material. The nature of this component, presumably a polysaccharide, is not yet known. It is interesting to note that, even after 24-48 h of self digestion with a change of buffer, microscopic inspection shows the walls to be intact. Some visible digestion is apparent after 4-5 days of incubation but the walls never completely disappear as do the ascus walls after sporulation.

Pattern of action of the endo-glucanase. The reducing sugars released into the supernatant solution were analysed by paper chromatography. Chromatograms were run for 20 h on Whatman No. 1 paper with the solvent system ethylacetate, pyridine, water in a ratio of 12:5:4. Sugar positions were detected by the alkaline silver nitrate method [25]. Standards of glucose, laminaribiose, laminaritriose, laminaritetraose along with the products of hydrolysis of laminarin and pachyman with a purified bacterial endo-β-1,3-glucanase were used. The pattern of action of fresh *S. versatilis* walls on laminarin was studied concurrently. After 90 min of autohydrolysis, spots with R_{glc} of 0.839, 0.636, 0.475 and 0.338 were visible. These corresponded to laminaribiose, -triose, -tetraose and -pentaose. A spot corresponding to glucose was also present. As autohydrolysis proceeded the glucose spot became more predominant along with spots corresponding to laminaripentaose and laminarihexaose. The latter two spots were very prominent after 12 h of autohydrolysis and remained as such even after 36 h. A similar pattern of products was obtained when walls were incubated in the presence of laminarin. In this case the oligosaccharides as well as glucose could be detected after 30 min of incubation.

Walls of *S. pombe* during autohydrolysis and upon incubation with laminarin gave the same release of oligosaccharides. However, glucose was not produced during the early stages of hydrolysis and the larger oligosaccharides did not show the unexplainable tendency to accumulate as in *S. versatilis* digests during the later stages. The spots representing laminaribiose, -triose and -tetraose were most prominent with this yeast.

These results suggest that the wall associated β-1,3-glucanase acts by the endo-mechanism, i.e. cleaves its substrate randomly. The early appearance of glucose in the products of the *S. versatilis* enzyme is a deviation from the conventional pattern of endo-glucanase action and this would tend to suggest that the walls of this yeast may be contaminated with a small amount of exo-β-glucanase. Further studies are required to substantiate such a possibility. At this time our endo-β-1,3-glucanase is not yet available in sufficient quantity or purity.

Attempts to elute endo-glucanase from the cell walls. Wall-associated enzymes such as the acid phosphatase and invertase of other species of yeast have been eluted from the walls by treatment with various reagents [7, 26, 27]. Although the glucanase associated with isolated walls of *S. versatilis* was solubilized during autohydrolysis it was thought that the application of various reagents might enhance the elution. Isolated walls were treated with the following reagents for periods up to 24 h: 0.5 M KCl; 1.0 M KCl; 0.05 M 2-mercaptoethanol; 0.05 M 2-mercaptoethanol-0.5 M KCl; 0.02 M dithiothreitol; 0.02 M dithiothreitol-0.5 M KCl. These compounds were dissolved in 0.05 M sodium succinate buffer pH 5.5. None of the treatments resulted in a release of wall glucanase greater than that by autohydrolysis. Next, two proteolytic enzymes, trypsin and subtilisin, were tested. Neither of these enhanced glucanase release, but, instead, caused inactivation.

It was felt that if the glucanase is attached to the wall by an enzyme-substrate interaction, substances which interfere or compete with this association might favour dissociation of such a complex.

Incubation of walls in the presence of high concentrations of soluble laminarin or glucono-δ-lactone, the latter possibly a competitive inhibitor of glucanase activity, did not enhance elution of the enzyme.

Purification of the wall-associated endo-β-1,3-glucanase. In view of the instability problems encountered by Barras [10] with the wall glucanase of *S. pombe* it was decided to attempt purification of the apparently more stable glucanase of *S. versatilis.* The results reported below represent our preliminary findings.

Freshly isolated walls of *S. versatilis* were permitted to undergo autohydrolysis in the presence of 0.05 M succinate buffer pH 5.0 at 30° C. After 24 h the walls were centrifuged and resuspended in fresh buffer for further incubation. This was repeated three times. The supernatant fractions containing the enzyme were combined, dialysed against 0.005 M succinate buffer pH 5.0 and concentrated by lyophilization. The powder was then reconstituted in distilled water and centrifuged to remove insoluble material. The enzyme remained soluble and no loss in activity was observed after all of these treatments.

Samples of the concentrated enzyme were then applied to a Sephadex G-100 column equilibrated with 0.05 M succinate buffer pH 5.0. The same buffer was used for elution and the enzyme came off the column as a single sharp peak (purification approximately 20-fold). After dialysis against 0.01 M sodium succinate buffer pH 5.0 the enzyme was adsorbed onto a DEAE-cellulose column equilibrated with the same buffer. The enzyme could be eluted as a single peak of activity with a linear gradient to 0.02 M sodium succinate-0.75 M NaCl buffer pH 5.0 (cf. Table III and text). The elution profiles of the enzyme during gel filtration and DEAE-cellulose chromatography were very similar to those of the exo-β-glucanase. Overall the enzyme was purified approximately 50-fold with only small losses in activity.

Samples of the purified enzyme were incubated with laminarin at pH 5.0 and the reaction products after various times were analysed by paper chromatography. The pattern of glucose and oligosaccharide release with laminarin was the same for the purified enzyme as was observed with the crude enzyme (see pattern of action above). It would seem that if the wall enzyme is contaminated with traces of the exo-β-glucanase the latter was not removed by the above purification steps. Further purification and specific tests for the presence of exo-β-glucanase are necessary to ascertain whether the presence of this enzyme is responsible for the rather early appearance of glucose among the reaction products.

G. *Wall-associated Glucanases in Other Yeasts*

As referred to in the introduction, there have been isolated reports on the occurrence of glucanases associated with *Saccharomyces* species [5, 6, 7, 9]. Abd-El-Al and Phaff [2] did not include cell wall preparations in their study of species of several other genera of yeast. It was therefore of interest to see if the occurrence of endo-glucanases might be more universal. Cell walls were isolated from several yeast species and assayed for the presence of β-1,3-glucanase. The procedure used was similar to that employed for the *Schizosaccharomyces* species. The results presented in Table IX show the association of various levels of β-1,3-glucanase activity with the walls of three other species studied quantitatively. For comparison the levels of wall enzyme in *S. versatilis* and in *S. pombe* are also shown. Chromatographic analyses of the products of wall autohydrolysis and the action of walls on laminarin revealed a release of oligosaccharides of the laminarin series. This indicates that the wall glucanases of these other yeasts also act by the endo-mechanism. When cell walls were incubated with laminarin a large increase in oligosaccharide release occurs as compared to autohydrolysis but with pustulan no such increase was noted. This experiment provides preliminary evidence that the walls of these yeasts do not possess activity on β-1,6-glucan. Isolated walls of commercial baker's yeast also exhibited an endo-β-1,3-glucanase activity. Isolated walls of *Pichia pastoris* underwent autohydrolysis (as measured by increase in reducing value) and hydrolysed laminarin, but the pattern of action was not studied in this case.

Thus all of the yeasts surveyed in the above manner show β-1,3-glucanase activity associated with the wall fraction and on the basis of our observations thus far, most or all of this represents an endo-β-glucanase.

TABLE IX

A comparison of the β-1,3-glucanases in the wall fractions and intracellular extracts of several yeast species. Assays were done with laminarin 5 mg/ml at pH 5.0 and 30°C.

Yeast	Isolated wall fraction	Dialysed intracellular extract
	Units/g of walls (D.W.B.)	Units/mg of protein
Sch. versatilis	4.3	0.630
Sch. pombe	0.33	0.002
Sacch. cerevisiae	0.225	0.041
Sacch. rosei	0.125	0.032
Kluyveromyces fragilis	1.73	0.187

Table IX also includes values for the level of intracellular glucanase in the different yeasts. In the case of *S. versatilis, S. cerevisiae* [2] and *K. fragilis* [2] these are exo-β-glucanases. Of interest is the fact that *S. versatilis* and *K. fragilis* have much higher levels of glucanases (both in the wall and intracellular) than the other yeasts. These two yeasts are known to release their spores rapidly from the ascus at maturity, whereas *Sacch. cerevisiae* and *S. rosei* do not release their spores. *Sch. pombe* forms an exception since its asci lyse when they are mature. The low levels of activity in this species may be connected with the instability of the glucanase complex in this yeast or with harvest of the cells during a stage in the cell cycle when the enzyme level is repressed.

H. *Action of Bacterial Glucanases on the Walls of* S. pombe

Results in the previous section have shown a β-1,3-glucanase associated with isolated walls of *S. pombe* and *S. versatilis*. Autohydrolysis of the walls of either yeast for extended periods never resulted in complete enzyme liberation and digestion of the walls. This observation could perhaps be explained by the fact that these yeasts have a high α-1,3-glucan content in their walls [1] and do not produce any apparent α-1,3-glucanase. This polysaccharide might possibly protect some of the insoluble β-glucan from being hydrolysed by the cell wall-associated glucanase and at the same time prevent quantitative leaching of the endo-β-glucanase from the walls. It was therefore of interest to have available a microbial α-1,3-glucanase to degrade pseudonigeran from cell walls.

Tanaka and Phaff [11] observed that if the bacterium *Bacillus circulans* WL-12 is grown on cell walls of *S. pombe* complete lysis of these walls takes place. Enzymes induced by growing this bacterium on *Sacch. cerevisiae* walls caused incomplete lysis and clearing. These results suggest the production of an α-1,3-glucanase by the bacterium when grown on cell walls of *S. pombe*.

Induction experiments. Experiments were done to determine if this is the case. *B. circulans* WL-12 was grown in yeast nitrogen base medium (Difco) containing *S. pombe* cell walls (0.5%) as the sole carbon source for growth. The medium was buffered with 0.1 M sodium phosphate at pH 6.5 [12]. Cultures were incubated with vigorous shaking at 30° C. Similar cultures were grown with baker's yeast cell walls (0.5%) and purified α-1,3-glucan (0.5%) as substrates. After 3 days of incubation the bacterial cells were removed by centrifugation and the culture filtrate dialysed against 0.01 M phosphate buffer pH 6.0. The filtrate was then assayed for the presence of three glucanase activities. The results are shown in Table X. Growth on *S. pombe* walls or on α-1,3-glucan caused the induction of an α-1,3-glucanase as well as β-1,3-glucanase and β-1,6-glucanase. Growth on baker's yeast cell walls did not result in the production of α-1,3-glucanase but on this substrate higher levels of β-1,3 and β-1,6-glucanases were produced. The low levels of α-1,3-glucanase produced in comparison to the other two enzymes can be attributed, at least in part, to a limitation of the assay procedure for an insoluble substrate (see Methods). Pseudonigeran (α-1,3-glucan) is insoluble at pH 6.0, whereas laminarin and pustulan are soluble. Another variable, not shown in Table X, is based on the observation that the relative concentrations of the various enzymes produced vary with culture age.

TABLE X
Production of extracellular glucanases by *Bacillus circulans WL-12* grown for 3 days at 30°C on different substrates. Assay substrate concentration 5 mg/ml; activity values in units/mg of protein at pH 6.0

Substrate	β-1,3-glucanase	β-1,6-glucanase	α-1,3-glucanase
Baker's yeast cell walls	0.43	3.24	0
Sch. pombe cell walls	0.138	1.25	0.131
Purified α-1,3-glucan	0.041	0.276	0.131

Pattern of action. The β-1,3 and β1,6-glucanases of *B. circulans* WL-12 have been shown to possess endo-hydrolytic actions on their respective substrates [12]. Next the α-1,3-glucanase was tested for its pattern of action. The filtrate was concentrated by lyophilization, reconstituted, and reacted with pseudonigeran. The products of hydrolysis were analysed by paper chromatography with a solvent of propane-l-ol-ethyl acetate-water (6:1:3). A control of acid hydrolysed pseudonigeran [17] was also chromatographed. A series of oligosaccharides was produced by the enzyme reaction showing that the α-1,3-glucanase possessed an endo-hydrolytic pattern of action.

Action on cell walls. Endo-β-1,3 and endo-β-1,6-glucanases produced by growth of *Bacillis circulans* WL-12 on baker's yeast cell walls have been extensively purified [12]. The purification procedure used here is a modification of the one cited [12] by including a column of

Sephadex G-100. Full details will be given in a later publication.

The purified bacterial β-glucanases were then examined for their abilities to digest purified cell walls of *S. pombe*. The techniques used were similar to those described for wall hydrolysis by the exo-β-glucanase (Table VII). The results and experimental details are given in Table XI. In contrast to the exo-β-glucanase, the endo-β-glucanases caused a steady release of reducing sugars from the wall suspensions. Endo-β-1,3-glucanase caused much more extensive hydrolysis of susceptible bonds than did endo-β-1,6-glucanase, indicating a greater proportion of linear β-1,3-linked chains than of β1,6-linked chains. When the two enzymes were used together there appeared to be more reducing sugar produced than by the individual enzymes because in the mixture each was used in one half the original concentration. Wall degradation by the enzymes did not appear to be enhanced by pretreatment with DTT. For comparison the table also includes the action of endo-β-1,3-glucanase on laminarin and pachyman and the action of endo-β-1,6-glucanase on pustulan.

Cell wall plates of *S. pombe* [12] showed only a very incomplete clearing in the presence of endo-β-1,3-glucanase or by a mixture of the two endo-β-glucanases. Endo-β-1,6-glucanase alone did not produce any visible clearing. Similar incomplete clearing was caused by crude preparations of the wall enzyme from *S. versatilis*. At this time insufficient purified α-1,3-glucanase was available to use this key enzyme for testing on the walls. Such experiments will be reported in a later publication.

III. Discussion

The work reported here was a continuation of earlier studies in this laboratory [2] on the occurrence and function of exo-β-glucanase in multilaterally budding yeasts. Since further work [3] revealed an endo-β-1,3-glucanase associated with two bipolarly budding species of *Hanseniaspora* it was of interest to study the association of glucanases with species of the fission yeast *Schizosaccharomyces*. An additional reason for considering this genus is the presence of high levels of α-1,3-linked glucan in the cell walls of its species [1].

Our work has shown that *Sch. versatilis* produces two kinds of β-glucanases in high concentration (i) an exo-β-glucanase in cell extracts and in the culture medium and (ii) an endo-β-1,3-glucanase, which appears to be associated only with the cell walls. The first of these two enzymes shows specificity for both β-1,3- and β-1,6-glucans, degrading them in an end-wise fashion, which

TABLE XI

Action of purified bacterial endo-β-1,3-, and endo-β-1,6-glucanases (1 unit/ml except as noted) on cell walls of *S. pombe* and on β-glucans. All substrates were used at a concentration of 5 mg/ml. Temperature 30°C. The values represent μmoles of glucose equivalents per ml.

Enzyme treatment	Incubation time (h)						
	0	1	3	8	16	26	48
β-1,3-glucanase	0.040	0.350	0.450	0.640	0.900	0.970	
β-1,6-glucanase	0.040	0.085	0.105	0.140	0.200	0.225	
β-1,3 + β-1,6-glucanase (0.5 unit per ml each)	0.040	0.320	0.410	0.580	0.730	0.900	1.05
β-1,3-glucanase[a]	0.040	0.240	0.275	0.360	0.500	0.600	0.880
β-1,6-glucanase[a]	0.040	0.087	0.100	0.125	0.165	0.200	0.275
β-1,3-glucanase on laminarin[b]	0.317	3.29					
β-1,3-glucanase on pachyman[b]	0.082	0.446					
β-1,6-glucanase on pustulan[b]	1.08	4.32					

[a] These walls had been pretreated with 0.02 M dithiotreitol for 24 h and washed free of this reagent.
[b] Enzyme concentrations were approx. 0.033 unit/ml.

confirms similar findings with this enzyme from other yeasts [2, 24]. The second enzyme is specific for β-1,3-glucans only. Thus far little detailed information has been published on the occurrence of endo-β-glucanases in yeasts [3, 5, 6, 7, 9]. For this reason we surveyed several other yeast species and found endo-β1,3-glucanases present in the walls of *Schizosacch. pombe, Sacch. cerevisiae, Sacch. rosei* (all in low concentration) and of *Kluyveromyces fragilis* (in high concentration).

Our results on the localization of endo-β-1,3-glucanase in yeast are not consistent with those of other investigators [7, 9]. Arnold [7] observed very low activities of endo-glucanase in cell extracts of *S. cerevisiae* prepared by shearing the cells in a French pressure cell. Matile and coworkers [8, 9] have demonstrated the existence of endo-β-1,3-glucanase within vesicles located inside the cells of baker's yeast and that the contents of these vesicles may be secreted into the cell wall. Our failure to demonstrate endo-β-1,3-glucanase in cell-free extracts could be due to several causes: (i) it is possible that the shearing action of a French pressure cell releases some wall-associated enzyme in contrast to a Braun homogenizer used by us, (ii) stage of the cell cycle at the moment of harvest [6, 9], and (iii) the possibility that endo-glucanase present in cell extract might become associated with the wall polysaccharides during the process of cell disruption. The question regarding the association of endo-β-1,3-glucanase with the cell wall *in vivo* might be resolved in studies with protoplasts of these yeasts. The lack or low level of action of exo-β-glucanase on isolated walls of the fission yeasts *in vitro* makes it difficult to suggest a role for this enzyme in the physiology of the yeast (cf. Table VII). The fact that the enzyme is excreted into the culture medium suggests that at least one of its functions is hydrolysis of β-1,3- or β-1,6-linked substrates in the medium. Storage substrates with β-linkages are not known to occur in yeasts as is known in *Euglena* [28]. Glucanases are thought to be involved in the budding process [9] in cell conjugation [22, 24] and in cell wall expansion [10, 29]. An endo-glucanase would seem to be a more likely candidate for these activities than an exo-glucanase. Our observation on the association of an endo-β-1,3-glucanase with the isolated wall fractions of all the yeast species so far studied tends to support this suggestion. Further investigations are needed in the direction of cellular localization of these enzymes and on how such enzymes might be regulated and controlled.

Thus far we have been unable to demonstrate any α-1,3-glucanase activity in *Schizosaccharomyces* spp. Based on the studies by

Zonneveld [30] on morphogenesis in *Aspergillus nidulans* and the high content of α-1,3-glucan occurring in the walls of the fission yeasts [1] such an enzyme might be expected to play a role in cell conjugation and other physiological activities.

References

1. Phaff, H. J. (1971). *In:* "The Yeasts—Physiology and Biochemistry of Yeasts", Vol. 2 (A. H. Rose and J. S. Harrison, eds). Academic Press, London and New York.
2. Abd-El-Al, A. T. H. and Phaff, H. J. (1968). *Biochem. J.*, **109**, 347.
3. Abd-El-Al, A. T. H. and Phaff, H. J. (1969). *Can. J. Microbiol.*, **15**, 697.
4. Bacon, J. S. D., Davidson, E. D., Jones, D. and Taylor, I. F. (1966). *Biochem. J.*, **101**, 36C.
5. Bacon, J. S. D., Gordon, A. H., Jones, D., Taylor, I. F. and Webley, D. M. (1970). *Biochem. J.*, **120**, 67.
6. Maddox, I. S. and Hough, J. S. (1971). *J. Inst. Brew.*, **77**, 44.
7. Arnold, W. N. (1972). *J. biol. Chem.*, **247**, 1161.
8. Matile, P., Cortat, M., Wiemken, A. and Frey-Wyssling, A. (1971). *Proc. natn. Acad. Sci. U.S.A.*, **68**, 636.
9. Cortat, M., Matile, P. and Wiemken, A. (1972). *Arch. Mikrobiol.*, **82**, 189.
10. Barras, D. R. (1972). *Antonie van Leeuwenhoek*, **38**, 65.
11. Tanaka, H. and Phaff, H. J. (1967). *In:* "Symposium on Yeast Protoplasts, Jena 1965", (R. Müller, ed.) Abhand. deutsch. Akad. Wissensch. Berlin, Klasse f. Medizin 1966, No. 6 Akademie-Verlag. Berlin.
12. Tanaka, H. and Phaff, H. J. (1965). *J. Bacteriol.*, **89**, 1570.
13. Gale, E. F. and Folkes, J. P. (1953). *Biochem. J.*, **53**, 483.
14. Hay, G. W., Lewis, B. A. and Smith, F. (1965). *In:* "Methods in Carbohydrate Chemistry", Vol. 5, p. 357. (R. L. Whistler, ed.). Academic Press, New York and London.
15. Tanaka, H., Ogasawara, N., Nakajima, T. and Tamari, K. (1970). *J. Gen. appl. Microbiol.*, **16**, 39.
16. Hoffmann, G. C., Simson, B. W. and Timell, T. E. (1971). *Carbohyd. Res.*, **20**, 185.
17. Johnston, I. R. (1965). *Biochem. J.* **96**, 659.
18. Hasegawa, S. and Nordin, J. H. (1969). *J. biol. Chem.*, **244**, 5460.
19. Bacon, J. S. D., Jones, D., Farmer, V. C. and Webley, D. M. (1968). *Biochim. biophys. Acta*, **158**, 313.
20. Spiro, R. G. (1966). *In:* "Methods in Enzymology", Vol. VIII Complex Carbohydrates (Neufeld, E. F. and Ginsburg, V., eds). Academic Press, New York and London.
21. Lowry, O. H., Rosebrough, N. J., Farr, A. L. and Randall, R. J. (1951). *J. biol. Chem.*, **193**, 265.
22. Brock, T. D. (1964). *J. Cell. Biol.*, **23**, 15A.
23. Fiske, C. H. and Subbarow, Y. (1925). *J. biol. Chem.*, **66**, 375.
24. Brock, T. D. (1965). *Biochem. biophys. Res. Commun.*, **19**, 623.
25. Treveleyan, W. E., Proctor, D. P. and Harrison, J. S. (1950). *Nature, Lond.*, **166**, 444.

26. Weimberg, R. and Orton, W. L. (1965). *J. Bacteriol.*, **90**, 82.
27. Sommer, A. and Lewis, M. J. (1971). *J. gen. Microbiol.*, **68**, 327.
28. Bull, A. T. and Chesters, C. G. C. (1966). *Adv. Enzymol.*, **28**, 325.
29. Johnson, B. F. (1968). *J. Bacteriol.*, **95**, 1169.
30. Zonneveld, J. M. (1972). *Biochim. biophys. Acta*, **273**, 174.

The Contribution of β-Glucanases to the Lysis of Fungal Cell Walls

J. S. D. BACON*
Department of Biochemistry, Macaulay Institute for Soil Research, Aberdeen, Scotland

Dedicated to the memory of David James Bell
(12 August 1905–3 April 1972)

The release of the protoplast is a result of the breaking of the cell wall. In theory, at least, this could simply consist of the splitting of the wall into a few large pieces, but in practice it seems more often to consist of the conversion of the water-insoluble wall into water-soluble material, a true lysis, or dissolving, of the wall. For this reason it is useful to consider why the cell walls of fungi are insoluble in water.

Solubility, as Pirie pointed out some thirty years ago [1], is a quality that must be defined in relation to the operation being carried out, and often for biochemists simply means failure to sediment noticeably in a particular centrifugal field. It has something to do with size as well as with interaction with solvent molecules.

A very simple answer to our question might be that the cell walls are composed entirely of material that is intrinsically water-insoluble, but the evidence so far obtained has shown that this is not the case in fungal cell walls. The predominant constituents are polysaccharides, whose building units are freely water-soluble [2].

We might therefore go to the other extreme and suggest that the insolubility of the fungal cell wall is due to its being composed of a few, or perhaps only one, giant polysaccharide molecule. Subdivision of this molecule could be achieved by the hydrolysis of an almost imperceptible number of bonds, considered in relation to the total hydrolysable linkages present, and would give readily soluble products. Such a wall, one feels, would be far too susceptible to attack by harmful organisms. It is still possible though that giant molecules of this kind could form part of a more complex wall structure.

* Present address: Rowett Research Institute, Bucksburn, Aberdeen, Scotland, U.K. AB2 9SB.

A degree of complexity is found even within the predominantly polysaccharide nature of the wall, there being usually a fibrillar element (cellulose or chitin) surrounded by amorphous (matrix) material. The full range of functions served by these walls, which constitute a high proportion of the dry weight of the organism, has yet to be recognized and so it is not surprising that they show considerable chemical complexity, far beyond what would be expected if only mechanical support was required.

If the cell wall consists of more than one structural element, evidently more than one enzymic system could be needed to lyse it, and it is the purpose of the present paper to discuss this question as it relates to cell walls in which glucan is a major component.

The earliest guesses concerning polysaccharide structure included the idea that Nature favoured the $(1 \rightarrow 4)$-linkage between sugar units; this was based on the structures of starch, cellulose, xylan and pectin then recently discovered. Subsequent research has shown that no such rule applies. Similarly our ideas about the solubility of polysaccharides have moved very far from the idea that β-linkages (as in cellulose and xylan) conferred insolubility. As the fine structure of more polysaccharides has become known theories have been developed to explain the great range of behaviour of polysaccharides in aqueous media, from total insolubility through gel formation to complete solubility (see, for example, the lecture by Rees [3]).

The use of dilute aqueous alkali, usually at 100°C, played a large part in early investigations, although even in the 1930's there were those who looked for less drastic means of separating polysaccharides, such as glycogen [4] , from living tissue. Hot aqueous alkali is a better solvent for polysaccharides than water. Its degradative action may be negligible (though not always), and it was an advantage in early investigations that it converted accompanying proteins and fats to products soluble in aqueous ethanol. This solvent could then be used to precipitate and purify the polysaccharide.

Because glycosidic linkages are more or less sensitive to acidic conditions, acids were used sparingly. Alkaline extraction methods proved adequate for the isolation and structural characterization of many polysaccharides, but very serious difficulties were met with polysaccharides that would not dissolve readily in (say) *N*-NaOH at 100° C. A classical example of this is the glucan of yeast, once thought to be a form of cellulose, but now known to be predominantly a β-$(1 \rightarrow 3)$-glucan. Rather desperate efforts to solubilize it were made [5] . We have ourselves struggled with this problem recently, without any great success [6] .

The use of hot aqueous alkali almost always dissolves some of the cell wall polysaccharide, and this is not necessarily precipitated when the pH of the extract is brought to 7. In the case of baker's yeast the cell wall mannan was early obtained in this way, but later work showed that alkaline conditions were not essential and it could be released by autoclaving whole yeast in citrate buffer at pH 7 [7].

A suitable combination of acid and alkali treatments will dissolve almost the whole of most fungal cell walls leaving only chitin or cellulose, unless melanin is present [8]. Both chitin and cellulose have linear molecules capable of close-packed arrangements that resist dissolution, but enzymes are known that will degrade them to the corresponding water-soluble monomers. We may therefore direct our attention for the moment to the less well characterized polysaccharides that are not extracted by hot aqueous alkali, and those that dissolve in it but are insoluble at pH 7, since both these types must clearly have to be solubilized somehow, if the wall is to be lysed.

These polysaccharides are often β-(1 → 3)-glucans which may be recognized by their characteristic infrared spectra [9]. They seem to be essentially amorphous in the yeast cell wall, giving no strong X-ray diffraction pattern. This impression is confirmed by the discovery by Kreger [10] that boiling in acid, which leads to the appearance of progressively thickening fibrils of "hydroglucan", does not simply remove matrix material from pre-existing fibrils; the X-ray diffraction pattern intensifies, and less hydroglucan is formed if the alkali-soluble glucan fraction is first removed. We have found that boiling in acid makes the cell wall, and the glucan extracted from it by alkali, more resistant to the action of β-(1 → 3)-glucanases [11]—a further indication that the native form of the polysaccharide is amorphous. Northcote [12] had earlier come to a similar conclusion from studies of the sorption of water by the glucan.

He concluded that the insolubility of the glucan was probably due to its molecular size, but it remains a possibility that firm associations might be formed between parts of smaller molecules, so constituting a continuous network throughout the wall.

Unfortunately, the very insolubility of the glucan has made chemical investigation difficult, and fine details of structure on which such an hypothesis could be based, are lacking. For the glucan of the cell wall of baker's yeast, the chemical evidence was often based on examination of only a small fraction of the total (Table I). Evidence that this was representative of the whole, or, for that matter, that it was homogeneous itself, is not available.

TABLE I. Yields of glucan and mannan from pressed baker's yeast.

Glucan: yields are expressed as mg/g after (a) alkali treatment and (b) subsequent acid treatment		
	(a)	(b)
Hassid, Joslyn and McCready [13]	—	24
Barry and Dillon [14]	26	—
Bell and Northcote [15]	16	5.5
Peat, Whelan and Edwards [16]	19	6.5
Misaki *et al.* [17]	23	8.4
Mannan: yields are expressed as mg/g after (a) initial extraction, by hot aqueous alkali or autoclaving at pH 7, and (b) purification as the copper complex		
	(a)	(b)
Haworth, Hirst and Isherwood [18]	21	13
Peat, Whelan and Edwards [7]	—	5.5
Edwards [19]	16	10
Theory:		
(i) wall, 25% cell; polysaccharide, 45% wall: 30 mg/g		
(ii) wall, 15% cell; polysaccharide, 30% wall: 12 mg/g		

Extraction of intact cells with hot aqueous alkali seems to remove very little of the β-(1 → 3)-glucan component, although cell walls isolated from mechanically broken cells yield a substantial proportion, even at room temperature [20 21]. The discrepancy has led us to postulate [6] that the intact cell wall constitutes a semi-permeable bag, through which the alkali-soluble glucan cannot escape. The fact that dimethylsulphoxide does not extract much glucan unless ultrasonics are applied could be explained in a similar way. We thus visualized an outer layer of alkali-insoluble and dimethylsulphoxide-insoluble glucan, different in character from the rest. We speculated that this might be associated with chitin, but Cabib has provided good evidence [22, 23] that at least in his strains of yeast chitin is confined to the bud-scar region. Our attempts to prove the existence of the layer by microscopy were inconclusive, but the production of "cell ghosts" by others [22] offers some support.

Very mild acid treatments, which had been used repeatedly in the past, following Bell and Northcote [15], to removed trapped glycogen from the alkali-treated cells, render most of the glucan soluble in hot alkali [24]. It seems likely that these acid treatments, although as mild as heating to 80° C at pH 5, split some glycosidic linkages and allow alkali degradation, to which β-(1 → 3)-glucans are particularly sensitive [25], to take place. The fact that glycogen, which should be stable to the acid treatment, is released suggests an increase in the permeability of the outer layer. A β-(1 → 6)-glucan is also released. The situation could be clarified by a careful examination of the changes in ultrastructure accompanying these chemical and enzymic treatments, but even as it stands the evidence leaves little doubt that the glucan of the yeast cell wall is heterogeneous. The structural analyses so far made by chemical techniques provide no explanation for this, and the history of investigations into starch structure indicates that a combination of enzymic and chemical studies is likely to be needed. In some fungi, e.g. *Polyporus* species [26], *Schizophyllum commune* [27], α-(1 → 3)-glucans accompany the β-(1 → 3)-glucans, and both show heterogeneity with respect to solution in aqueous alkali or dimethyl sulphoxide. The α-(1 → 3)-glucan, originally recognized as a distinct type by Kreger [28], was studied most carefully by Johnston [29], who obtained it from the cell walls of *Aspergillus niger* [30], and it has been found also in *Cryptococcus* species [31]. Duff was the first to suggest its nature [26], but he found structural investigation by the methylation technique very difficult. He later spent some efforts in attempts to fractionate the alkali-soluble material from *Polyporus betulinus.* A convenient large-scale preparation from the flesh of *Polyporus,* in which it is the major component, would help the search for convenient sources of glucanase capable of attacking fungal walls containing this polysaccharide, e.g. *Schizosaccharomyces* [32].

Where the cell wall consists, as in these cases, of several water-insoluble polysaccharides, there is no difficulty in visualizing the need for several enzymes to lyse it. In the cell wall of most yeasts however, there is present in addition to β-(1 → 3)-glucans, a mannan, much of which can be obtained in a freely water-soluble form. The mannan obtained by extraction with aqueous alkali is an artifact. Material of much larger molecular weight is released from cell walls by enzymic action [33], but this also is freely soluble. Some mannan passes into solution when cells are broken [34] (this includes the "periplasmic" enzymes, invertase and acid phosphatase), but the greater part remains firmly in the wall. Extraction of the walls in the dry state with ethylene diamine produces a water-soluble mannan with some peptide attached; its molecular weight is relatively low (76,000) and is only marginally reduced by treatment with aqueous alkali [35]. It too would appear to be an artifact, because the material released by glucanase action has a higher molecular weight [11, 33]. If the mannan occurs as complexes with molecular weights of the order of 10^6 one could conceive, that though water-soluble, it was surrounded by a glucan network, disruption of which would set it free, but an idea as simple as this must at once be rejected if one considers the difference in response of *Cytophaga johnsonii* to cell wall preparations and to living cells of *S. cerevisiae, S. carlsbergensis* and *S. fragilis.* In each case the presence of cell walls induces the formation of extra-cellular β-(1 → 3)-glucanase, but whole cells induce none. Further, yeast cells killed by a minimal heat treatment do not induce glucanase [11]. The conclusion seems inescapable that glucan is not exposed on the surface of the yeast cell wall. Lampen [36] has proposed that part of the mannan is present on the outside of the cell wall, a continuous layer (a giant molecule) being formed by phosphodiester cross-linkages. This layer could prevent glucanase action so long as the wall was intact, but cell breakage would expose the inner surface and the glucan component. The remainder of the mannan could then be trapped in the glucan mesh.

This explanation is not entirely satisfactory. Lytic action on living cells [37] and on cell wall preparations [38] is facilitated by thiols, which will act in a pre-treatment. This effect is connected with the release of invertase from living cells by thiol action [39], and is most simply explained as an increase in the porosity of the wall. Although there is no satisfactory direct evidence, the thiol is supposed to act upon disulphide bridges between the peptide portions of mannan-peptide complexes. The thiol thus by action on the mannan component disaggregates the wall structure sufficiently to allow large enzyme molecules (consisting mainly of mannan) to escape, but leaves the bulk of the mannan component fixed in the wall. The outer layer postulated by Lampen would have to be susceptible to thiol action.

It would seem simpler to imagine that a glucan network is surrounded by a matrix composed of mannan molecules held together by a variety of ways [40], in some places by close association between the side groups of polysaccharide molecules. Whether the latter could include associations between mannans and glucans cannot be stated on the basis of our present knowledge. The integrity of the wall structure would thus be threatened by any one of a variety of agents, but total dissolution might require several.

Further complexity would be added if co-valently linked glucan-mannan complexes exist. The evidence for these is still lacking although since the early experiments of Falcone and Nickerson [41] with mild alkaline extractions, the results of chemical and enzymic treatments have often called for this interpretation [42, 43]. Our own experiments, done in the presence of thiol, with a

β-(1 → 6)-glucanase from *C. johnsonii* have some relevance. Cell walls treated with this enzyme in the complete absence of β-(1 → 3)-glucanase lost almost all of their mannan, suggesting that the mannan may be held in the wall because it is linked to the β-(1 → 6)-glucan fraction. If this is the case it raises the question of the fate of this glucan when mannan is extracted from intact cells. Is it present, for example, in the material extracted by autoclaving at pH 7?

Peat, Whelan and Edwards [7] obtained from partial acid hydrolysates of their mannan preparations a disaccharide fraction consisting only of glucose residues linked (1 → 6), but from its optical rotation containing both α- and β-isomers.

Little other evidence is available, but it may be significant that the yields of mannan from whole yeast often fall short of those expected (Table I). The final purification is by precipitation with copper from alkaline solution, and losses may be incurred there; failure to precipitate has been ascribed to association with glucan in the case of cell wall extracts [41]. Heating at 80° C with 3% NaOH extracts all the mannan in 1 h, subsequent extractions doing no more than wash the bulky insoluble residue. We could find no evidence that glucan insoluble at pH 7 was present in the extract [6], and recently we have looked for soluble β-(1 → 6)-glucan, but detected none. This may indicate merely that the linkage between mannan and glucan is alkali-labile.

In the course of this symposium we shall hope to hear more of the structure of yeast mannans, and particularly of those obtained by the milder methods of extraction. As with the glucan, there is doubt whether the alkali-treated preparations used for structural analysis are typical of the whole mannan component.

Some clarification of the situation might be gained if enzymes could be found capable of degrading the mannan extensively *in situ*. There is no evidence that snail digestive juice, or most of the other lytic enzyme preparations used for preparing protoplasts degrade the mannan to the point at which mannose or mannose-containing oligosaccharides can be detected. Dr H. W. Morgan, working in the Department of Microbiology of this Institute, has isolated several organisms that can participate in the breakdown of typical alkali-extracted mannan. The first, resembling that used by Ballou and his colleagues [44], achieves a substantial, though partial degradation, but it has no action at all on mannan *in situ* in the cell walls of a number of yeasts. He has found two other organisms capable of completing the degradation of the residue left by the first, but having no detectable action on the starting material. These two organisms produce at the same time considerable extracellular β-(1 → 3)-glucanase activity, and it has not proved possible to remove this from the mannanase. Tests on intact cells or cell wall preparations, are therefore confused by the presence of the glucanase.

The same is true of the phosphomannanase preparation described by McLellan, McDaniel and Lampen [45]. This enzyme has a very limited action upon *S. cerevisiae* mannan, but this could plausibly convert insoluble mannan into soluble fragments. It is unfortunate that the presence of glucanase makes it impossible to be sure that modification of glucan plays no part in the process.

If there are uncertainties about the action of enzymes that have been subjected to purification procedures, they cannot be compared with those created by the use of snail digestive juice, which is known to contain numerous activities [46, 47], several of which might be relevant to its action on yeast cell walls. Other crude mixtures of enzymes have been used, and there is no doubt that it would be preferable to use the minimum of highly purified enzymes, and so avoid unnecessary damage to the protoplast.

Our own interest in this subject began with the isolation by Dr D. M. Webley of a strain of *C. johnsonii* that was present abundantly on the surface of grass roots [48]. When grown in the presence of yeast cell walls this organism releases into the culture medium enzymes capable of very extensive lysis of yeast cell walls *in vitro*; when grown without the cell walls it has no action on them. It therefore seemed that some, or perhaps all, the enzymes needed for lysis were inducible. Later work showed that a glucan fraction from the yeast cell wall would induce a lytic culture filtrate.

The question therefore arises: is a β-glucanase the only enzyme that is needed for the lysis of yeast cell walls?

Certainly all the enzyme preparations used to lyse cell walls, or to form protoplasts, contain β-(1 → 3)-glucanase, but the *C. johnsonii* culture fluid is the least active in this respect (Table II). Snail digestive juice has several thousand times as much activity; it has little effect on cell walls when diluted to match the *Cytophaga* fluid. This in itself might raise suspicions that some other enzyme is involved. The Cytophaga produces a great deal of β-(1 → 6)-glucanase, in which snail juice is relatively deficient, but we could obtain no evidence that this enhanced the action of the β-(1 → 3)-glucanase [11]. Enzyme preparations devoid of β-(1 → 6)-glucanase activity, such as those from actinomycetes studied by Jones and Webley [53] or the commercial laminarinase preparation tested by Baird and Cunningham as a protoplasting agent [52], will lyse walls.

TABLE II. Laminarinase activities used to prepare protoplasts. So far as can be ascertained the values refer to 1 ml of protoplasting medium

Source	Units of laminarinase used	Authors
Snail digestive juice[a]	15	Nagasaki, *et al.* [49]
	3.8	Ottolenghi [50]
Bacillus circulans glucanase	5.5	Nagasaki, *et al.* [49]
B. circulans phosphomannanase	1.0	McLellan, *et al.* [45]
plus snail digestive juice	3.0	
Basidiomycete QM 806 exo-glucanase	5.0	Bauer, *et al.* [51]
Cytophaga enzyme (Glaxo)	0.015	Baird and Cunningham [52]
Cytophaga johnsonii culture fluid (unfractionated)	0.01	Dr W. L. Cunningham, personal communication
C. johnsonii "lytic enzyme"	0.03	
C. johnsonii "non-lytic enzyme"	0.06	

[a] Activity taken as 150 units/ml. One unit liberates reducing sugar equivalent to 1 μ mole of glucose per minute.

Chitinase also could not be implicated, as might be expected if chitin were confined to the bud-scar region [22].

Fractionation of the *C. johnsonii* culture fluid has given a possible explanation for its greater effectiveness on a β-(1 → 3)-glucanase basis. This is that it contains two kinds of endo-β-(1 → 3)-glucanase activity in approximately equal amounts. One of these is much more effective in lysing cell walls. The other is the "ordinary" type of β-(1 → 3)-glucanase, very widely distributed in microorganisms and plants, and probably the main agent of snail juice action.

The more effective enzyme is not confined to *C. johnsonii*, having been noticed by Chesters and Bull in fungal culture filtrates [54]. It is difficult to detect in the presence of the other, because its action ceases at an earlier stage of glucan degradation; its end-products are thus further degraded by the ordinary endo-enzyme. It might be detected by a kinetic study, which would show a faster liberation of reducing products early in the reaction, but the best chance of recognizing it is to find some means of separating the two enzymes.

We have attempted this for a commercial Cytophaga preparation enzyme (L.1, B.D.H. Ltd.), but while the pattern of separation of β-(1 → 3)-glucanase activity on DEAE cellulose was similar to that for *C. johnsonii*, endo-β(1 → 3)-glucanase of the "non-lytic" type was distributed through all the fractions. Fractionation of snail digestive juice by gel-filtration on Biogel polyacrylamide gel (P-100), although revealing at least three endo-β-(1 → 3)-glucanases gave no indications of the lytic type. We have tested the two endo-glucanases from *C. johnsonii* on a variety of substrates ranging from glucan extracted from the cell wall to living yeast [11]. In the latter tests we were under the mistaken impression that all strains of *S. cerevisiae* should yield protoplasts in a suitable medium, provided that a thiol was present. Baird and Cunningham have since shown that this is not the case, and have been able to produce protoplasts from susceptible strains by the use of the commercial Cytophaga preparation (Table II). Dr Cunningham has very kindly tested a *C. johnsonii* culture filtrate on some susceptible and non-susceptible strains with similar results. He has also found that our "lytic enzyme" is more effective at forming protoplasts than the non-lytic.

The difference between the two enzymes is therefore evident whether they are tested on glucan preparations, cell walls, or living yeast. The lytic enzyme is more active on all substrates, and it seems that a systematic study of the production and purification of this enzyme would be well justified.* If its action on living yeast is

*See Doi *et al.* in *Jap. J. Biochem.*, (1971), **70**, 711; (1973), **73**, 667.

retained in spite of further purification it should be a particularly innocuous agent for protoplast formation.

Before any explanation of the greater effectiveness of the lytic enzyme could be offered one would need to know more about its specificity. There is no reason to believe that it will be more effective against every fungal cell wall glucan. It has been tested so far only against cell walls of *Fusarium culmorum.* These walls disappear from cultures of *C. johnsonii,* but the separated culture fluid does not dissolve them completely. A concentrated solution of the lytic enzyme did not achieve complete lysis. It may therefore be the case that each cell wall requires a particular glucanase, or combination of glucanases.

Such a situation is well illustrated by the cell walls of some sclerotia-forming fungi. We have studied particularly *Sclerotinia sclerotiorum* and *Sclerotium rolfsii* [55, 56]. Dr D. Jones has isolated two fungi that are parasitic on the sclerotia of *S. sclerotiorum* [57]. When grown on a medium containing crushed sclerotia both produce culture fluids rich in β-(1 → 3)-glucanase activity, which will lyse the cell walls. The culture fluid of *C. johnsonii* has practically no action on them.

Fractionation of the enzymes in the lytic culture fluids shows that the predominant β-(1 → 3)-glucanase produced by both parasitic fungi (*Coniothyrium minitans* and *Trichoderma viride*) is an exo-enzyme [58]. Enzymes of this type have been purified by American and Japanese workers [59, 60], and are known to attack a glucan present in sclerotia. This glucan, named "sclerotan" by Kitahara and Takeuchi [61], is much more gelatinous than the β-glucans we have so far considered. It forms a major part of the sclerotial cell walls, and a lesser but still dominant part of the hyphal walls. The infra-red spectrum resembles that of the other fungal β-(1 → 3)-glucans, but has small differences which are probably ascribable to some crypto-crystalline arrangement of the molecules [56].

Although sclerotan has many substituted glucose residues, it is completely degraded by the exo-β-(1 → 3)-glucanase from Basidiomycete QM 806 [62]. The enzyme ignores the fact that every third glucose residue has another β-glucose residue attached at position 6 and so produces a mixture of gentiobiose and glucose in molar ratio about 1 : 2 [62, 63]. The glucan resists the action of endo-β-(1 → 3)-glucanases.

The exoglucanases we have examined from *C. minitans* and *T. viride* do not achieve complete degradation of the sclerotan preparations we have tested; degradation practically ceases at 35%

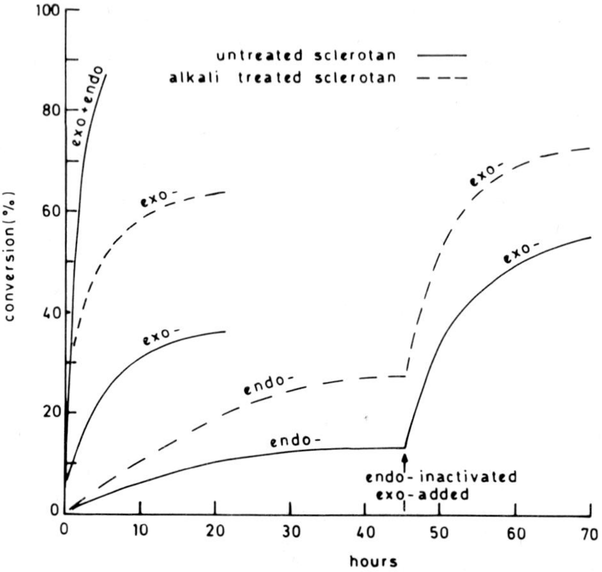

Fig. 1. Cooperative action of glucanases on sclerotan. The sclerotan was prepared from autoclaved mycelium of *Sclerotium rolfsii.* [58]. The endo-enzyme was inactivated by heating to 100° C for 15 min. Both enzymes were prepared by ion-exchange chromatography: endo-β-(1 → 3)-glucanase from *C. johnsonii* on DEAE-cellulose; exo-β-(1 → 3)-glucanase from *Coniothyrium minitans* on DEAE-Sephadex.

conversion to glucose and gentiobiose [58]. If some endo-β-(1 → 3)-glucanase is added breakdown goes to completion (Fig. 1).

Failure to achieve complete degradation has been found when the QM 806 enzyme acts upon the alkali-extracted glucan from the Basidiomycete itself; here about 35% remained in the form of polysaccharide [64].

It is clear from these examples that it is very likely that complete degradation of the variety of β-glucans in fungal walls is rarely likely to be achieved by a single glucanase, but there is good evidence that mixtures of glucanases will suffice to lyse many of the cell walls so far investigated. Bearing in mind that changes in the cell shape, budding processes, and perhaps growth itself require the cell to loosen or modify its own wall, it is perhaps significant that glucanases, but not mannanases or chitinase, have been found either fixed to the wall or in the periplasmic space [65, 66]. This could indicate that the wall is so constructed that the state of the glucan determines its mechanical properties and integrity.

Acknowledgements

I wish to thank Miss Sheila Clark, Miss Lorraine Stefani and Mr. A. H. Gordon for assistance with the new experiments described. It is a pleasure to have the opportunity to record my appreciation of the excellent facilities provided by successive Directors of the Macaulay Institute, and the invaluable contributions of many colleagues during the eight years in which this problem was my major preoccupation.

References

1. Pirie, N. W. (1940). *Biol. Rev. Camb. Phil. Soc.*, **15**, 377.
2. Bartnicki-Garcia, S. (1968). *Ann. Rev. Microbiol.*, **22**, 87.
3. Rees, D. A. (1972). *Biochem. J.*, **126**, 257.
4. Bell, D. J. (1935). *Biochem. J.*, **29**, 203.
5. McAnally, R. A. and Smedley-Maclean, I. (1937). *Biochem. J.*, **31**, 72.
6. Bacon, J. S. D., Farmer, V. C., Jones, D. and Taylor, I. F. (1969). *Biochem. J.*, **114**, 557.
7. Peat, S., Whelan. W. J. and Edwards, T. E. (1961). *J. Chem. Soc.*, 29.
8. Jones, D. (1970). *Trans. Br. mycol. Soc.*, **54**, 351.
9. Michell, A. J. (1970). *Aust. J. Biol. Sci.*, **23**, 345.
10. Kreger, D. R. (1967). *Abh. dt. Akad. Wiss. Berl. Kl. Med.*, **6**, 81.
11. Bacon, J. S. D., Gordon, A. H., Jones, D., Taylor, I. F. and Webley, D. M. (1970). *Biochem. J.*, **120**, 67.
12. Northcote, D. H. (1953). *Biochim. biophys. Acta*, **11**, 471.
13. Hassid, W. Z., Joslyn, M. A. and McCready, R. M. (1941). *J. Am. Chem. Soc.*, **63**, 295.
14. Barry, V. C. and Dillon, T. (1943). *Proc. R. Ir. Acad.*, Sect. B, **49**, 177.
15. Bell, D. J. and Northcote, D. H. (1950). *J. Chem. Soc.*, 1944.
16. Peat, S., Whelan, W. J. and Edwards T. E. (1958). *J. Chem. Soc.*, 3862.
17. Misaki, A., Johnson, J., Kirkwood, S., Scaletti, J. V. and Smith, F. (1968). *Carbohydrate Res.*, **6**, 150.
18. Haworth, W. N., Hirst, E. L. and Isherwood, F. A. (1937). *J. Chem. Soc.*, 784.
19. Edwards, T. E. (1965). *In:* "Methods in Carbohydrate Chemistry" (Whistler, R. L., ed.). Vol. 5, p. 176. Academic Press, New York and London.
20. Roelofsen, P. A. (1953). *Biochim. biophys. Acta*, **10**, 477.
21. Eddy, A. A. and Woodhead, J. S. (1968). *FEBS Letters*, **1**, 67.
22. Cabib, E. and Bowers, B. (1971). *J. biol. Chem.*, **246**, 152.
23. Cabib, E. and Farkas, V. (1971). *Proc. natn. Acad. Sci., U.S.A.* **68**, 2052.
24. Bacon, J. S. D., Davidson, E. D., Jones, D. and Taylor, I. F. (1966). *Biochem. J.* **101**, 36C.
25. Corbett, W. M. and Kenner, J. (1955). *J. Chem. Soc.*, 1431.
26. Duff, R. B. (1952). *J. Chem. Soc.*, 2592.
27. Niederpruem, D. J. and Wessels, J. G. H. (1969). *Bacteriol. Rev.*, **33**, 505.
28. Kreger, D. R. (1954). *Biochim. biophys. Acta*, **13**, 1.
29. Johnston, I. R. (1965). *Biochem. J.*, **96**, 659.
30. Johnston, I. R. (1965). *Biochem. J.*, **96**, 651.

31. Jones, D., Bacon, J. S. D., Farmer, V. C. and Webley, D. M. (1969). *Soil Biol. Biochem.*, **1**, 145.
32. Bacon, J. S. D., Jones, D., Farmer, V. C. and Webley, D. M. (1968). *Biochim. biophys. Acta*, **158**, 313.
33. Eddy, A. A. and Longton, J. (1969). *J. Inst. Brewing*, **75**, 7.
34. McMurrough, I. and Rose, A. H. (1967). *Biochem. J.*, **105**, 189.
35. Sentandreu, R. and Northcote, D. H. (1968). *Biochem. J.*, **109**, 419.
36. Lampen, J. O. (1968). *Antonie van Leeuwenhoek. J. Microbiol. Serol.*, **34**, 1.
37. Anderson, F. B. and Millbank, J. W. (1966). *Biochem. J.*, **99**, 268.
38. Bacon, J. S. D., Milne, B. D., Taylor, I. F. and Webley, D. M. (1965). *Biochem. J.*, **95**, 28C.
39. Kidby, D. K. and Davies, R. (1970). *J. gen. Microbiol.*, **61**, 327.
40. Bacon, J. S. D. (1970). *In:* "Essays in Cell Metabolism" (Bartley, W., Kornberg, H. L. and Quayle, J. R., eds) p. 45. Wiley, London.
41. Falcone, G. and Nickerson, W. J. (1956). *Science, N.Y.* **124**, 272.
42. Korn, E. D. and Northcote, D. H. (1960). *Biochem. J.*, **75**, 12.
43. Cawley, T. N., Harrington, M. G. and Letters, R. (1972). *Biochem. J.*, **129**, 711.
44. Jones, G. H. and Ballou, C. E. (1968). *J. biol. Chem.*, **243**, 2442.
45. McLellan, W. L., McDaniel, L. E. and Lampen, J. O. (1970). *J. Bacteriol.*, **102**, 261.
46. Holden, M. and Tracey, M. V. (1950). *Biochem. J.*, **47**, 407.
47. Myers F. L. and Northcote, D. H. (1958). *J. exp. Biol.*, **35**, 639.
48. Webley, D. M., Duff, R. B., Bacon, J. S. D. and Farmer, V. C. (1965). *J. Soil Sci.*, **16**, 149.
49. Nagasaki, S., Neumann, N. P., Arnow, P., Schnable, L. D. and Lampen, J. O. (1966). *Biochem. biophys. Res. Commun.*, **25**, 158.
50. Ottolenghi, P. (1966). *C. r. Trav. Lab. Carlsberg*, **35**, 363.
51. Bauer, H., Bush, D. A. and Horisberger, M. (1972). *Experientia*, **28**, 11.
52. Baird, J. K. and Cunningham, W. L. (1971). *Biochem. J.*, **125**, 32P.
53. Jones, D. and Webley, D. M. (1968). *Plant Soil*, **28**, 147.
54. Chesters, C. G. C. and Bull, A. T. (1963). *Biochem. J.*, **86**, 31.
55. Jones, D. (1970). *Trans. Br. mycol. Soc.*, **54**, 351.
56. Jones, D., Farmer, V. C., Bacon, J. S. D. and Wilson, M. J. (1972). *Trans. Br. mycol. Soc.*, **59**, 11.
57. Jones, D. and Watson, D. (1969). *Nature, Lond.*, **224**, 287.
58. Bacon, J. S. D., Gordon, A. H. and Jones, D. (1972). *Biochem. J.*, **129**, 27P.
59. Ebata, J. and Satomura, Y. (1963). *Agr. Biol. Chem. (Tokyo)*, **27**, 478.
60. Huotari, F. I., Nelson, T. E., Smith, F. and Kirkwood, S. (1968). *J. biol. Chem.*, **243**, 952.
61. Kithahara, M. and Takeuchi, Y. (1961). *Nippon Nogeikagaku Kaishi*, **35**, 468, 474.
62. Johnson, J., Kirkwood, S., Misaki, A., Nelson, T. E., Scaletti J. V. and Smith, F. (1963). *Chemy Ind.*, 820.
63. Buck, K. W., Chen, A. B., Dickerson, A. G. and Chain, E. B. (1968). *J. gen. Microbiol.*, **51**, 337.
64. Bush, D. A. and Horisberger, M. (1972). *Carbohydrate Res.*, **22**, 361.
65. Abd-El-Al, A. T. and Phaff, H. J. (1968). *Biochem. J.*, **109**, 347.
66. Abd-El-Al, A. T. and Phaff, H. J. (1969). *Can. J. Microbiol.*, **15**, 697.

PART II

Structure and Morphogenesis

Cell Wall Genesis in a Natural Protoplast: The Zoospore of *Phytophthora palmivora*

S. BARTNICKI-GARCIA

Department of Plant Pathology
University of California
Riverside, California, U.S.A.

I. Introduction

These International Symposia have been concerned with the production and behavior of artificial protoplasts from walled organisms, mainly yeasts but also molds and higher plants. The present article departs somewhat from this tradition and deals with zoospores, a kind of *natural* protoplast produced by the aquatic phycomycetes. In particular, we have been studying the zoospores of *Phytophthora palmivora* and their ability to form a cell wall *de novo*.

A simplified asexual life cycle of this notorious plant pathogen is shown in Fig. 1. The fungus develops the typical coenocytic mycelium of a phycomycete. Later, in response to illumination with visible, or near ultraviolet light, the mycelium differentiates into abundant sporangia [1]. The sporangia, which are deciduous in this species, remain quiescent until they are immersed in water [2]. The sporangium then undergoes its final differentiation; a host of elaborate subcellular changes takes place and the cytoplasm is ultimately cleaved into numerous uninuclear bodies; each one becomes a biflagellate zoospore [3]. Following the dissolution of the papillar plug at the distal end of the sporangium, the zoospores swim out into the medium. The zoospores are capable of swimming for relatively long periods but sooner or later they undergo encystment. Motility ceases, the flagella are shed or retracted [4], and the distinctly pyriform zoospore rounds off and fabricates a cell wall—the cyst wall. Some time later the cyst germinates by emitting a slender germ tube, and, if the nutritional conditions are suitable, the tube will continue to grow and branch into a mycelium.

Zoospores lack a cell wall and can therefore be regarded as protoplasts of natural origin. Their encystment constitutes an amenable system for experimental studies on *de novo* synthesis of cell walls. In many respects the process of zoospore encystment, i.e. cell wall generation by a natural protoplast, may be equated to the process of cell wall regeneration by artificial protoplasts, a subject that has received considerable attention in recent years (see reviews by Villanueva [5]; Nečas [6]). Comparisons between natural and artificial protoplasts may prove to be mutually useful and will be made whenever pertinent.

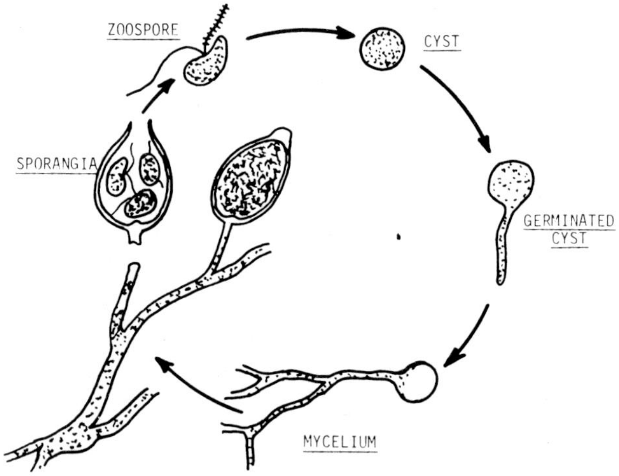

Fig. 1. Asexual life cycle of *Phytophthora palmivora* (not drawn to scale).

From a practical viewpoint, the zoospore system described herein affords the following advantages for biochemical investigations on cell wall genesis: (i) ease of preparation of large, uniform populations of protoplasts; (ii) no need to introduce potentially deleterious agents (lytic enzymes, osmoticants, and miscellaneous reagents) used in the preparation of artificial protoplasts; (iii) facility to generate a cell wall in reproducible, highly synchronous fashion, in response to well defined stimuli.

Large populations of *Phytophthora palmivora* zoospores may be prepared simply by flooding sporulated cultures with distilled water for 20-30 min [2, 7]. These zoospore populations may be encysted synchronously as described before [7].

II. Are Zoospores Devoid of a Cell Wall?

The degree of nudity of artificial protoplasts has been a persistent point of discussion among protoplast experts. The question of whether or not artificial protoplasts have remnants of the original wall need not solely result in a semantic argument; it also creates a fundamental dispute on the ability of protoplasts to regenerate a wall *de novo*, without the assistance of pre-existing wall fragments (see references 5, 6, 8).

Electron microscopy of thin-sectioned zoospores, conventionally fixed with $KMnO_4$ or OsO_4, do not show any evidence of a cell wall. This is true of the zoospores of *Phytophthora* spp. [9-11] as well as the zoospores of other phycomycetes [12-14] where the plasmalemma seems to be the outermost boundary of the cell. Barring the existence of some wall polymers on the cell surface, invisible in the electron micrographs, we must conclude that zoospores are true protoplasts. Hence their ability to generate a cell wall *de novo* would support in principle the claims by Villanueva [5] and Nečas [6] that artificial, *bona fide* protoplasts of fungi are capable of regenerating a cell wall *de novo*.

III. Synchronous Encystment: Kinetics of *de novo* Wall Formation

Although zoospores of *P. palmivora* may spend hours or even days swimming [15] before they spontaneously settle down and encyst, it is possible to induce them to encyst in a rapid and highly synchronous manner by subjecting them to vigorous agitation or other stimuli [7] (Fig. 2).

We routinely encyst zoospores by agitation in a Vortex mixer [7]. The encystment process may be quantitatively followed by counting the number of encysted zoospores after addition of an equal volume of 2M KOH. The alkali causes unencysted zoospores to disintegrate completely whereas the encysted ones retain a spherical wall, visible by phase contrast microscopy, with some particles inside. The kinetics of synchronous zoospore encystment is shown in Fig. 3. After 2-3 min agitation essentially all zoospores have formed an alkali-resistant cell wall [7]. Wall synthesis may be followed by measuring the amount of alkali insoluble glucan in the cell suspension as shown in Fig. 4. Accordingly, wall synthesis proceeds linearly for about 10 min after the beginning of agitation and then it stops [16]. Presumably by this time the cysts have reached maturity. Cyst wall formation may also be observed microscopically with M2R calcofluor, a fluorescent brightener that acts as a selective wall stain [27]. Thus Tsao [27] found that the zoospores of *Phytophthora* spp. do not exhibit any fluorescence but 2-4 min after motility has ceased the cells begin to fluoresce and in 8-14 min they attain

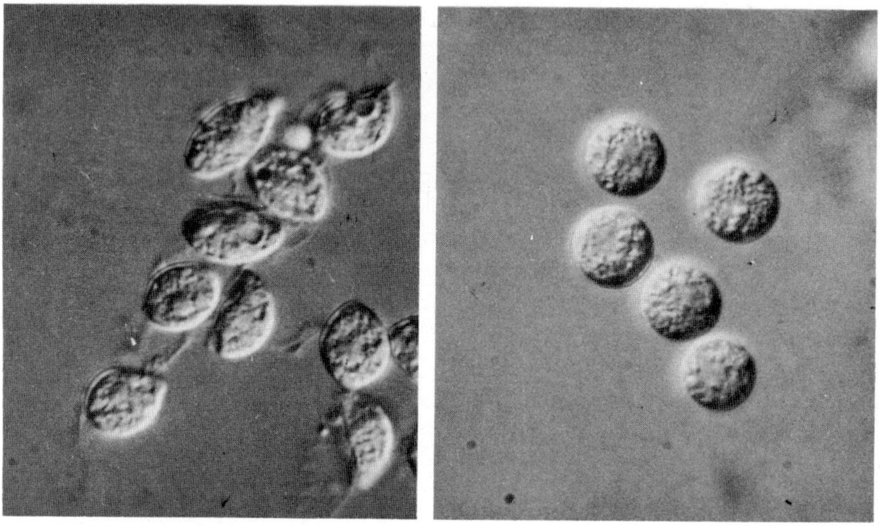

Fig. 2. Synchronous encystment of zoospores of *Phytophthora palmivora*. Zoospores before agitation (left) and resulting cysts after 2 min agitation (right). Cells fixed with glutaraldehyde and photographed with interference contrast optics. Magnification ×1300.

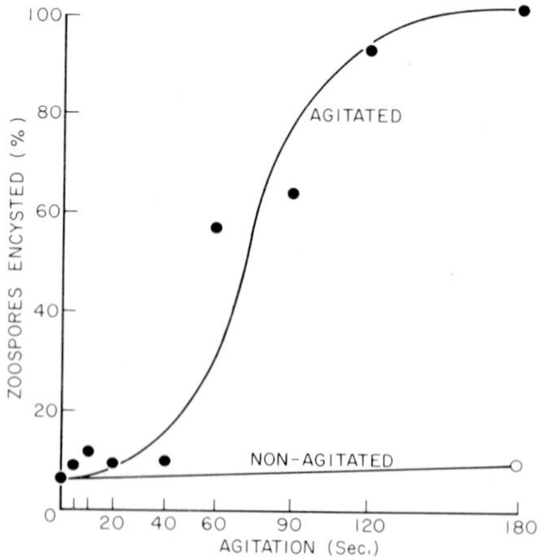

Fig. 3. Effect of agitation on zoospore encystment in *Phytophthora palmivora* [7].

maximum fluorescence. These microscopic observations are in good agreement with the chemical determinations shown in Fig. 4. Earlier, Streiblová *et al.* [17] had successfully used a similar method (with the fluorescent dye primulin) to follow the course of wall regeneration in yeast protoplasts.

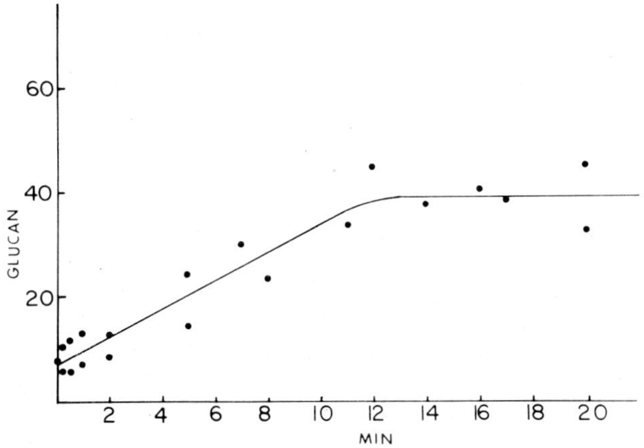

Fig. 4. Synthesis of alkali-insoluble cyst-wall glucan by a population of zoospores of *P. palmivora* undergoing synchronous encystment. Values from 0 to 2 min are all from continuously agitated cell suspensions. Longer incubations correspond to cell suspensions agitated for 2 min and then incubated stationarily (Wang, M. C. and Bartnicki-Garcia, S., unpublished results).

The rapidity with which the zoospores form a wall contrasts with the much longer times needed for wall regeneration by artificially produced protoplasts. For instance, Kopecká *et al.* [18] found that it took 30-60 min of growth before protoplasts of *Saccharomyces cerevisiae* prepared with snail enzymes began forming fibrils; protoplasts prepared by autolysis required 4-6 h. This slow synthesis of wall by the normally fast-growing yeast cells is not surprising. The removal of the wall and other manipulations most certainly disturbed, if not injured, the protoplasts. Furthermore the artificial protoplasts may find it difficult to build wall material onto the free protoplast surface and must gradually recover their capacity to manufacture cell walls efficiently [8].

IV. Microfibrillar Nature of the Cyst Wall

The wall synthesized by the encysting zoospore can be isolated by sonic vibration or by extraction with ethanolic alkali [19]. In either case, the purified wall from mature cysts appears as a thin, collapsed envelope made of long, tightly interwoven, microfibrils (Fig. 5). The overall impression is of a continuous perfectly uniform envelope of even texture and thickness. The microfibrils are about 15 nm in diameter; their length however could not be measured but segments spanning several μm could readily be discerned on the wall surface. Possibly some of the microfibrils may run around the entire cell circumference. Both the inner and the outer surface of the cyst wall are distinctly microfibrillar (Fig. 6).

Shadow-cast specimens of isolated cyst walls of *P. palmivora* show no evidence of any permanent amorphous substance associated with

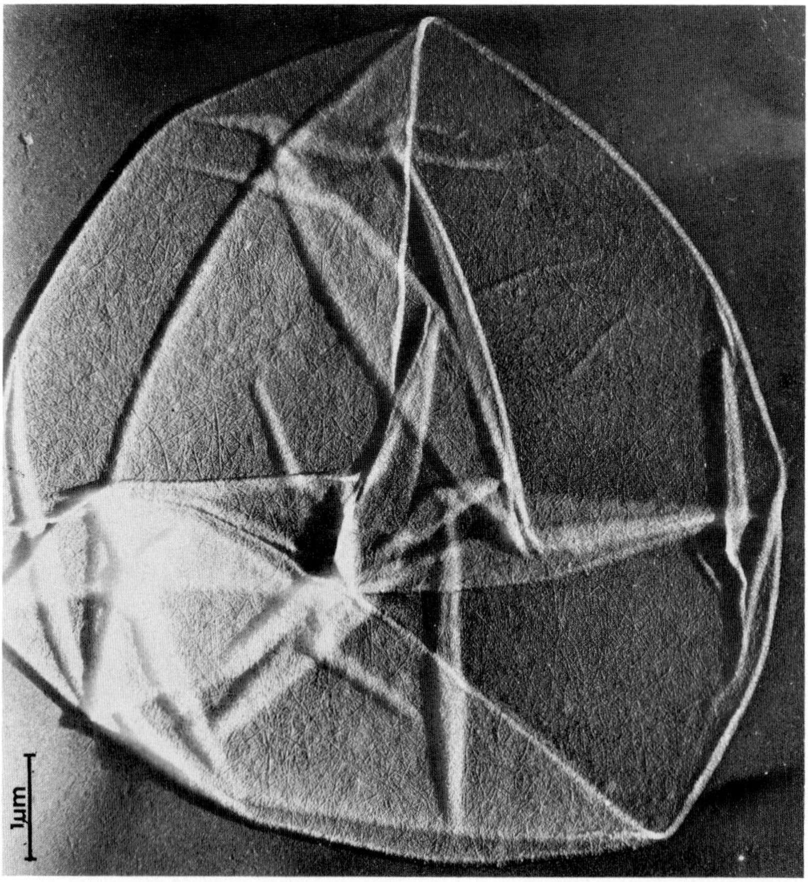

Fig. 5. Cyst wall of *Phytophthora palmivora* prepared by extraction of whole cysts with ethanolic KOH [19].

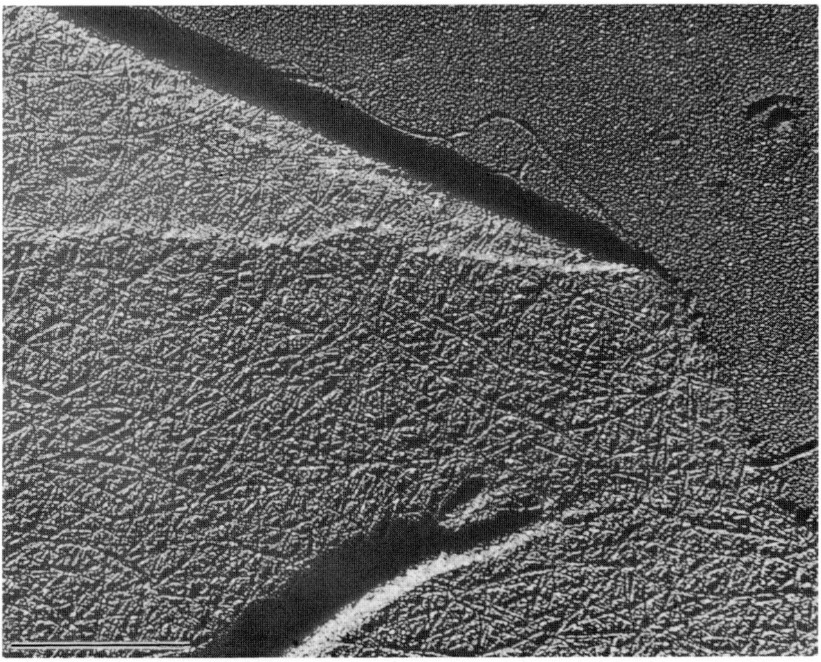

Fig. 6. Texture of a cyst wall fragment of *P. palmivora* prepared by sonication [19]. The folding permits observation of inner (top) and outer (bottom) surface textures. Bar = 0.5 μm.

the microfibrils. However, examination of carbon replicas of intact cysts reveals some irregular patches of amorphous substance (Fig. 7) partially covering the microfibrillar surface of the cyst [20]. This amorphous material is probably removed during purification of the cyst wall. The significance of this amorphous material is unclear. It could be extraneous matter adsorbed by the wall or intracellular material released during fixation and drying. It could be the remnants of a temporary matrix associated with the formation of the wall microfibrils, or it could be part of the cementing material that allows encysted zoospores to adhere strongly to solid surfaces [21].

V. Chemical Composition of the Cyst Wall

Cyst walls of *P. palmivora,* prepared by sonic disruption, were found to consist mainly of β-glucans (92.9%) and some protein (2.9%) [19]. Chemical fractionation revealed two different types of glucans: cellulose which represents about 20% of the total glucan, and a

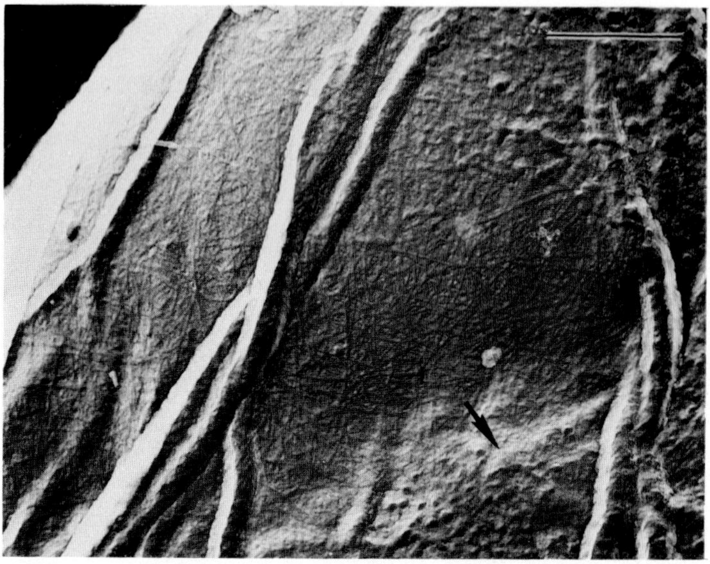

Fig. 7. Surface texture of a whole cyst of *P. palmivora*. Carbon replica of mature cyst fixed with glutaraldehyde [20]. Arrow indicates amorphous patch. Bar = 0.5 μm.

branched β-glucan of complex structure with 1-3, 1-6 and 1-4 links [16, 19].

Notwithstanding the highly regular appearance of the wall microfibrils, the X-ray diffraction pattern of isolated cyst walls shows only a broad diffraction band at about 4.0 Å [19]. The lack of sharp reflections could be an indication of the presence of masking substances not readily evident in the electron micrographs or, perhaps, the relative lack of periodicity in the molecular structure of the cyst wall microfibrils compared with those of cellulose or chitin [22].

VI. Site of Synthesis of Cyst Wall Polysaccharides

One of the major questions of cell wall biogenesis concerns the site of synthesis of wall polysaccharides. Two main alternatives may be considered: (i) intracellular elaboration of polysaccharides, presumably inside vesicular elements derived from the Golgi apparatus, or an equivalent vesicle generator; (ii) *in situ* formation of wall polymers either on the outer surface of the plasmalemma or at some distance from it, within the cell wall fabric. There may be of course intermediate schemes wherein the wall polymers would be partially synthesized in the cytoplasm but the final assembly would take place outside the plasmalemma.

A. *Cytological Evidence*

Thin-section studies of zoospores of *Phytophthora parasitica* [10], *P. palmivora* [11, 23], as well as those with a closely related organism, *Pythium aphanidermatum* [24], reveal the existence of a number of large vesicles with granular or "fibrillar" contents distributed around the cell periphery in close proximity to the plasmalemma (Fig. 8). These vesicles have been termed "fibrillar vacuoles" [11] or "peripheral vesicles" [24]. Significantly, at the beginning of encystment, the peripheral vesicles fuse with the plasmalemma and their contents form a coat (wall?) around the cell surface [11,24]. Also consumed during the initial phase of encystment are numerous flat vesicles that lie directly under the plasmalemma [11]. In addition, Hemmes and Hohl [11] recognized

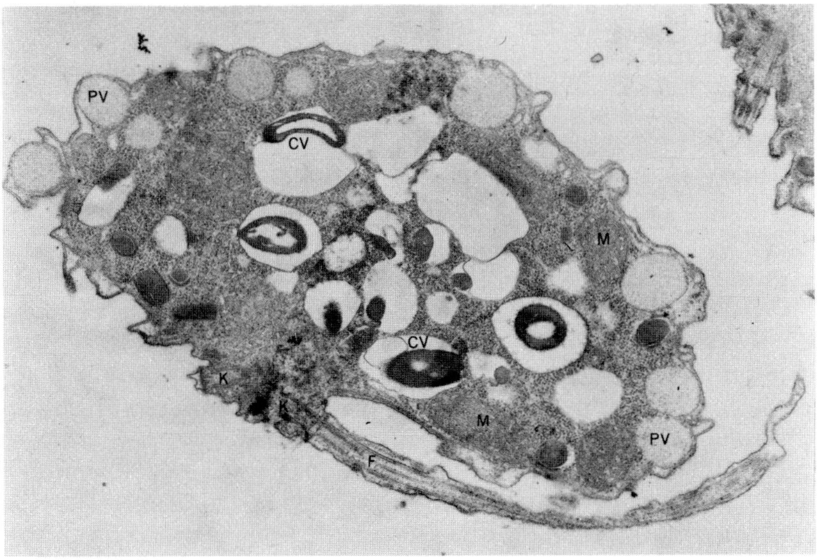

Fig. 8. Internal structure of a zoospore of *Phytophthora pamivora* fixed with glutaraldehyde and OsO_4, and stained with uranyl acetate [23]. The section runs along a longitudinal median plane. PV: peripheral vesicle; CV: vesicle with crystalline "finger print" contents; M: mitochondrion; K: kinetosome; F: flagellum.

a second stage during which materials from dictyosome-derived vesicles are added to the initial cyst wall.

All these studies strongly suggest that vesicular elements move to the zoospore surface materials needed for cyst wall formation. However, biochemical studies (see below) indicate that the materials secreted by these vesicles cannot be prefabricated wall microfibrils.

B. *Biochemical Evidence*

Chemical analyses detected little, if any, preformed cyst wall glucans in the zoospores of *P. palmivora* [7]. The zoospores do contain a large amount of β-glucans but these are soluble cytoplasmic glucans of low molecular weight. By contrast the cyst wall glucans, both the cellulosic and the non-cellulosic ones, are highly insoluble polymers.

Two kinds of cytoplasmic β-glucan have been characterized from the zoospores of *P. palmivora*: neutral and acidic [25]. The former is a highly homogeneous β-glucan of about 5000 M.W. while the latter is a family of phosphoglucans structurally related to the neutral glucan [25].

Although these findings preclude the existence of preformed microfibrils in the zoospore, they do not rule out the possibility that the soluble β-glucans may serve as "subunits" for the assembly of the wall microfibrils, particularly since a portion of the cytoplasmic glucan is consumed during encystment with the concomitant appearance of cyst wall glucan [7]. This possibility, however, seems unlikely; recent methylation studies [16] show considerable structural differences between the cytoplasmic glucan and the cyst wall glucans (Table I). The cytoplasmic glucan is predominantly

TABLE I
Methylation analyses of cyst-wall glucans (unfractionated) and soluble neutral glucan from the zoospore of *P. palmivora* [16]

O-Methyl glucose derivatives	Structural significance	Glucan: Cyst wall	Glucan: Zoospore cytoplasm
		%	%
2,3,4,6-tetra-	terminals	17.4	6.5
2,4,6-tri-	linear 1 → 3	15.9	89.9
2,3,6-tri-	linear 1 → 4	48.0	0.0
2,6-di-	branch 3 & 4	5.1*	0.0
2,4-di-	branch 3 & 6	8.9	3.5
2,3-di-	branch 4 & 6	4.9*	0.0

*Tentative identification

$1 \rightarrow 3$ linked whereas the cyst wall glucans have a large proportion of $1 \rightarrow 4$ links, some of them obviously belong to cellulose but others are associated with the non-cellulosic portion. It is therefore unlikely, that intracellular soluble glucan molecules are directly polymerized into the insoluble microfibrillar glucans. At most, the cytoplasmic glucan might be a primer accounting for a minor portion of the non-cellulosic wall glucan, the rest, if not all of the cyst wall glucan, must be synthesized anew during encystment.

Probably the various secretory vesicles supply the enzymes and precursors needed for wall glucan biosynthesis and microfibrillar assembly at the cell surface.

VII. Enzymic Readiness of the Zoospore

Zoospore encystment in *Blastocladiella emersonii* [26] or *P. palmivora* [11] is insensitive to cycloheximide, thus indicating that synthesis of new proteins is not needed during encystment; the enzymic machinery required for cyst wall formation must therefore be present in the zoospore and need only be set in motion. Such readiness is consistent with the rapidity of the encystment process. It is also supported by the finding of β-1,3- and β-1,6-glucan synthetase activity in the swimming zoospore of *P. palmivora* [28]. Such enzyme incorporates glucosyl units from uridine diphosphoglucose into an alkali-insoluble glucan. Likewise the motile zoospore of *Blastocladiella emersonii* contains chitin synthetase activity [29].

The enzymic sufficiency, however, does not extend beyond encystment; the subsequent developmental stage, i.e. germ tube formation, is inhibited by cycloheximide indicating a requirement for new enzyme synthesis [26].

VIII. On the Synthesis of Wall Microfibrils and Wall Matrix: Experimental and Natural Uncoupling

As indicated before, the isolated cyst wall of *P. palmivora* is solely a microfibrillar envelope. However, the germ tube wall, that later arises from the cyst wall, has a more elaborate structure [19]. In addition to an inner microfibrillar layer (continuous with the cyst wall fabric), the germ tube wall has a permanent outer layer of non-fibrillar β-glucan (Fig. 9). This dual structure is typical of hyphal walls [19]. Clearly, the first wall made around the zoospore, or natural protoplast, has a simpler structure compared to subsequent walls

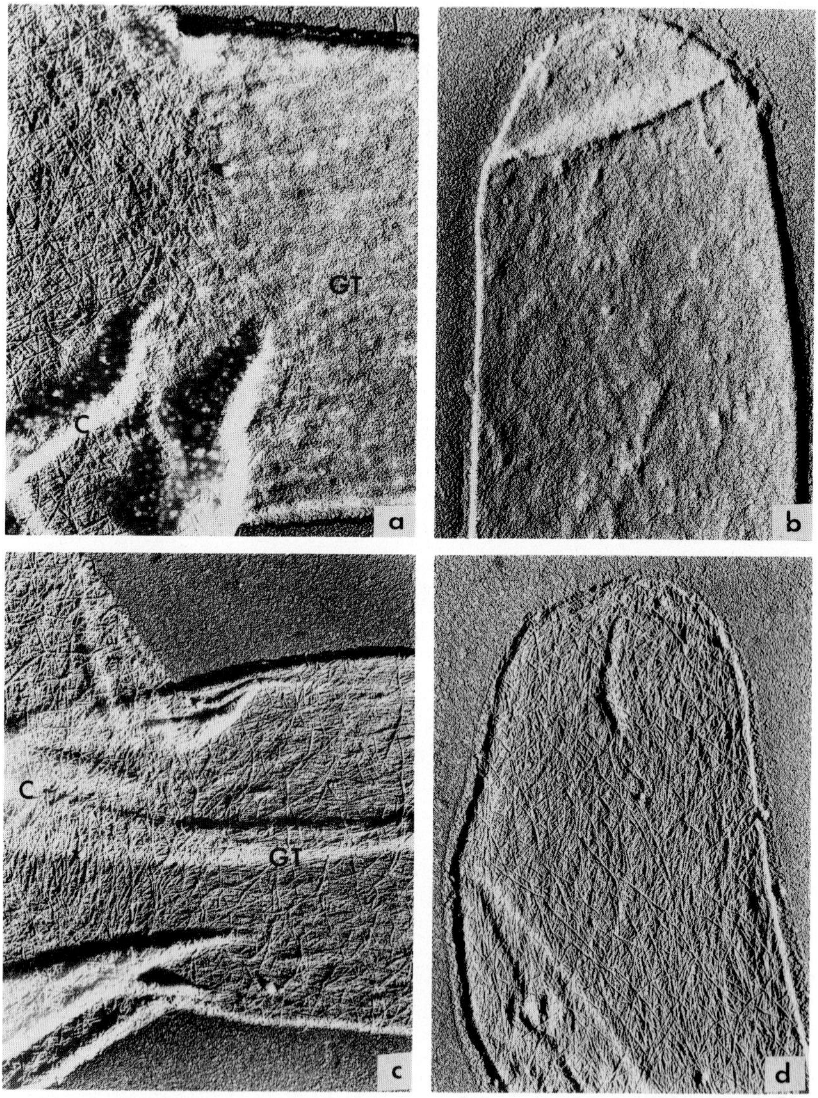

Fig. 9. Dual texture of the germ tube wall of *P. palmivora* [19]. (a) and (b): outer texture at the base and apex of the germ tube, respectively. (c) and (d): inner texture revealed by digestion with exo-β-1,3-glucanase. C: cyst wall; GT: germ tube wall.

made by the fungus. On this point, an interesting parallel can be seen with artificial protoplasts. The regenerating protoplasts of *Saccharomyces cerevisiae* produce, more or less readily, the microfibrillar component of the wall; however, the development of a normal, complete wall (microfibrils + matrix) is more sensitive and necessitates special conditions and may often be delayed [6]. Moreover, it is possible to inhibit (uncouple), experimentally, the synthesis of the "matrix" (non-fibrillar component) by treatment with cycloheximide [30, 31]. Thus protoplasts of *Sacch. cerevisiae* incubated with this antibiotic can only regenerate a microfibrillar wall [30] reminiscent of the cyst wall of *P. palmivora.* To stress the analogy between natural and artificial protoplasts, we may consider that in the encysting zoospore the synthesis of non-fibrillar components has been naturally inhibited.

To conclude, artificial as well as natural protoplasts seem to have a durable enzymic capacity for microfibrillar synthesis. But the formation of the non-fibrillar or matrix component of the wall, whenever present, requires continuous (*de novo*) protein synthesis.

IX. Morphogenetic Implications

The conversion of zoospores of *P. palmivora* into cysts involves a marked change in cell shape: from a pyriform protoplast with a lateral groove to a spherical walled cell (Fig. 2). This is a curious reversal of the situation with artificial protoplasts where the protoplast is characteristically spherical and the walled forms are not [5]. The peculiar shape of the zoospore is probably dictated by an internal microtubular cytoskeleton which becomes dismantled at the time of encystment [20]. The cyst wall formed around the zoospore protoplast probably plays a minor, if any, role in the rounding-off process. The cyst wall may be viewed as a relatively simple microfibrillar sac formed around an already existing cell—a "morphogenetically passive" wall which probably has little control on the shape of the cell. By contrast, the germ tube wall not only has a much more elaborate architecture (Fig. 9) but, perhaps because of this greater complexity, it also plays a more active role in cell morphogenesis—as this "morphogenetically active" wall grows, it dictates the cylindrical shape of the germ tube. An interesting analogy exists in reverting yeast protoplasts where "the morphology of the cell wall synthesized *de novo* follows the shape of the protoplast" [6]. These initial walls are chemically incomplete or defective [32,33]. Eventually, the yeast recovers its capacity to make a complete cell wall and, with it, the ability to undergo normal morphogenesis [6].

X. Summary

The fungal zoospore ostensibly lacks a cell wall and can therefore be regarded as a true protoplast of natural origin. The acquisition of a cell wall by the encysting zoospore affords an amenable system for studying the mechanism of cell wall genesis in fungi. Zoospore encystment may be equated to the process of cell wall regeneration by artificially produced protoplasts, a process which has received a great deal of attention in recent years.

By vigorous agitation, zoospores of *Phytophthora palmivora* undergo a rapid and highly synchronous encystment. In 2 min nearly all zoospores fabricate an alkaline-resistant cyst wall. The synthesis of this wall proceeds linearly for about 10 min. Like the walls of other morphological stages of this fungus, the cyst wall is made principally of β-glucans. The cyst wall, however, has a unique, simple architecture consisting of a thin envelope of tightly interwoven microfibrils of uniform diameter. There is no evidence of a permanent amorphous matrix associated with the microfibrils. This contrasts with the more complex architectural organization of the hyphal wall which has also a microfibrillar skeleton but this is heavily coated by an outer layer of non-fibrillar glucan.

Electron microscopy of thin sections of zoospores reveals large peripheral vesicles which are consumed during encystment. There is, however, no chemical evidence that the zoospore contains pre-fabricated cyst-wall microfibrils. More likely, the microfibrillar polysaccharides are assembled largely, if not entirely, *de novo* during encystment. Cell-free extracts of zoospores contain particulate enzymes capable of transferring glucosyl residues from uridine diphosphoglucose to insoluble glucans. The swimming zoospore is equipped with the enzymatic machinery needed for encystment. Agitation, or any other stimulus triggering encystment, somehow initiates the operation of this machinery.

Acknowledgement

Experimental work from my laboratory described herein was supported in part by research grants from the National Institutes of Health (AI-06205) and the National Science Foundation (GB-32165).

References

1. Aragaki, M. and Hine, R. B. (1963). *Phytopathology*, **53**, 854.
2. Aragaki, M. and Hine, R. B. (1964). *Phytopathology*, **54**, 1431.
3. Hohl, H. R. and Hamamoto, S. T. (1967) *Am. J. Bot.*, **54**, 1131.
4. Reichle, R. E. (1969). *Arch. Mikrobiol.*, **66**, 340.
5. Villanueva, J. R. (1966). *In:* "The Fungi" Vol. 2, p. 3 (Ainsworth, G. C. and Sussman, A. S., eds), Academic Press, New York.
6. Nečas, O. (1971). *Bact. Rev.*, **35**, 149.
7. Tokunaga, J. and Bartnicki-Garcia, S. (1971). *Arch. Mikrobiol.*, **79**, 283.
8. Bartnicki-Garcia, S. and McMurrough, I. (1971). *In:* "The Yeasts" (Rose, A. H. and Harrison, J. S., eds), Vol. 2, p. 441. Academic Press, London and New York.
9. Ho, H. H., Zachariah, K. and Hickman, C. J. (1968). *Can. J. Bot.*, **46**, 37.
10. Reichle, R. E. (1969). *Mycologia* **61**, 30.
11. Hemmes, D. E. and Hohl, H. R. (1971). *J. Cell Sci.*, **9**, 175.
12. Cantino, E. C., Lovett, J. S., Leak, L. V. and Lythgoe, J. (1963). *J. gen. Microbiol.*, **31**, 393.
13. Fuller, M. S. and Reichle, R. (1965). *Mycologia*, **57**, 946.
14. Hill, E. P. (1969). *J. gen. Microbiol.*, **56**, 125.
15. Bimpong, C. E. and Clerk, G. C. (1970). *Ann. Bot.*, **34**, 617.
16. Wang, M. C. and Bartnicki-Garcia, S. Unpublished observations.
17. Streiblová, E., Svoboda, A. and Nečas, O. (1967). *In:* "Symposium über Hefeprotoplasten" (Müller, R. ed.), p. 91. Akademie-Verlag, Berlin.
18. Kopecká, M., Čtvrtniček, O. and Nečas, O. (1967). *In:* "Symposium über Hefeprotoplasten". (Müller, R., ed.), p. 73. Akademie-Verlag, Berlin.
19. Tokunaga, J. and Bartnicki-Garcia, S. (1971). *Arch. Mikrobiol.*, **79**, 293.
20. Desjardins, P. R., Wang, M. C. and Bartnicki-Garcia, S. (1973). *Arch. Mikrobiol.* **88**, 61.
21. Sing, V. O. and Bartnicki-Garcia, S. (1972). *Phytopathology*, **62**, 790.
22. Roelofsen, P. A. (1959). "The Plant Cell Wall". Gebrüder Borntraeger. Berlin-Nikolassee.
23. Lindsey, S. and Bartnicki-Garcia, S., unpublished results.
24. Grove, S. N. (1970). *Am. J. Bot.*, **57**, 745. And personal communication.
25. Wang, M. C. and Bartnicki-Garcia, S. (1973). *J. biol. Chem.* **248**, 4112.
26. Lovett, J. S. (1968). *J. Bact.*, **96**, 962.
27. Tsao, P. H. (1969). *Int. Bot. Congr.*, p. 221.
28. Wang, M. C. and Bartnicki-Garcia, S. (1971). *Bact. Proc.*, p. 61.
29. Camargo, E. P., Dietrich, C. P., Sonneborn, D. and Strominger, J. L. (1967). *J. biol. Chem.*, **242**, 3121.
30. Nečas, O., Svoboda, A. and Kopecká, M. (1968). *Expl Cell Res.*, 53,291.
31. Elorza, M. V. and Sentandreu, R. (1969). *Biochem. biophys. Res. Commun.*, **36**, 741.
32. Eddy, A. A. and Williamson, D. H. (1959). *Nature, Lond.*, **183**, 1101.
33. Garcia-Mendoza, C. and Novaes-Ledieu, M. (1968). *Nature, Lond.*, **220**, 1035.

Cell Wall Regeneration of Protoplasts of *Histoplasma capsulatum*

L. M. CARBONELL, F. GIL and F. YEGRES
Center of Microbiology and Cell Biology,
Instituto Venezolano de Investigaciones Cientificas,
Caracas, Venezuela

Regeneration of protoplasts from yeast [1, 2] and from mycelia [3] have been reported both in solid and in liquid media. A review of this subject appeared recently [4].

Recently, protoplasts of fungi pathogenic to humans have been obtained using enzymatic [5] and non-enzymatic methods [6]. An electron microscope study of protoplast formation in *Histoplasma capsulatum* using a non-enzymatic method [7] appeared recently. This paper presents our observations on the regenerating cell wall of *H. capsulatum,* and the identification of the first fibers to appear during this process.

I. Materials and Methods

The procedure to obtain protoplasts from the yeast phase of strain G-184B (supplied by Dr. M. D. Berliner, Harvard School of Public Health, Boston, Mass.) and strains SwB (supplied by Dr. J. E. Domer, Tulane University, New Orleans, La.) of *H. capsulatum* has been described elsewhere [7]. Maximum production of protoplasts occurred at 48 h for strains G-184B, and at 24 h for strain SwB.

To separate protoplasts from the yeast, the following procedure was used. The protoplast-yeast suspension was poured aseptically into a glass centrifuge tube of 50 ml size and centrifuged at 1500 × g for 30 min. Abundant protoplasts with few yeasts were found in the uppermost 5 ml of the supernatant. This layer was transferred to another centrifuge tube and 1/5 volumes of 0.8M mannitol was added. After centrifuging for 30 min at 1500 × g, the yeasts were found in the sediment and the protoplasts in the supernatant. In order to obtain protoplast pellets, 0.8 mannitol was added to the liquid medium containing the protoplasts (3:1 v/v) and centrifuged at 2000 × g for 30 min. The pellet was resuspended in 1 ml of mannitol and used for the regeneration experiments.

Regeneration of the protoplasts was done in both solid and liquid media. The solid medium was prepared with 2% brain heart infusion (Difco), 0.8 mannitol and 1.25% of agar. The liquid medium was prepared similarly but without agar. Regeneration in liquid medium was done in Leyton tubes containing 1 ml of medium and two drops of protoplast suspension. Regeneration in solid medium was done in glass vials stoppered with gauze, by mixing 0.5 ml of the solid media and one drop of the protoplast suspension at 43° C, the lowest temperature at which the agar remains liquid. Each vial was put into a centrifuge tube of 10 ml size containing 1 ml of distilled water, and stoppered with gauze in order to prevent desiccation of the material. All regeneration experiments were done at 37° C.

After a given incubation time (2, 18, 24, and 72 h), the regenerated material in the solid medium was cut into small blocks and fixed in 2% glutaraldehyde in 0.1M cacodylate buffer pH 7.2 containing 0.8M mannitol. The material regenerated in liquid was concentrated by centrifugation, fixed with glutaraldehyde as above, and embedded in 2% agar. These two materials were prepared for electron microscope examination as described elsewhere [7]. Since the osmotic fragility of the regenerating protoplast is much less than that of fresh and growing protoplasts, the following procedure was used to study the morphology of the microfibrils appearing in the first steps of synthesis of the cell wall. The incubated material from liquid or solid media was diluted with water in order to burst protoplasts that had not regenerated. This material was put in an Omnimixer (Onm 200, Ivan Sorvall) for 2 min at maximum speed, and centrifuged. The pellet was ground by hand in a mortar with liquid nitrogen. Material was centrifuged at low speed, and the sediment was examined with the electron microscope, after shadowing or negative staining. Shadowing was done with platinum at a 30° angle; negative staining with 4% phosphotungstic acid at pH 6.8.

II. Results

When the sections were observed with the electron microscope there were no remnants of cell wall on the surface of the fresh protoplasts. Description of the fresh and growing protoplasts was made elsewhere. At an early stage of regeneration, the cell membrane of the protoplast became wavy. In solid medium there was a centripetal invagination of the cell membrane which divided the protoplast into three to six cytoplasmic masses (Fig. 2). Some of these masses had one nucleus while others had none (Fig. 2). During the first 24 h,

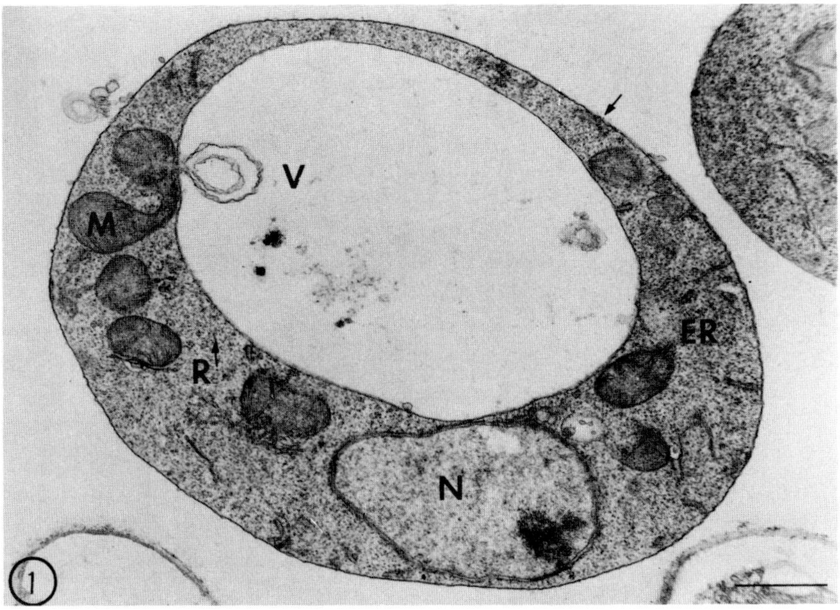

Fig. 1. Fresh protoplast G-184B. There are no remnants of cell wall attached to the cell membrane. The cell membrane is clearly outlined (arrow). Note in the cytoplasm mitochondria (M) with electron dense matrixes, nucleus (N) with clearly delimited nuclear envelope, a big out of center vacuole (V) with membranes and electron dense material inside, scarse endoplasmic reticulum (ER) and ribosomes (R). x60.000. Bar = 0.25 μm for all figs unless otherwise specified.

fibrils which formed a loose network began to appear outside the cell membrane, appearing first at the site of invagination. Later, the same type of fibers appeared at the external surface of the protoplasts (Fig. 3). Vesicles were observed outside of the cell membrane and at the beginning of the formation of the cell wall they were entwined with the fibrils (Fig. 3).

After 24 h, the thickness of the cell wall increased and there was a deposition of amorphous material which obscured the morphology of the newly formed fibrils. At this stage, distinct fibrils were seen only at the periphery of the cell wall. Later the cytoplasmic masses disengaged themselves producing a yeast-like structure. Some of the cytoplasmic masses before disengagement, produced buds indistinguishable from normal buds. Not all of the cytoplasmic masses produced yeasts (Fig. 4).

When the invagination of the cell membrane occurred, the characteristics of the cytoplasmic masses, which had a nucleus, were

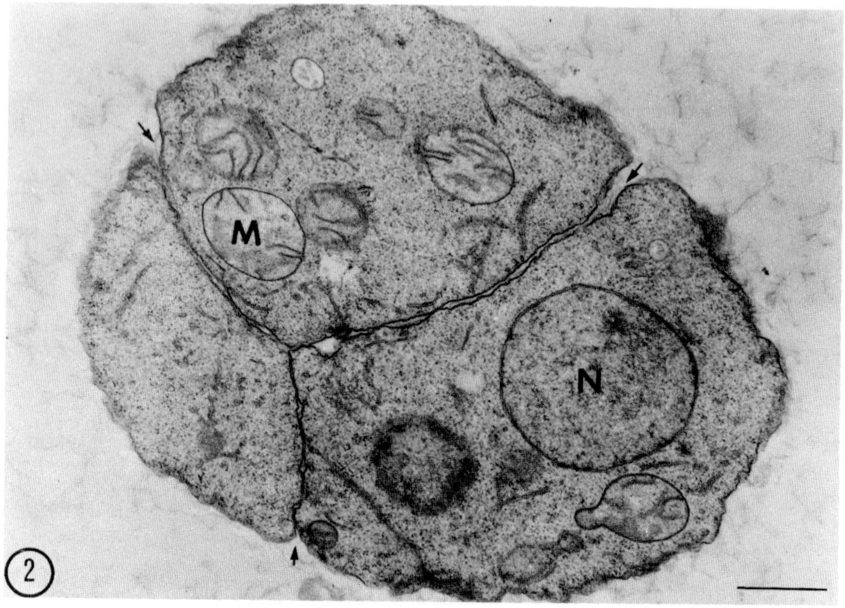
M
N
2

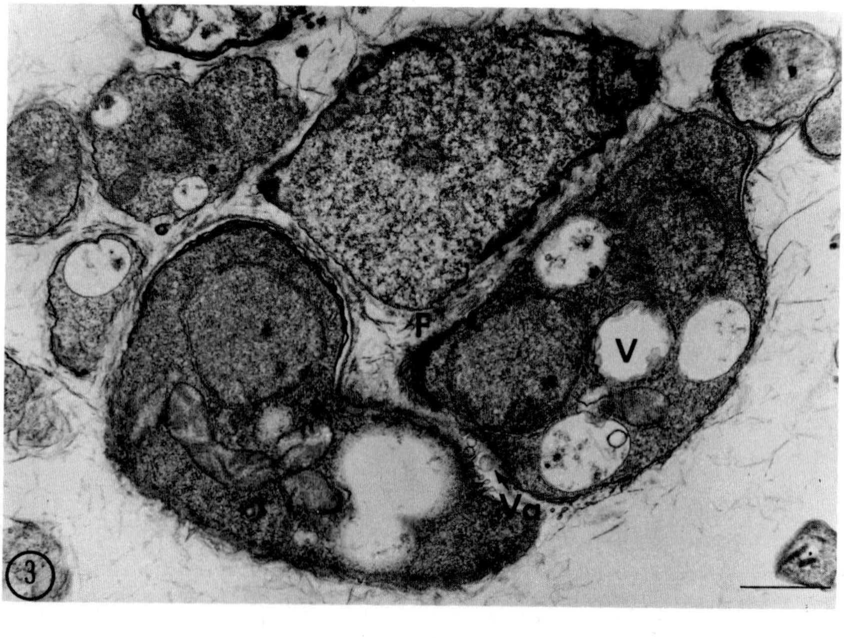
P
V
Va
3

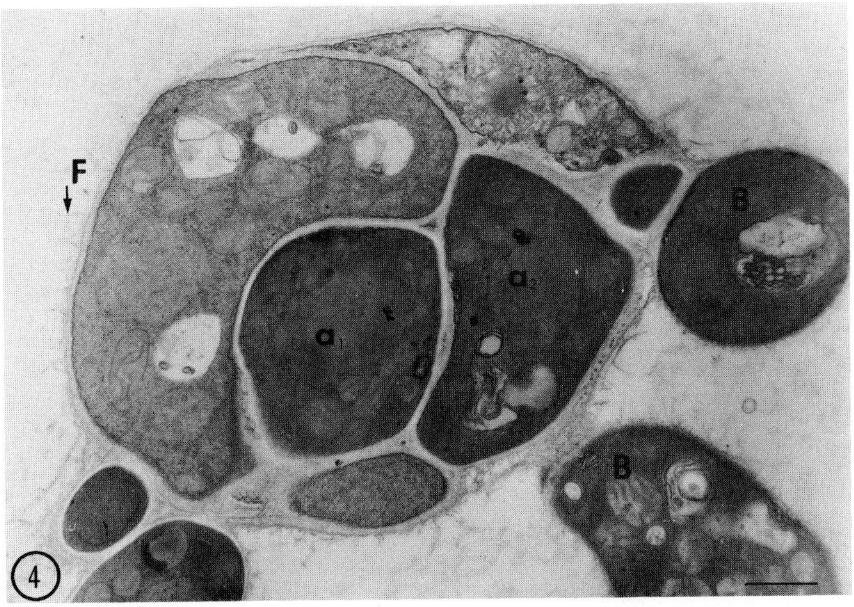

Fig. 4. The cell wall is formed and fibrils (F) could be only identified at the surface of the cell wall due to the production of the amorphous matrix. Some of the cytoplasmic masses are indistinguishable from a yeast (a_1 a_2). One of them is budding (B). Regeneration in solid medium. *H. capsulatum* G-184G protoplast. x50.000.

similar to the cytoplasm of fresh protoplasts, with low electron density of the cytoplasm and a clear outline of the nuclear envelope (Fig. 1). As the cell wall was synthesized, the density of the cytoplasm increased due mainly to an increasing number of ribosomes. When the regenerated protoplast became a yeast, the density of the cytoplasm was so high that it obscured the organelles within (Fig. 4). Mesosomes appeared only when the yeast was formed. The big, out of center vacuole (Fig. 1) which was observed in the fresh protoplast

---

Fig. 2. Early stage of regeneration of the yeast protoplast G-184B. Observe the invagination of the cell membrane (arrows). The cytoplasmic masses contain nucleus (N) and/or mitochondria (M). Regeneration in solid medium. x60.000.

Fig. 3. Observe the fibrils (F) between the cytoplasmic masses. The amorphous matrix has not been yet fully synthesized allowing a clear contour of the fibrils. The cytoplasmic masses have increased electron density as compared with those in Fig. 2. Small vacuoles are seen (V). Observe the vacuoles (Va) between the cell membrane and the cell wall. Regeneration in solid medium. *H. capsulatum* G-184B protoplast. x60.000.

disappeared. A few small vacuoles empty, or containing electron dense material, or membranes, were seen in the regenerating protoplast and in the newly formed yeast (Fig. 4).

In some cases wall production was not preceded by invagination of the cell membranes. In this case the wall regenerated at the periphery of the protoplast (Fig. 5). Multiple nuclei were found in these regenerated protoplasts. Some protoplasts instead of budding (Fig. 6) produced a multinucleated septated hyphae.

Regeneration in liquid medium very seldom shows the invagination of the cell membrane. Observations at 10 h revealed that the cell wall began to form as a net of loosely bound fibers, with no amorphous matrix between them. The electron density of the cytoplasm increased and the fibrils were so loosely distributed that the cell wall became quite thick 0.5 μm (Fig. 7). When the protoplasts regenerated in a solid medium, the synthesis of the amorphous

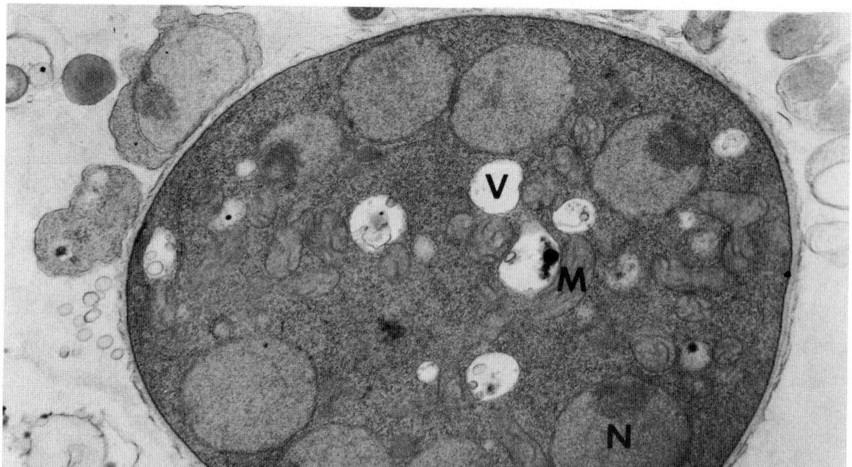

Fig. 5. Regenerated protoplast G- 184B in which there was an absence of cytoplasmic masses. Observe the abundant number of nuclei (N), the high electron density of the cytoplasm, the small vacuoles (V) and the mitochondria (M). Regeneration in solid medium. x50.000.

Fig. 7. Regeneration of the cell wall in liquid medium. Observe the clear outline of the fibres due to the absence of the amorphous matrix. The fibers are not tightly bound which increases the thickness of the cell wall. N: nucleus, CW: cell wall. *H, capsulatum* G-184B x50.000.

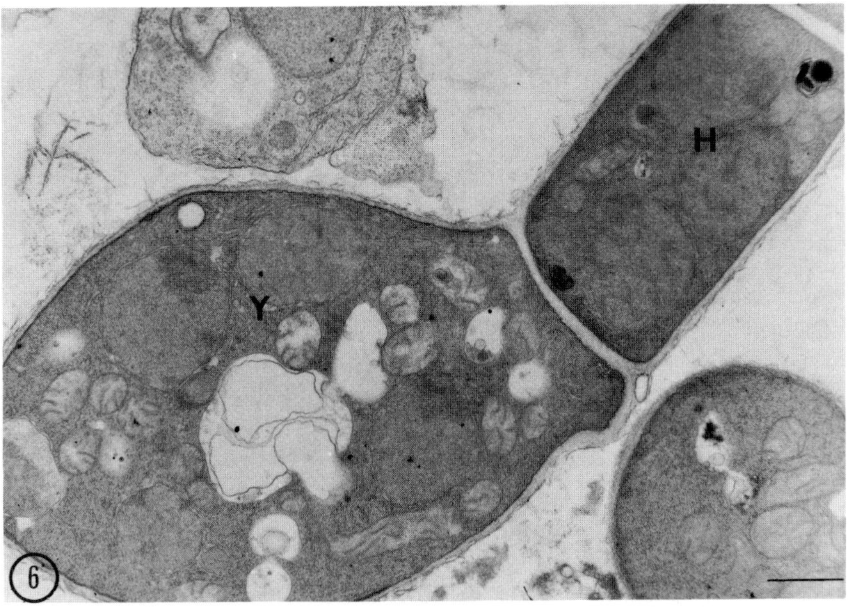

Fig. 6. Regenerated protoplast G-184B with all the characteristics of the yeast (Y). Instead of budding it produced a septated hyphae (H). Regeneration in solid medium. x50.000.

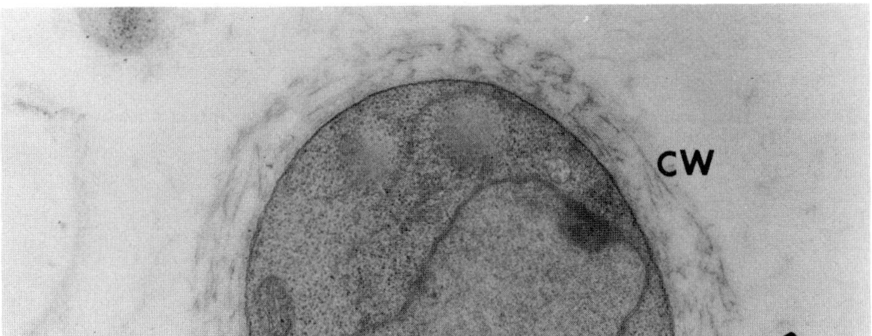

matrix obscured the morphology of the newly synthesized microfibrils while in liquid medium the microfibrils remained clearly outlined. The nature of the newly formed fibers depended upon the strain used.

Two kinds of fibers appeared in strain G-184B, and only one kind in strain SwB when they were regenerated in liquid medium. One was a thick microfibril of undefined length with undulating borders and measuring between 12 to 17 nm in diameter. This fiber was previously identified as an α-1,3-glucan fiber [8]. The other fibril was much longer and slender, measuring 2 to 3 nm across. Bundles of these, measuring from 10 to 15 nm in thickness, were irregularly interwoven resembling the pictures of frozen etched regenerated protoplasts of *S. cerevisiae* [9]. These thinner fibers have been identified as chitin fibers (Figs 8, 9) [8, 10]. The thin fibers of chitin can be identified only with negative staining. With the metal-shadowing technique only the bundles were observed. Strain G-184B contains both α-1-3-glucan and chitin; strain SwB contains only the chitin fibers.

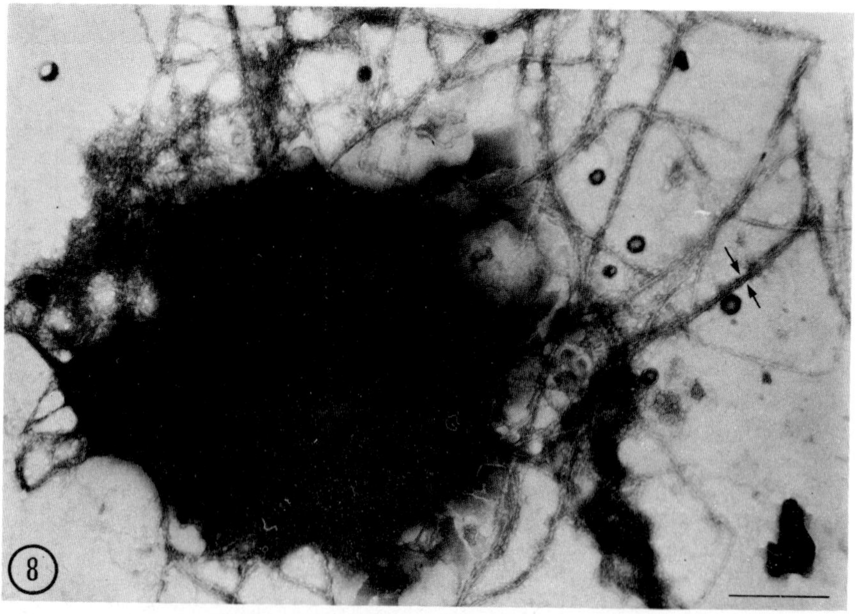

Fig. 8. Early stage of regeneration of *H. capsulatum* SwB protoplast. Observe the thick bundle which is formed by thinner elements (arrows). These microfibrils are identified as chitin fibers. Negative Stain. 19 h regeneration in liquid medium. x75.000.

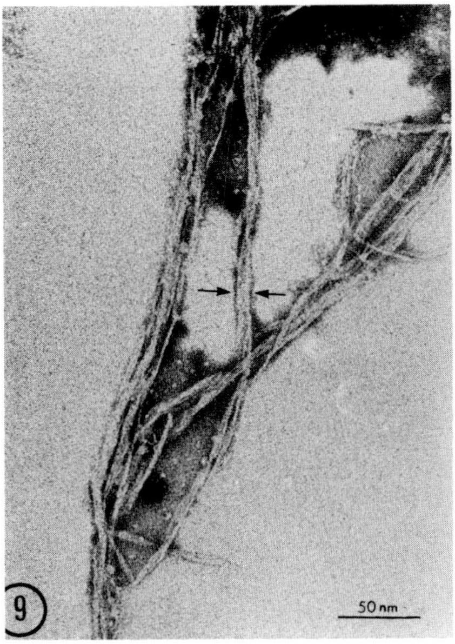

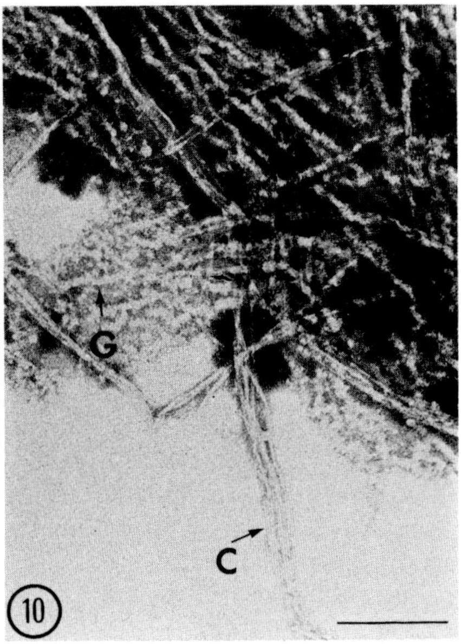

Fig. 9. Chitin fibers in a regenerated *H. capsulatum* SwB protoplast. Observe the thin microfibril (arrow) and the bundle of microfibrils (arrows). Negative stain. Regeneration in liquid medium. x250.000.

Fig. 10. Regeneration in liquid medium of cell wall of *H. capsulatum* G-184B protoplast. Observe the two types of fibers: the chitin microfibril (c), forming bundles and with smooth edges and the much thicker glucan fibers (g) not forming bundles and with a slightly dented edge. Negative stain. x75.000.

Discussion

The loose fibrillar network that formed around the protoplast when liquid medium was used, did not resemble normal yeast morphogenesis. The reversion of *H. capsulatum* protoplasts that took place in solid medium was more nearly "normal". Under these conditions septation by invagination of the cell membrane of the protoplast took place, and microfibrils and amorphous matrix were synthesized producing a complete cell wall. Svoboda and Nečas obtained similar results with *S. cerevisiae* when using agar [11] and gelatin gel [12]. The vesicles extruding from the protoplast [7], were seen in increasing numbers in the regenerated protoplast. They are located, as are the lomasomes [13], between the cell membrane and the inner surfaces of the newly formed cell wall. The presence of lomasomes in the yeast and in the regenerated protoplast indicates that they may play an important role in building the cell wall as demonstrated in *Penicillium vermiculatum* [14].

Domer [15] divided the strains of *H. capsulatum* into two groups: chemotypes I and II. Domer's data strongly suggested that there was no α-glucan in the chemotype I group, and we have confirmed this in the yeast form of *H. capsulatum* strains SwA, SwB, and Bonn (kindly supplied by Dr. J. E. Domer as representatives of group I).

α-1,3-glucan is present in the yeast form of *H. capsulatum* G-184B which belongs to the chemotype II. α-1,3-glucan and chitin fibers have been morphologically identified in the cell wall of the yeast phase of *H. capsulatum* G-184B and of *Paracoccidioides brasiliensis* [8]. Only one type of fiber (chitin) was found in chemotype I (Carbonell unpublish data). In SwB protoplasts the first fibers formed were chitin fibers. The two types of fibers that appeared in the regenerated protoplasts of *H. capsulatum* G-184B (Chemotype II) were identified morphologically as α-1,3-glucan and chitin. The glucan fibers were more abundant than the chitin ones. The morphological data are consistent with the chemical analysis of the walls of the two strains. Thus, SwA wall contained chitin and no α-glucan, while G-184B wall contained large amounts of both chitin and α-glucan. (Kanetsuna, unpublished data). β-glucan is the other major component of the yeast phase walls of *H. capsulatum*. In strain G-184B β-glucan accounts for 40% of the wall glucan (Kanetsuna, and Carbonell, in press); in strain SwB, 68% of the total glucan as determined by exo-β-1,3-glucanase hydrolysis, (Kanetsuna unpublished data). The ultrastructural identification of this β-glucan was difficult, since we did not have a pure chitinase. We believe its ultrastructure may be similar to that of chitin fibers, but with fibers scattered at random, of shorter lengths, and not forming bundles as does chitin.

Nećas [4], has commented on the possibility that if a protoplast is left some time in an inhibitory medium for the synthesis of the cell wall, the synthetic processes making wall material are allowed to run to waste for a time without the feedback action of the wall, the normal cell cycle and cytokinesis are irreversibly impaired. In a budding yeast such as *S. cerevisiae* the length of time for this process is 6 h [4], with our material 24 and 48 h, depending on the strain. If *H. capsulatum* protoplasts were harvested from the inhibitory medium after 24 or 48 h, the regeneration rate was sharply reduced. To explain these facts it is important to take into account the difference in methods used to produce the *S. cerevisiae* protoplasts (snail enzyme) and the *H. capsulatum* protoplasts ($MgSO_4$, 1.5M), as these agents have different metabolic effects on the protoplast itself.

The number of nuclei varied from one to five in the fresh and growing protoplast [7]. When the protoplast regenerated and cytoplasmic masses were formed by the invagination of the cell membrane, one or two nuclei were enclosed in each cytoplasmic membrane mass which later formed a yeast. When there was no formation of cytoplasmic masses, accumulation of nuclei occurred as seen in Fig. 6. The fate of these nuclei and abnormal yeasts is not known. In *S. cerevisiae* the synchronization of karyokinesis with cytokinesis in regenerated walled protoplasts containing a larger number of nuclei did not take place in the first generation, but in the second. The reduction of the nuclear number per cell was due to the greater probability that one nucleus, rather than several passed into the bud before it was closed by the septum [4].

References

1. Nećas, O. and Svoboda, A. (1967). *Fol. biol, (Praha)*, **13**, 379.
2. Havelková, M. (1969). *Fol. Microbiol.*, **14**, 155.
3. Peberdy, J. F. and Gibson, R. K. (1971). *J. gen. Microbiol.*, **69**, 325.
4. Nećas, O. (1971). *Bact. Rev.*, **35**, 149.
5. Berliner, M. D. and Reca, M. E. (1969). *Mycopathol. Mycol. Appl.*, **37**, 81.
6. Berliner, M. D. and Reca, M. E. (1970). *Science*, **167**, 1255.
7. Carbonell, L. M., Berliner, M. D. and Gil, F. (1972). *J. Bact.*, **110**, 731.
8. Carbonell, L. M. (1970). *J. Bact.*, **101**, 636. University J. E. Purkyne, Brno.
9. Nećas, Kopecká, M. and Brichta, J. *In* Yeast Protoplasts, p. 33 (eds. O. Nećas and A. Svoboda). University J. E. Purkyné, Brno. (1970).
10. Moreno, R. E., Kanetsuna, F. and Carbonell, L. M. (1969). *Arch. Biochem. Biophys.*, **130**, 212.
11. Svoboda, A. (1966). *Exp. Cell Res.*, **44**, 640.
12. Nećas, O. (1961). *Nature, Lond.*, **192**, 580.
13. Carbonell, L. M. and Rodriguez, J. (1968). *J. Bact.*, **96**, 533.
14. Wilsenach, R. and Kessel, M. (1965). *J. gen. Microbiol.*, **40**, 401.
15. Domer, J. E. (1971). *J. Bact.*, **107**, 870.

Yeast Septum Formation as a Model System for Morphogenesis

ENRICO CABIB, VLADIMIR FARKAS and RODNEY E. ULANE

Laboratory of Biochemistry and Metabolism
National Institute of Arthritis, Metabolism and Digestive Diseases

BLAIR BOWERS

Laboratory of Bochemistry, National Heart and Lung Institute
National Institutes of Health, Bethesda, Maryland, U.S.A.

Despite the time and effort devoted to studies of cellular differentiation, we are still far from the goal of understanding the biochemical basis of morphogenesis. Success in attaining this goal will depend largely on the finding of adequate model systems. Preferably, these systems should be endowed with certain properties that will facilitate the experimental attack at both the ultrastructural and the molecular level. In the first place, the object whose formation is to be studied should be of a specific and simple chemical composition, consisting, if possible, of a single substance; yet it should have a definite shape, recognizable at the level of resolution of the electron microscope. It would also be desirable that the structure in question occupy a precisely defined region of the cell and that its formation be restricted to a specific phase of the cell cycle. In this way it would be possible to study the synthesis of the structural component and its regulation independently from the total process of cell growth. In what follows we shall try to show that the primary septum formed during cell division in budding yeasts satisfies the above-mentioned conditions.

I. Septum Formation in Yeast

The different stages of septum formation in a budding yeast are shown schematically in Fig. 1, (A to E). In (B) the primary septum between mother cell and bud begins to appear, pushing the plasma membrane centripetally. Although the primary septum appears in electron micrographs as electron lucent material [1] this area has

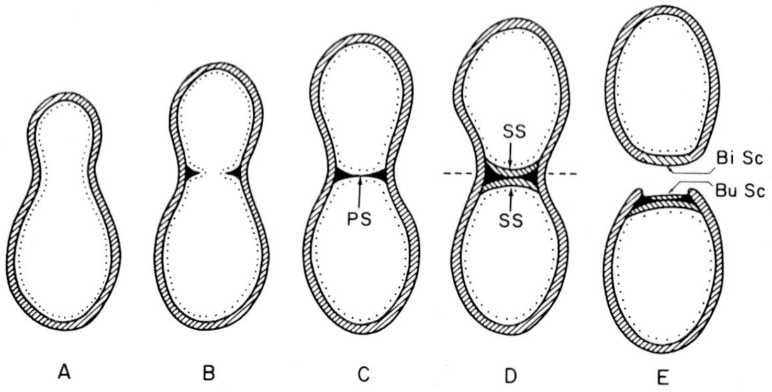

Fig. 1. Septum formation in budding yeast. The dotted line represents the plasmalemma. The cross-hatched areas are the cell walls. The secondary septa are distinguished by a cross hatching in the opposite direction, although they may well have the same composition as the cell wall. *PS*, primary septum; *SS*, secondary septa; *Bu Sc*, bud scar; *Bi Sc*, birth scar.

been darkened in Fig. 1 for clarity of presentation. The inward growth of the primary septum finally closes the gap between the two cells while the membranes pinch off on both sides (Fig. 1C) [2]. In later stages, secondary septa are laid down from both sides, in juxtaposition to the primary septum, as shown in Fig. 1D [1, 2]. Finally, the two cells separate in such a way that most or all of the primary septum remains with the mother cell (Fig. 1E). This asymmetrical separation accounts for the well known structural difference between bud scar and birth scar. Our own studies on the chemical composition and internal structure of the bud scar [3] strongly suggest that a chitin disc with a raised ridge occupies the position in which the primary septum ends up in the bud scar (Fig. 1E). We therefore propose that the primary septum is made up of chitin, whereas secondary septa would have a composition similar to that of the normal cell wall.* In agreement with this hypothesis, we have shown, using a synchronized yeast culture, that chitin synthesis takes place at a precise time in the cell cycle, that is, before cell division occurs [4].

On the strength of the previous discussion, it would appear that the yeast primary septum satisfies the conditions outlined above for a good morphogenetic model. It has a well-defined localization in the cell, a simple and specific chemical composition, and its formation is

* Glucan, a normal component of the cell wall, is also associated with the bud scar, as previously reported [3].

restricted to a limited portion of the cell cycle. In order to understand how such localization in space and time is attained, it was necessary to study this process at the molecular level.

II. The Enzymic System for Chitin Synthesis

We have reported previously [5] that a particulate fraction isolated from a yeast protoplast lysate catalyses the reaction shown in Equation 1, where $(AG)_n$† stands for chitin.

$$\text{nUDP-AG} + \text{particle} \rightarrow \text{nUDP} + (\text{AG})_n - \text{particle} \qquad (1)$$

The acetylglucosamine acceptor is endogenous and has not yet been identified. In addition to an absolute requirement for divalent cation, the presence of free *N*-acetylglucosamine is necessary for maximal activity.

A striking and unexpected observation was the almost complete loss of activity in the particles recovered after exposure to a very mild sonic treatment. Nevertheless, incubation of the particles with the supernatant fluid (AF) from the sonic-treated extract restored most or all of the activity (Table I). An even more dramatic activation of chitin synthetase took place after treatment with trypsin (Table I). These and similar results [4] led to the conclusion that the particulate preparation contained chitin synthetase in an inactive or zymogen form, and also included an "activating factor" (AF), capable of transforming zymogen into active enzyme. Sonic treatment released AF from the particles into the supernatant fluid, from which AF could be precipitated with ammonium sulfate [4]. Figure 2 shows schematically the procedure currently used to prepare zymogen and AF. Further experiments [4] showed that AF was thermolabile and that the activation of chitin synthetase was linearly dependent on the amount of AF added and on incubation time. Thus, AF manifests the typical properties of an enzyme.

The finding that both trypsin and AF could activate zymogen suggested that AF might also be a protease. Proteolytic activity was indeed detected in the AF preparations, but this fact could not be taken as proof that such an activity was responsible for the transformation of zymogen. In order to solve this problem an inhibitor of the chitin synthetase system was used. Previous work [6] had shown that yeast extracts contain a heat-stable, nondialysable, trypsin-

† The abbreviations used are AG, *N*-acetylglucosamine; AF, chitin synthetase activating factor.

TABLE I
Separation of activating factor from chitin synthetase and reconstitution of the system

Enzyme preparation	Incubation before assay	*N*-acetyl glucosamine incorporated (nmoles)
Crude enzyme	—	5.2
Zymogen	—	1.3
Zymogen	60 min at 30°C	0.52
Zymogen	60 min at 30°C + 10 μl AF	3.1
Zymogen	60 min at 30°C + 20 μl AF	5.1
Zymogen	15 min at 30°C + 10 μg trypsin	17.0

Crude enzyme, zymogen (sonic-treated, washed particles) and AF (supernatant fluid from sonic-treated suspension) were prepared from *Saccharomyces cerevisiae* S288C, as previously described [4, 5] (see also Fig. 2 below). After incubation with AF or trypsin, the activation was stopped with an excess of AF inhibitor (see below) or soybean trypsin, respectively. The amount of active chitin synthetase was then determined as already reported [4, 5].

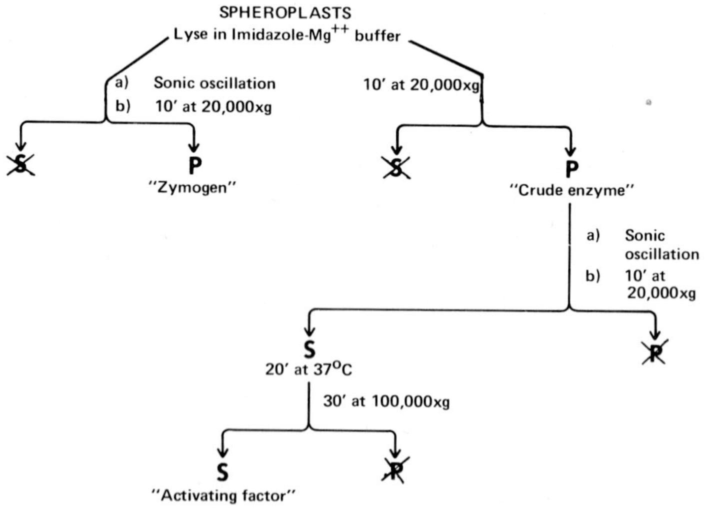

Fig. 2. Method of isolation of zymogen and AF. S, supernate; P, pellet. Crossed-out fractions were discarded. For further details and for purification of AF by ammonium sulfate precipitation, see Cabib and Farkas [4].

sensitive inhibitor of chitin synthetase. This heat-stable protein was purified to near-homogeneity from *Saccharomyces carlsbergensis* [6]. With a slightly modified method, a similar inhibitor was isolated from *S. cerevisiae* S288C, and freed from practically all of the remaining impurities by preparative disc gel electrophoresis (Fig. 3).

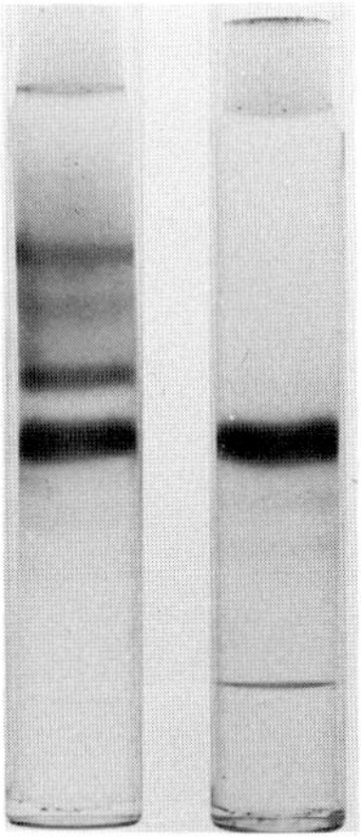

Fig. 3. Disc gel electrophoresis pattern of AF inhibitor, purified from *S. cerevisiae* S288C as previously reported [4, 6], before (*A*) and after (*B*) further purification by preparative disc gel electrophoresis. Gels were stained with Coomassie blue. The two minor bands in *B* account for only a few percent of the stainable material and seem to be artifacts, since they appear again after reisolation and new electrophoresis of the main band.

The resolution of the chitin synthetase system into zymogen and AF allowed testing of the inhibitor on the different steps of the reaction. It was found [4] that although inhibitor prevented the activation of zymogen by AF, it was without effect on the enzyme already activated, that is, the inhibitor acted only on AF. The inhibition of AF was practically stoichiometric [4]. The availability of the purified inhibitor made it possible to test the hypothesis that AF was a protease, by measuring the effect of the inhibitor on both activation of zymogen and the proteolytic action of AF preparations. As shown in Fig. 4, the inhibition curve is the same for both activities, a result which strongly supports the conclusion that AF is a protease.

From the foregoing results, a rather complicated picture emerges for the regulation of the chitin synthetase system, as summarized in Fig. 5. According to this scheme, the zymogen form can be transformed into active enzyme by AF, a process which is blocked by the

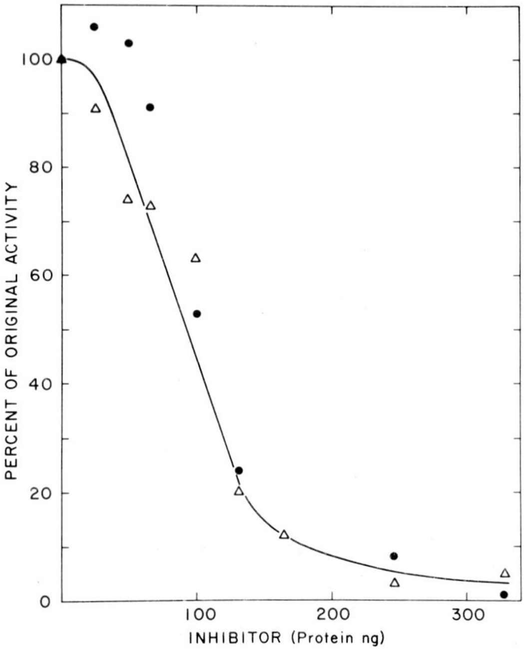

Fig. 4. Titration of AF with inhibitor in chitin synthetase (●) and protease (△) assay. The activation of zymogen was measured as previously reported [4]. Protease was determined with the insoluble substrate "Azocoll", from Calbiochem (protein from cowhide, complexed with a dye). The incubation mixture contained 21 mg/ml of Azocoll, 0.1 M imidazole chloride at pH 7.4, AF (0.54 μg of protein) and variable amounts of purified inhibitor, as indicated. The activity was determined from the increase in absorbance at 520 nm of a filtered reaction mixture.

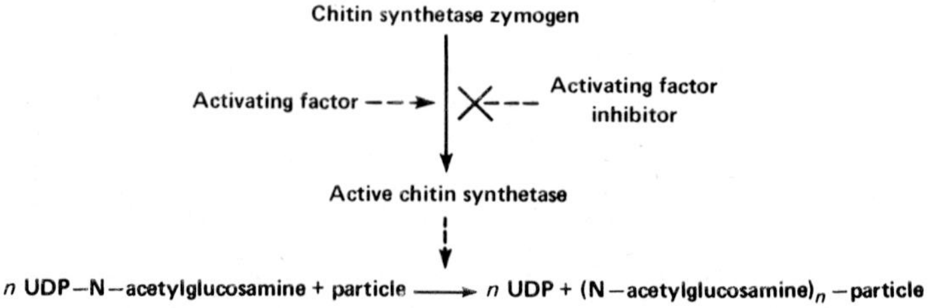

Fig. 5. Scheme of the chitin synthetase enzymatic system.

AF inhibitor. The effect of AF appears to be proteolytic in nature and may result in the removal of a peptide which otherwise masks the active site in zymogen. The activated enzyme can now catalyse the synthesis of chitin. In an attempt to understand how this system might function *in vivo,* the intracellular localization of its different components had to be studied.

III. Intracellular Distribution of the Components of the Chitin Synthetase System and a Hypothesis About Their Interaction

After osmotic lysis of yeast protoplasts, both the chitin synthetase zymogen and AF are found in a particulate fraction. Nevertheless, under a very mild sonic treatment AF becomes soluble, whereas the zymogen remains bound to particles. This result suggested that the zymogen is attached to a membrane, whereas AF is soluble, but enclosed in a fragile vesicle easily disrupted by sonic oscillation.

The firmness of the attachment of zymogen to an insoluble element is illustrated by the results of Table II. Even after treatment with 0.5% cholate, two-thirds or more of the zymogen activity is recoverable in the pellet; further sonic treatment in the presence of cholate did not affect the results. Recovery was even higher using Tris buffer at pH 8 rather than imidazole buffer at pH 6.5. It is worthy of notice that the activity of mannan synthetase [7] which is present in the same preparation, is almost completely abolished by the cholate treatment, as also happens with the small amount of active chitin synthetase which is measured in the absence of AF treatment. The latter effect may be due to removal by cholate of residual traces of AF from the zymogen preparation.

Proof that zymogen and AF reside in different organelles required separation of these organelles under mild conditions, with exclusion of sonic oscillation. Our speculations concerning the enclosure of AF within a vesicle led us to attempt the isolation of yeast vacuoles. The published procedures were unsatisfactory in our hands. The method of Indge [8] was not successful because of heavy contamination with other organelles and that of Matile and Wiemken [9] because of the low yield. Nevertheless, metabolic lysis of protoplasts according to Indge [10], followed by a centrifugation in a discontinuous Ficoll gradient similar to that used by Matile and Wiemken [9] led to the isolation of clean vacuoles in good yield (Fig. 6, A and B). The vacuole-free pellet obtained from the gradient was again submitted to gradient centrifugation but with different concentrations of Ficoll, mainly to eliminate residual cells and intact protoplasts. The

TABLE II
Treatment of zymogen with cholate

Treatment of zymogen	Chitin synthetase assay: N-acetyl glucosamine incorporated (nmoles) No AF	Chitin synthetase assay: N-acetyl glucosamine incorporated (nmoles) AF	Mannan synthetase assay: Mannose incorporated (nmoles)
Experiment 1			
None	0.43	11.1	29.6
Cholate (imidazole)	n.d.	7.35	0.96
Experiment 2			
None	0.7	8.2	
Cholate (imidazole)	n.d.	4.7	
Cholate, sonic oscillation (imidazole)	n.d.	4.4	
Cholate (Tris)	n.d.	7.4	

To 0.5 ml of a zymogen suspension in 0.05 M imidazole chloride pH 6.5, containing 2 mM $MgSO_4$ (protein content 7-10 mg/ml), 0.125 ml of 2.5% sodium cholate was added. The mixture was kept 20 min in ice and then centrifuged for 20 min at 100,000 x*g*. The pellet was washed three times with 1.5-ml portions of 0.05 M imidazole chloride at pH 6.5, containing 2 mM $MgSO_4$, and finally resuspended in the same buffer up to a total volume of 0.5 ml. Where indicated, 0.05 M Tris-chloride, pH 8, was substituted for the imidazole-Mg^{2+} mixture during cholate treatment. The imidazole-Mg^{2+} buffer was, however, used for the washings. In one experiment the cholate-containing suspension was submitted to sonic oscillation, using the microtip of a Branson sonifier Model S75 for 5 s at setting 1, before centrifugation and washing. Chitin synthetase activity, after 30 min incubation in the presence or absence of excess AF, was measured as previously reported [4, 5]. Mannan synthetase was determined as indicated by Behrens and Cabib [7]. The data are given in nmoles of sugar incorporated per 20 μl of enzyme suspension. n.d., not detectable.

fractions obtained, called "membrane fractions", remained heterogenous (Fig. 6C). The vacuole fraction was greatly enriched in AF (Table III) and contained very little zymogen. On the other hand, the membrane fractions contained zymogen and were completely devoid of AF. It is thus clear that AF and zymogen are located in different organelles.

The intracellular distribution of the AF inhibitor was also studied. As shown in Table IV, more inhibitor was obtained by extracting

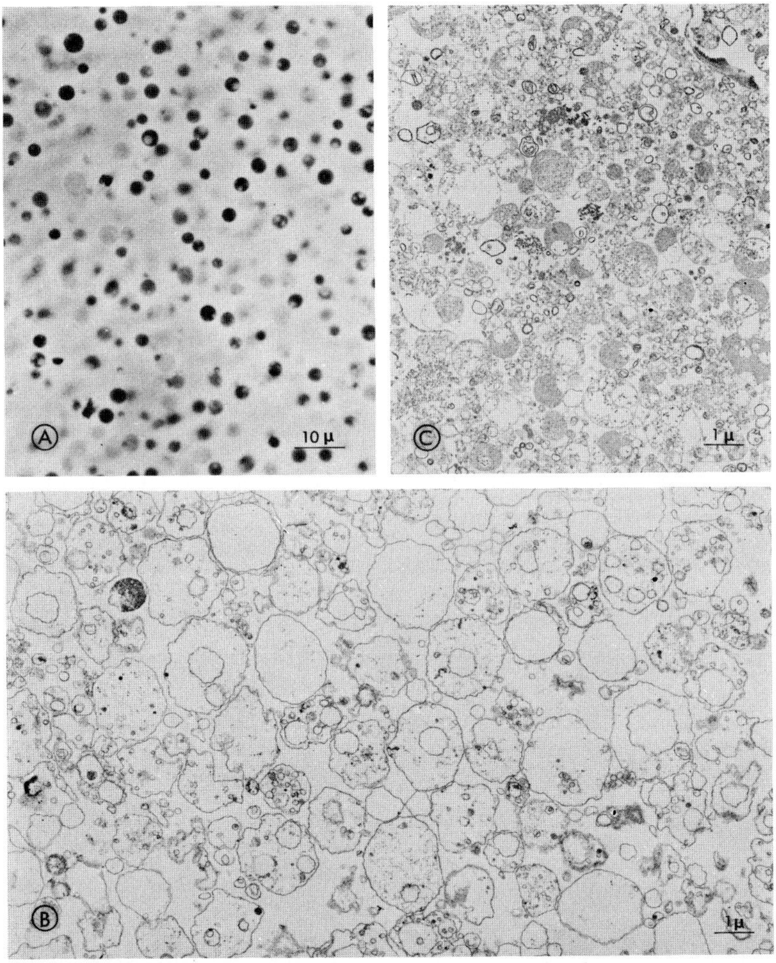

Fig. 6. Purified subcellular fractions. For the method of isolation see the legend to Table III. *A*, vacuole fraction under phase contrast microscopy. *B*, the same fraction in a thin section micrograph. Fixation was carried out in 3% glutaraldehyde-0.1 M phosphate buffer, pH 6.8, followed by Epon embedding. The sections were stained with uranyl acetate and lead citrate. *C*, "Membrane" fraction. Electron microscopy performed as for *B*.

protoplasts than intact cells, a result which eliminates the periplasmic region as a possible site of inhibitor localization. After centrifugation of a spheroplast lysate, most of the inhibitor was found in the supernatant fluid. It should be mentioned that a number of protoplast membranes reseal after osmotic lysis. Thus, it is possible that some of the inhibitor may be trapped inside of them and thus appear

TABLE III
Separation of zymogen and AF[a]

	Recovery per cent			Specific activity	
	Zymogen	AF	Ratio	Zymogen	AF
First gradient					
Crude particles	100	100	1	1	1
Vacuole fraction	2.6	45	17	—	26
Pellet	59	n.d.[c]	—	0.8	—
Second gradient					
Band 1	21[b]	n.d.[c]	—	2.3	—
Band 2	16.5[b]	n.d.[c]	—	0.86	—

[a] Protoplasts were obtained from *S. cerevisiae* X2180 (diploid isogenic with S288C) and subjected to metabolic lysis according to Indge [10]. The pellet obtained by centrifugation of this lysate for 5 min at 6000 ×*g* is called "crude particles". After homogenization in a Teflon-glass Potter homogenizer, the crude particles were fractionated in a discontinuous gradient similar to that used by Matile [9], except that the concentrations of Ficoll, from bottom to top, were 10, 7.5 and 6%, and the buffer used was imidazole at pH 6.5, containing 8% mannitol and 6 mM citrate. The centrifugation was for 45 min at 90,000 ×*g*. The meniscus layer (vacuole fraction) and the pellet were recovered and washed with 0.05 M imidazole, pH 6.5, containing 8% mannitol and 2 mM $MgSO_4$.

The pellet from the first gradient was refractionated in a second discontinuous gradient, with Ficoll concentrations of 40, 30, 20 and 10%, from bottom to top (the last layer contained the sample). The gradient was centrifuged for 1 h at 90,000 ×*g*, and the bands at the interphase between 10 and 20% Ficoll (Band 1) and between 20 and 30% Ficoll (Band 2) were recovered and washed with buffer as above.

Zymogen was measured in each fraction after treatment with trypsin [4]. AF was extracted from the different particles by sonic treatment as already reported [4]. The amount of zymogen and AF in the crude particles is taken arbitrarily as 100 and their specific activity as 1. The actual specific activities in the crude particles were 1.2 munit/mg protein for zymogen and 1.8 munit/mg protein for AF.

[b] Pellet taken as 100%.

[c] n.d. = not detectable.

in the pellet fraction. The tentative conclusion from these results is that the inhibitor, or at least most of it, is free in the cytoplasm.

On the basis of the foregoing evidence concerning the components of the chitin synthetase system and their distribution in the cell, a rationalization of the manner in which they might interact to produce the localized formation of chitin at a precise time in the cell cycle is depicted in Fig. 7. When precautions are taken to separate

TABLE IV
Intracellular distribution of inhibitor

	Treatment	Inhibitor content units/g yeast (wet wt)
Intact yeast	Heated, dialysed	11.4
Protoplasts	Heated, dialysed	15.2
Lysate supernate	Heated, concentrated, dialysed	8.4
Lysate pellet	Heated, dialysed	2.1

All samples were suspended or dissolved in 0.05 M imidazole chloride, pH 6.5, containing 2 mM $MgSO_4$. Heating for all the samples was for 4 min at 100° C followed by centrifugation and overnight dialysis of the supernatant fluid against water. The lysate supernate was obtained by centrifuging for 10 min at 20,000 x *g* the suspension obtained by lysing protoplast in buffer [5]. After heating as explained above and centrifuging off the denatured material, the supernatant fluid was concentrated in a rotary evaporator and dialysed. The lysate pellet was obtained in the same experiment, washed twice with the imidazole-Mg^{2+} buffer, heated, centrifuged and dialysed as described above. For the definition of units see Cabib and Farkas [4].

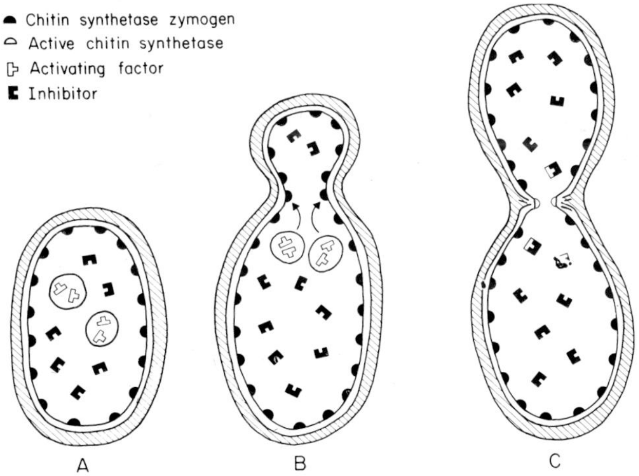

Fig. 7. Hypothetical mechanism of septum initiation. For details see text. (Reprinted from [4], with permission of the publisher.)

AF and prevent activation, practically all chitin synthetase is isolated in the zymogen form. Zymogen is very firmly bound to some structural element of the cell. Its amount remains essentially constant during cell growth [4] and throughout the cell cycle (Farkas and Cabib, unpublished results). We conclude that zymogen is present in large excess and probably bound to the plasmalemma (see Fig. 7A). For reasons explained above, AF is shown inside vesicles which may or may not be the cell vacuoles but fractionate together with the vacuoles. The AF inhibitor is assumed to be dissolved in the cytoplasm. At the time of primary septum initiation, the vesicles would carry AF to the site where chitin synthesis must occur (Fig. 7B), the membrane of the vesicle would coalesce with the plasmalemma and activation of chitin synthetase would occur (Fig. 7C). Thus, only a small part of total chitin synthetase (a few percent, considering the size of the bud scars) would be active in the culture at any given time. It is essential for this hypothesis that AF be carried inside an organelle in order to guarantee the localization of its action. In this scheme it is assumed that the inhibitor acts as a safety mechanism, i.e., as a trapping agent for any molecule of AF that may reach the cell lumen (see Fig. 7C). Activation of zymogen at any other than the desired place would thus be prevented. Clearly, several features of this scheme are hypothetical and need confirmation. Localization of zymogen in the plasmalemma, and proof that AF is the *physiological* agent for activation of zymogen, remain to be established. Other aspects are wholly unknown, such as the mechanism that would direct the AF- carrying vesicles to the right spot at the right time. On the other hand, there is little doubt that the final answer to this problem, indeed to any problem in morphogenesis, can only come from a simultaneous approach at the cellular, subcellular and molecular level, as has been attempted in the present study.

Acknowledgements

The authors are grateful to Dr. G. Ashwell, Dr. K.-P. Huang and Dr. W. B. Jakoby for a critical reading of the manuscript.

References

1. Marchant, R. and Smith, D. G. (1968). *J. gen. Microbiol.*, **53**, 163.
2. Sentandreu, R. and Northcote, D. H. (1969). *J. gen. Microbiol.*, **55**, 393.
3. Cabib, E. and Bowers, B. (1971). *J. biol. Chem.*, **246**, 152.
4. Cabib, E. and Farkas, V. (1971). *Proc. natn. Acad. Sci., U.S.A.*, **68**, 2052.
5. Keller, F. A. and Cabib, E. (1971). *J. biol. Chem.*, **246**, 160.
6. Cabib, E. and Keller, F. A. (1971). *J. biol. Chem.*, **246**, 167.
7. Behrens, N. H. and Cabib, E. (1971). *J. biol. Chem.*, **243**, 502.
8. Indge, K. J. (1968). *J. gen. Microbiol.*, **51**, 441.
9. Matile P. and Wiemken, A. (1967). *Arch. Mikrobiol.*, **56**, 148.
10. Indge, K. J. (1968). *J. gen. Microbiol.*, **51**, 433.

On the Nature of the Fibrillar Nets Formed by Protoplasts of *Saccharomyces Cerevisiae* in Liquid Media

D. R. KREGER AND M. KOPECKÁ

Department of Botany, University of Groningen, The Netherlands
Department of Biology, Medical Faculty, J. E. Purkyne University, Brno, Czechoslovakia

When protoplasts of *Saccharomyces cerevisiae* are cultured in liquid medium the regeneration of the new wall remains confined to the formation of a wide-meshed net-work of long microfibrils, as shown in Fig. 1 [1]. Since culture in solid media, particularly in 30% gelatin, results in synthesis of a complete wall [2], the formation of the fibrillar nets has tentatively been explained by a presumed partial removal of wall polymers, precursors and synthesizing enzymes by the liquid medium. With a view to the microfibrillar structure of the nets it is in particular the amorphous mannan fraction of the wall that has been supposed to be removed [3].

The present paper is a brief account of work undertaken to elucidate the chemical nature of the nets and thus to provide necessary data for a better insight into their meaning in terms of wall synthesis. The full data will be published elsewhere.

Kopecká *et al.* [1] have already reported that the nets are easily soluble in dilute alkali but are resistant against dilute mineral acid at 100° C. They were considered to be composed of a glucan, and this seemed to be confirmed when glucose was found as the only product of hydrolysis detectable [4]. However, neither the solubility of the nets in dilute alkali nor their conspicious microfibrillarity corresponds to the properties of the main glucan moiety of the normal wall. For the latter is usually isolated by its insolubility in hot dilute alkali and then does not show a microfibrillar structure [5,8]. Nevertheless, prolonged boiling of the walls in dilute mineral acid yields a product which by its ready solubility in dilute alkali and notable microfibrillarity [5] has indeed properties similar to those of the nets. These microfibrils, unlike those of the nets, are normally of a short rod-like type (Fig. 2) and may aggregate to form strings [5].

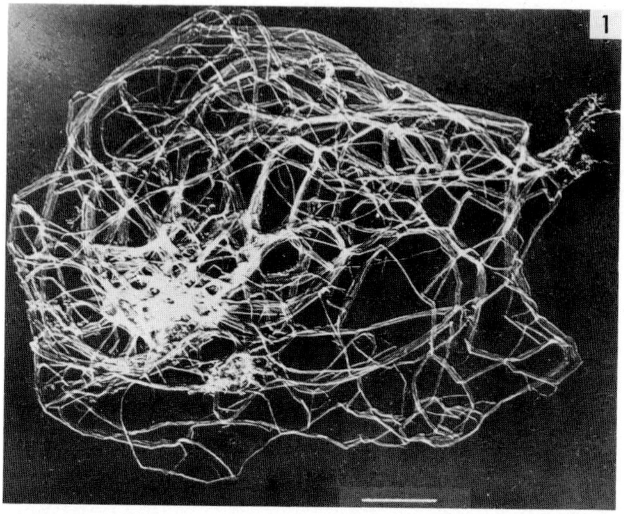

Fig. 1. Microfibrillar net formed by a protoplast of *S. cerevisiae.**

Fig. 2. Short microfibrils of micro-crystalline hydroglucan. (Courtesy Dr. A. L. Houwink.)

They are composed of a material that several authors have found to consist of linear chains of β-1,3-linked D-glucose residues, and has been termed hydroglucan [5].

In fact hydroglucan represents the acid resistant branches of the main glucan molecule of the normal wall. According to the present concept this molecule has a backbone chain of β-1,6-linked glucose

*The bar on all Figures corresponds to 1 μm.

residues with β-1,3-linked branches [6,7] and is accompanied in the native wall by a minor fraction of linear chains of 1,6-linked β-glucan [8,9], the latter being soluble in acid.

The general features of the nets therefore suggested that the protoplasts synthesize only the branches of the normal wall glucan but not the 1,6-linked backbone. To verify this, X-ray diffraction seemed to be the most suitable method because hydroglucan produces a characteristic X-ray diagram [5]. If the nets were found to produce the same diagram the identity of at least part of the net fibrils with hydroglucan could be taken for granted, and this was the basic thought from which the present work originated.

Two additional points have been given attention: first, when chitin was found to be present in the nets, whether it is formed *de novo* or comes from partly digested old walls; second, whether structures like the net fibrils may be detected in some way in normal walls.

I. Results

A number of resulting X-ray powder diagrams from which the nature of the nets may be derived as far as the crystalline components are concerned is given in Fig. 3–6. The diagrams are presented here in the form of their radial density curves. An X-ray powder diagram is normally a pattern of concentric rings, and each of the curves represents the density distribution along a radius of such a ring pattern. The centres of the patterns lie on the vertical line left of each series of curves. The numbered short lines at the bottom of each series indicate the diameters of the diffraction rings or, which is the same, the places of the peaks in the density curves for diagrams taken from hydroglucan and chitin under the same conditions as those from the nets. The upper series of numbers refer to the lines indicating hydroglucan peaks, the lower series to those indicating the predominant chitin peaks. Since part of the latter overlap with hydroglucan peaks, one line may have two numbers. The numerical values of the crystal spacings corresponding to the peaks of the curves will be given elsewhere.

II. The X-ray Diagrams from Native Nets

Figure 3 presents the diagrams from four specimens of nets from different cultures of protoplasts. It is clear that they all show a number of diffraction peaks, partly small and super-imposed on a broad peak from a diffuse diffraction ring. The first conclusion

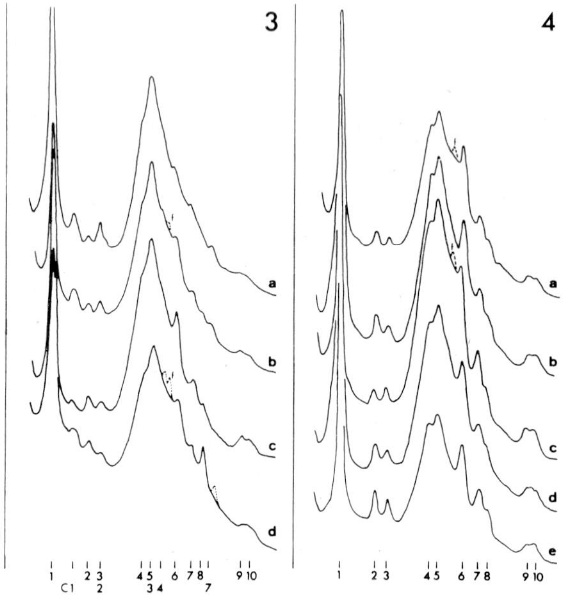

Fig. 3. X-ray density curves from samples of fibrillar nets from protoplasts.

Fig. 4. *a*, *b* and *c*: Curves from same samples as in Fig. 3 *a*, *b* and *c* but after extraction with cold 30 % HCl; *d*: Fibrillar nets after incubation with chitinase; *e*: Paramylon.

therefore is that the net fibrils contain crystalline material. In this respect they differ from the native walls since in the diagram of the latter the small peaks are lacking, as is demonstrated by curve *a* of Figure 6.

To identify the nature of the crystalline components of the nets we may put a ruler along the corresponding peaks in their curves and examine whether the positions of the peaks correspond to either those of hydroglucan or of chitin as indicated on the bottom. If we do so, it appears that the place of nearly every peak in the curves matches one of the numbered lines. The diagrams of the nets therefore strongly suggest that they contain chitin and hydroglucan in a crystalline form.

Conclusive support for the above suggestion may be derived from the curves in Fig. 5. Curve *a* represents the diagram from pure chitin and curve *c* from a mixture of the two containing 10% dry wt of chitin. It is evident that the curve from the mixture corresponds to those from the nets although its peaks are more pronounced. In

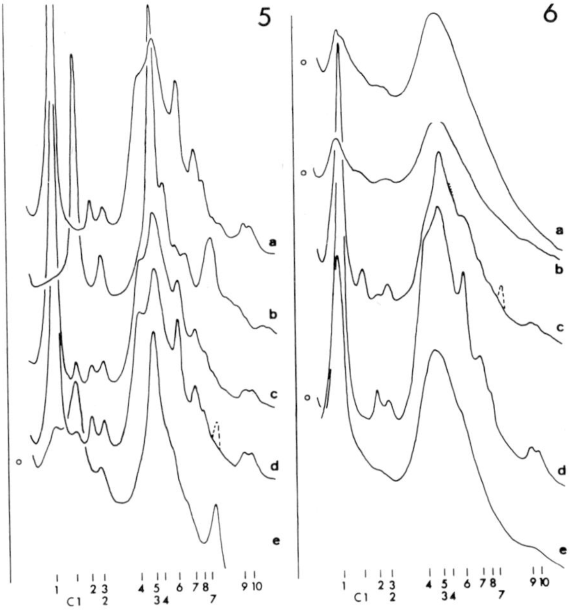

Fig. 5. X-ray density curves from: *a*: Hydroglucan; *b*: Chitin; *c*: Mixture of hydroglucan and chitin containing 10% of chitin; *d*: Hydroglucan residue from HCl-boiled walls of baker's yeast (chitin not removed); *e*: Chitin precipitate from 30% HCl extract from nets.

Fig. 6. X-ray density curves from: *a*: Native walls from baker's yeast; *b*: Walls partly digested by helicase; *c*: Same as *b*, but after treatment with 2% HCl 100°; *d*: Precipitate from alkali extract from nets, treated with boiling 2% HCl; *e*: Clusters formed after acid and alkali extraction of nets.

addition, the size of the chitin peak C1 in the curve from the mixture is nearly the same as in curves *a* and *b* from the net samples. This indicates that the chitin content of these samples must be around 10%. We may infer from an approximating calculation on this basis (to be discussed elsewhere) that the C1 peaks in curves *a, b, c* and *d* from the net samples correspond to chitin percentages about 15, 12, 3 and 6 respectively.

III. Extraction of the Nets

A. *Dissolution of chitin*

Some further data have been obtained by extraction of the nets. Extraction with cold 30% hydrochloric acid, or treatment with chitinase, removes chitin. When using hydrochloric acid chitin may be reprecipitated from the solution by dilution or neutralization of the acid. That the material obtained from the nets in this manner is indeed chitin is shown by its X-ray diagram represented by curve *e* of Fig. 5 which shows the same peaks as curve *b* of the same figure from authentic chitin.

Furthermore, the chitin extraction leaves the glucan component in a more nearly pure form which results in X-ray diffraction rings standing out more clearly, as is shown by the curves in Fig. 4. Curves *a, b* and *c* are from the same samples as are curves *a, b* and *c* respectively of Fig. 3, but after extraction with 30% hydrochloric acid, and curve *d* is from one of the net samples after incubation with chitinase. It is evident that in all of these curves the chitin C1 peak has disappeared as expected. In addition, the glucan peaks 4 and 5 on top of the main density peak appear with a better resolution, which enables a better comparison to be made with the curve from pure hydroglucan given as Fig. 5, *a*.

This comparison reveals that the peaks 4 and 5 are closer to each other in the glucan from the nets than in hydroglucans from fungi investigated so far. However, these hydroglucans, unlike the net glucan are crystallized *in vitro,* and this might be the reason of the small discrepancy. Another product that, like hydroglucan, is known to consist of linear chains of 1,3-linked β-glucan but which, like the net glucan, has crystallized *in vivo*, happens to be present in the form of the paramylum grains of Euglena's [10,11]. Its X-ray density curve is given as curve *e* of Fig. 4. All its crystallinity peaks, inclusive number 4 and 5, appear to have exactly the same position as those of the net glucan. Therefore, the net glucan, as far as its crystal modification is concerned, should be called paramylum.

B. *Extraction by Hot Dilute Acid and Alkali*

There is another feature by which the nets deviate from normal walls. The latter, as we know, dissolve nearly completely by successive treatments with boiling 2% hydrochloric acid and hot dilute alkali, leaving only a small fraction composed mainly of chitin [5,8]. We found that net samples treated similarly leave a remarkable kind of more or less microfibrillar clusters as are shown in Fig. 7. Such specimens normally contain interspersed small granules which may form aggregates. Because the granules are absent in specimens treated with cold 30% hydrochloric acid they are supposed to be chitin. The clean clusters yield an X-ray diagram with only two diffuse rings, as is shown by curve *e* of Fig. 6. The positions

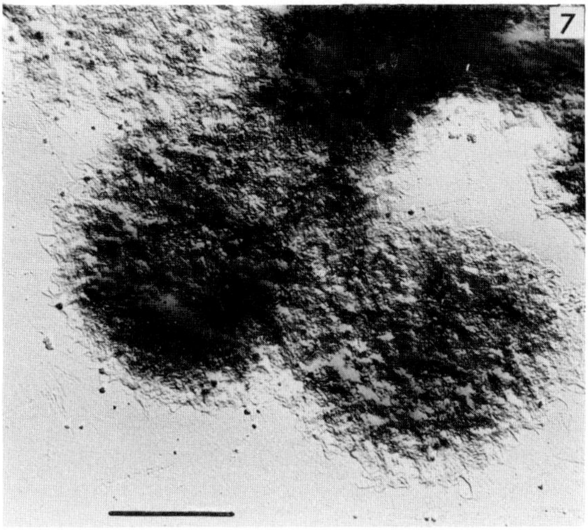

Fig. 7. Clusters remaining after successive treatments of nets with 2% HCl and 3% NaOH at 100°C.

of these rings cover the main intensity distribution of β-1,3-glucan, which suggests that the clusters contain β-1,3-glucan chains that cannot crystallize because of some kind of ramification. But the ramified product seems to be different from that in native walls since it is not easily hydrolysed to form hydroglucan. Our most recent data about the clusters indicate that they represent about 20% of the nets, but on alkali extraction alone 60% of the nets is left in this form, which would indicate that hydroglucan in the paramylum form represents only 40%. Further research on the clusters is in progress.

The alkali soluble fraction of the nets may be reprecipitated and then is microfribrillar under the electron microscope (Fig. 13). This material produces a rather diffuse X-ray diagram as does hydroglucan after precipitation from alkaline solution. But in both cases the precipitate may be recrystallized by boiling with 2% HCl, and then the net glucan, as expected, yields an X-ray diagram (Fig. 6, *d*) corresponding to that of hydroglucan (Fig. 5*a*).

IV. Formation of Chitin *de novo*

In order to preclude the possibility that chitin from partly digested cell walls of the original yeast culture could affect the outcomes of analysis of the nets, the protoplasts were separated from these old walls, in which a high degree of of separation could be attained. In addition, a sample of fresh protoplasts was lysed with distilled water before they had been cleaned from old wall material, and this material was subsequently collected from the suspension and X-rayed. The resulting X-ray diagram, shown by curve *b* of Fig. 6, exhibits the same diffuse rings as that of native walls (Fig. 6, *a*) and only a trace of a diffuse chitin C1 reflection. Residual wall material therefore cannot be the origin of the chitin peaks observed in the diagrams of the net samples, even if the above separation had been omitted.

Further proof that chitin is formed *de novo* came from a sample of nets grown on purified protoplasts which, in order to remove any chitin that might have been left on their surface, had undergone an additional incubation with chitinase before their being cultured for net production. The X-ray diagram of these nets (Fig. 3, *d*) also showed a clear chitin C1 peak.

Indications of an increased chitin content, though less conclusive than those above with regard to its formation *de novo*, have been reported before for protoplast nets of *S. carlsbergensis* [12] and regenerated walls of *Candida utilis* [13].

Concerning the chitin in the nets it may finally be noted that its crystallinity as witnessed by the C1 peak in the curves of Fig. 3 is a feature not shown by the chitin of the native wall. Evidence of the poor crystallinity of chitin in the native wall may be derived from absence of the C1 peak in the X-ray diagram from walls partly digested by helicase (Fig. 6, *b*) rather than from its absence in the diagram from the native walls themselves. For, in the latter, the chitin content is too small to give any observable X-ray diffraction even in case of full crystallinity. In the former, however, it has accumulated sufficiently—by selective digestion of other wall components by the helicase—to produce a clear peak if it were crystalline. This difference in chitin content between normal walls and partly digested walls appears from the difference in magnitude of the C1 peaks in the X-ray density curves from the specimens in question after treatment with boiling 2% HCl, by which chitin is brought to crystallization; compare the Cl peak in Fig. 6, *c* from partly digested walls after HCl treatment with that in Fig. 5, *d* from normal walls after this treatment. Evidence of the poor crystallinity of chitin in native walls of *S. cerevisiae* has also been given by Beran *et al.* [14].

V. Microfibrils in the Native Walls

An answer to the question of the occurrence of microfibrils similar to those of the nets also in native walls naturally requires electron micrographs of such walls revealing microfibrils. Despite the difficulty of the latter normally being obscured by amorphous wall components—which has led some authors to deny their presence—three electron micrographs of native wall material are shown here exhibiting clearly a microfibrillar pattern. The first one (Fig. 8) is taken from Houwink and Kreger [5] and presents a shadowed wall fragment with a fine microfibrillar network on the inside of the wall.

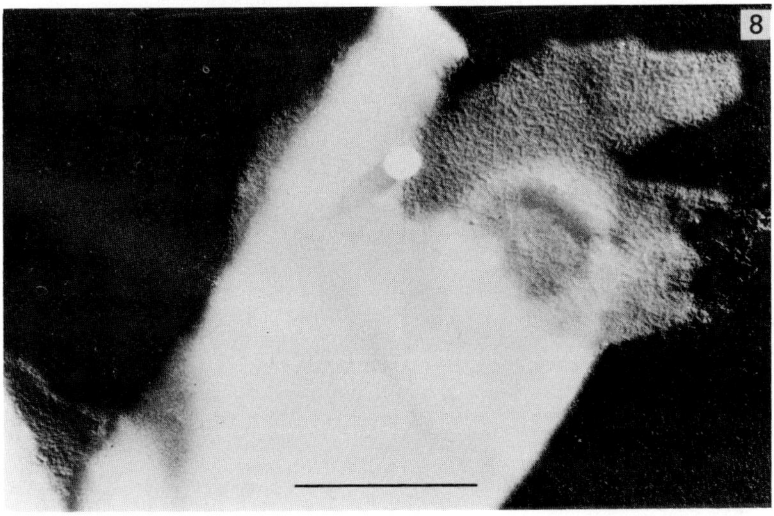

Fig. 8. Fibrillar network on inside of native wall.

There is also evidence of the presence of microfibrils in the native wall obtained by freeze-etching [15,16], but most elucidative we think are two photographs we obtained recently. One is taken from a specimen of wall fragments treated so as to attain so called negative staining (Fig. 9); the other one (Fig. 10) is from a shadowed preparation of a cell wall treated with an endo-β-1,3-glucanase, the latter prepared and kindly made available by Prof. H. J. Phaff.

These electron micrographs, resulting from three different techniques, correspond in exhibiting fine nets with meshes around 30–60 nm wide. But they do not show the type of microfibrils of the protoplast nets with their band formations and long unramified stretches. Two further electron micrographs, like the other ones made by the second author, may serve to underline the difference (Figs 11,12). They respectively show a net contrasted by negative staining and a shadowed preparation of a string of nets.

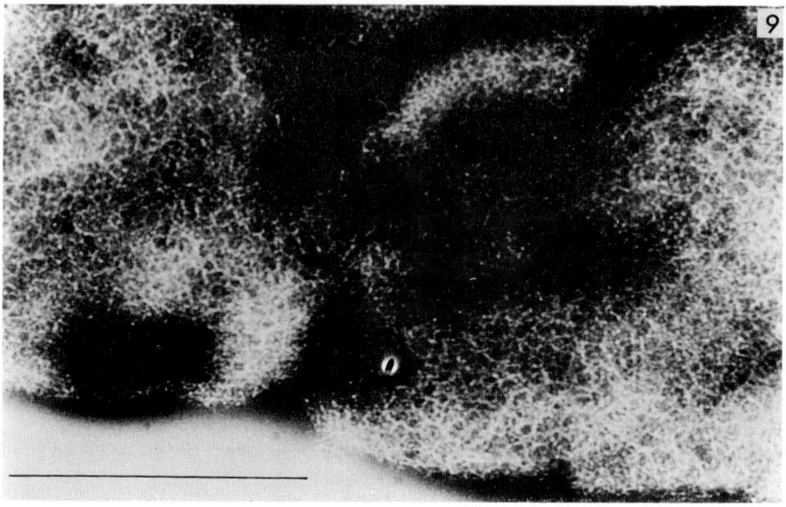

Fig. 9. Negative staining of fragment of native wall.

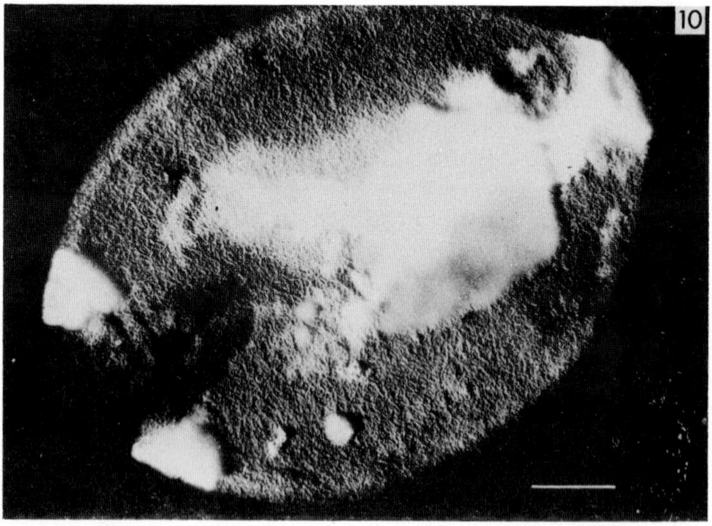

Fig. 10. Native wall after incubation with endo-β-1,3-glucanase.

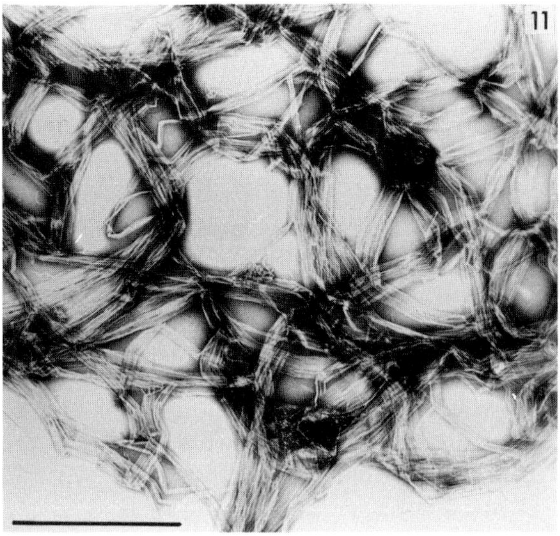

Fig. 11. Negative staining of fibrillar net from protoplast.

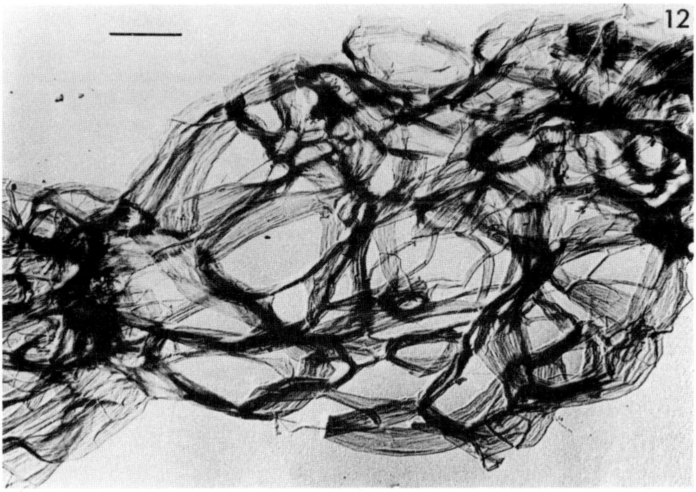

Fig. 12. Part of a string of nets with strong band formation of microfibrils.

VI. Discussion

The X-ray diffraction data have shown that the long microfibrils of the protoplast nets contain a high proportion of alkali soluble β-1,3-glucan in a crystalline form, which has not been observed in native walls. Neither do the latter possess the characteristic type of long microfibrils of the nets, as appeared from electron microscopy of special preparations of the walls. The remarkable length of the unramified stretches of the microfibrils produced by the protoplasts suggests that the β-1,3-glucan chains they contain are of a much higher degree of polymerization than are most likely the branches of the glucan of the native wall; this in view of the shortness of the hydroglucan microfibrils which in fact represent aggregates of these branches. A high degree of polymerization of the chains of the net glucan is also suggested by the observation of long microfibrils in the net glucan after dissolution and reprecipitation (Fig. 13).

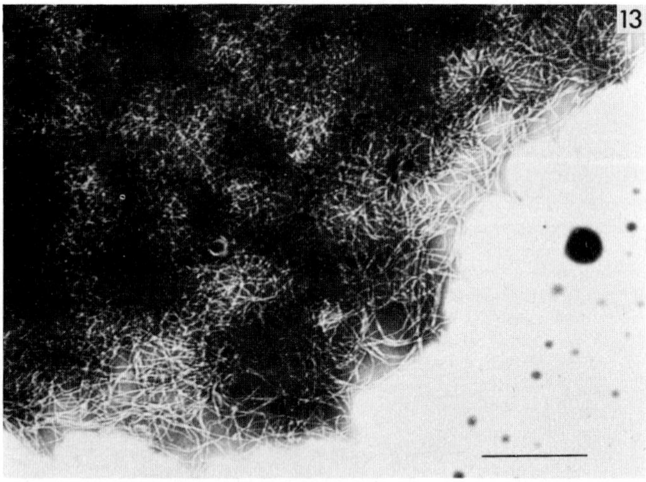

Fig. 13. Long microfibrils of glucan precipitate in alkali extract from nets.

Conclusive evidence concerning the length of the chains of the net glucan comes from molecular weight determinations on the alkali soluble fraction of the nets by Eddy and Woodhead [17]. The weight they found corresponds to a degree of polymerization around 600, whereas for hydroglucan and the branches of the native wall glucan a mean value of 28 has been reported [9, 18].

A plausible explanation of these data seems to be that the

protoplast membrane possesses a β-1,3-linking mechanism, but that the 1,6-linking mechanism required for synthesis of the ramified glucan of the normal wall is not, or is only partly, bound to the membrane and therefore carried away from its surface by the liquid medium. Consequently, a shortage of 1,6-linking activity will arise and, hence, the membrane may build 1,3 chains of considerably increased length. Furthermore we must presume an organizing mechanism in the plasmamembrane that in combination with mere physical-chemical properties of the molecules gives rise to the particular features of organization of the chains: the uniform direction over long stretches, the specific crystal modification and perhaps the conspicuous band formation.

Similarly, the crystallinity of the chitin in the nets, by which it deviates from its condition in the native wall, could be explained by the postulate that in normal walls extracellular complexing enzymes are present preventing the chains from crystallization as soon as they are formed, but which are carried away by the liquid medium. Then the chitin, apparently formed by a membrane bound enzyme, may aggregate into crystalline structures.

The increased chitin to glucan ratio of the nets, finally, might indicate that not all of the 1,3-linking activity is bound to the membrane. Dispersion of an extracellular or loosely bound part by the liquid medium would then result in an increased chitin to glucan ratio in the nets.

VII. Summary

The protoplast nets of *S. cerevisiae* have been found to contain, in contrast to native walls, free chains of β-1,3-linked D- glucose residues aggregated to form crystalline structures. These structures form part of the long microfibrils constituting the nets. The crystalline packing of the chains is slightly different from that of hydroglucan and corresponds to that of paramylum.

The nets contain chitin which is formed *de novo* by the protoplasts and, in contrast to the chitin in the native walls, is crystalline. The chitin to glucan ratio of the nets is normally higher than in native walls.

The net structure is disturbed by dilute alkali, which dissolves about 40%, the residual part consisting of more or less microfibrillar clusters and chitin. The X-ray diagram of the clusters is indicative of the presence in them of β-1,3-glucan chains, but their true nature has not yet been established.

The constitutional and structural differences between the nets and the glucan-chitin moiety of normal walls have tentatively been explained by differential removal of 1,6 and 1,3 linking enzyme and removal of a chitin complexing factor from their natural locus of action by the liquid medium.

References

1. Kopecká, M., Čtvrtníček, O. and Nečas, O. (1967). *In:* Symposium über Hefeprotoplasten, p. 73. (ed. Müller R.) Akademie-Verlag, Berlin.
2. Nečas, O. (1961). *Nature, Lond.*, **192**, 580.
3. Nečas, O. and Svoboda, A. (1967). *In:* Symposium über Hefeprotoplasten, p. 67. (ed. Müller, R.) Akademie-Verlag, Berlin.
4. Svoboda, A. and Nečas, O. (1970). *In:* Yeast Protoplasts, p. 211 (eds Nečas, O. and Svoboda, A.) Medical Faculty, J. E. Purkyně University, Brno.
5. Houwink, A. L. and Kreger, D. R. (1953). *Antonie van Leeuwenhoek*, **19**, 1.
6. Misaki, A. and Smith, F. (1963). *Abs. Am. Chem. Soc.*, 144th Meeting, 14.
7. Manners, D. J. and Patterson, J. C. (1966). *Biochem. J.*, **98**, 19.
8. Bacon, J. S. D., Farmer, V. C., Jones, D. and Taylor, I. F. (1969). *Biochem. J.*, **114**, 557.
9. Manners, D. J. and Masson, A. J. (1969). *FEBS Letters*, **4**, 122.
10. Kreger, D. R. and Meeuse, B. J. D. (1952). *Biochim. biophys. Acta*, **9**, 699.
11. Clarke, A. E. and Stone, B. E. (1960). *Biochim. biophys. Acta*, **44**, 161.
12. Eddy, A. A. and Williamson, D. H. (1959). *Nature, Lond.*, **183**, 1101.
13. Novaes-Ledieu, M. and Garcia-Mendoza, C. (1970). *J. gen. Microbiol.*, **61**, 335.
14. Beran, K., Řeháček, J. and Seichertová, O. (1970). *In:* Yeast Protoplasts, p. 171 (eds Nečas, O. and Svoboda, A.) Medical Faculty, University J. E. Purkyne, Brno.
15. Moor, H. and Mühlethaler, K. (1963). *J. Cell Biol.*, **17**, 609.
16. Streiblová, E. (1968). *J. Bact.* **95**, 700.
17. Eddy, A. A. and Woodhead, J. S. (1968). *FEBS Letters*, **1**, 67.
18. Barry, V. C. and Dillon, T., (1943). *Proc. R. I. Acad.* **49B**, 177.

Control of Yeast/Mold-dimorphism in *Mycotypha*

B. SCHULZ and G. KRAEPELIN
Institute of Botany, Technical University,
33 Braunschweig, GFR

Yeast/Mold-dimorphism in fungi is the ability of one organism to grow either as yeast or mycelium. The two morphological growth-types are generally referred to as M- (mycelial) and Y- (yeastlike) form. The factors affecting such a change in morphology vary. For example: the dimorphic *Mucor rouxii,* well studied by Bartnicki-Garcia and co-workers [1], aerobically grows as mycelium while anaerobically and with increased CO_2-tension it grows in the budding yeastlike form. Some pathogenic fungi, such as *Paracoccidioides brasiliensis* and *Blastomyces dermatitidis* switch their morphology in response to temperature change, growing at 37° C as Y- and at 20°C as M-form [2].

A particularly obliging object to study the control of dimorphism is *Mycotypha* (Mucorales), whose Y-/M- morphology may be triggered not just by one or two, but by many factors—including anaerobiosis/aerobiosis, CO_2, substrate or temperature. Under appropriate conditions even true dimorphs are produced, in other words cells which simultaneously are Y- and M- form as shown in Fig. 1.

In our experiments we used 3 species of *Mycotypha* characterized by their different tendency to grow as Y- and M-form. Under standard conditions (liquid glucose-peptone-yeast extract medium, pH 5, 28° C, vigorously shaken) *M. microspora* (CBS 230.32) uniformly grows as M-form, *M. "spirifera"* (a new species isolated in our laboratory and provisionally included in CBS as *M. microspora* 186.68) grows as a mixture of Y- and M-form, and *M. africana* (CBS 122.64) primarily as Y-form.

It has frequently been suggested that fermentation is coupled with Y-form and respiration with M-form. An affirmation of this hypothesis was the attainment of stable *Mycotypha* mutants following UV-treatment. Four such mutants of *M. spirifera* were obligatory Y-form; their Q CO_2/O_2 as well as cytochrome spectra indicating a defectiveness in oxidative metabolism. Cytochrome spectra of the wild type Y- and M-form were also compared. They differed in that

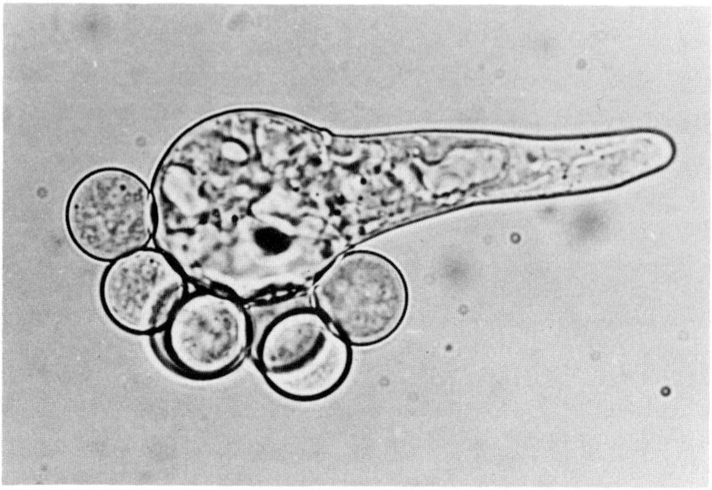

Fig. 1. Dimorph of *M. spirifera* showing simultaneous M-outgrowth and budding (the diameter of the primary cell is 30 μm).

Y-form, even after adding dithionite, did not show separated α-bands of cytochrome b and c, but a broad band at about 555 nm, also characteristic of strongly fermenting true yeasts.

If the relationship fermentation/Y-form and respiration/M-form is indeed true, electron microscopic sections should reveal typical respiratory characteristics for M-cells and fermentative characteristics for Y-cells. This was indeed the case. M-form was typified by many well developed mitochondria and a rather thin cell wall (Fig. 2), while in Y-cells the number and density of mitochondria was considerably lower (Fig. 3). As can be seen, the Y-cell wall is thicker. Moreover, numerous elongated ER-membranes are located at the periphery just beneath the cell membrane of Y-cells, as is known for strongly fermentative yeasts.

These cytological differences are poignantly shown in the two dimorphs in Figs 4 and 5. There is a concentration of mitochondria in the mycelial outgrowth as compared to the budding part. The peripheral ER-network is particularly prominent in the Y-progeny. The electron micrographs of the three morphological cell types shown here are typical of the approximately 200 sections examined. They again suggest that a predominance of fermentation leads to Y-morphology, while respiratory activity leads to M-development.

As previously noted, Y-form cell wall is consistently thicker than that of the M-form. Kanetsuna and Carbonell [3] made similar

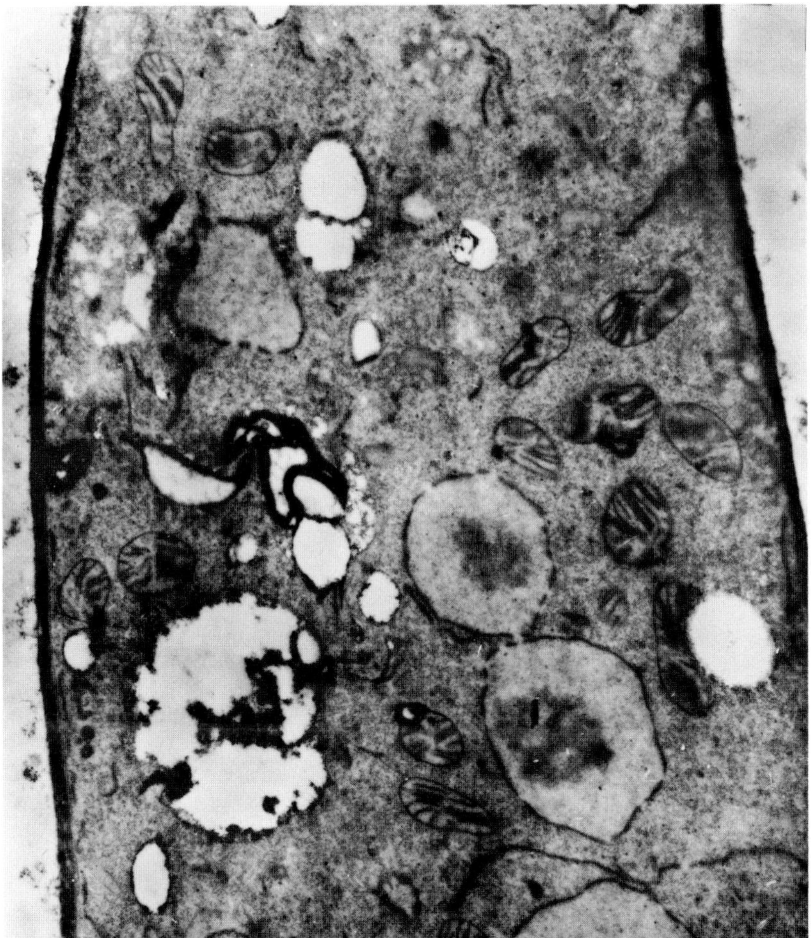

Fig. 2. Mycelial-form from *M. africana*. Note the abundance of mitochondria, absence of peripheral ER-membranes and rather thin cell wall. (Fixation glutaraldehyde/potassium permanganate; embedding Epon; magnification 16.000 x.)

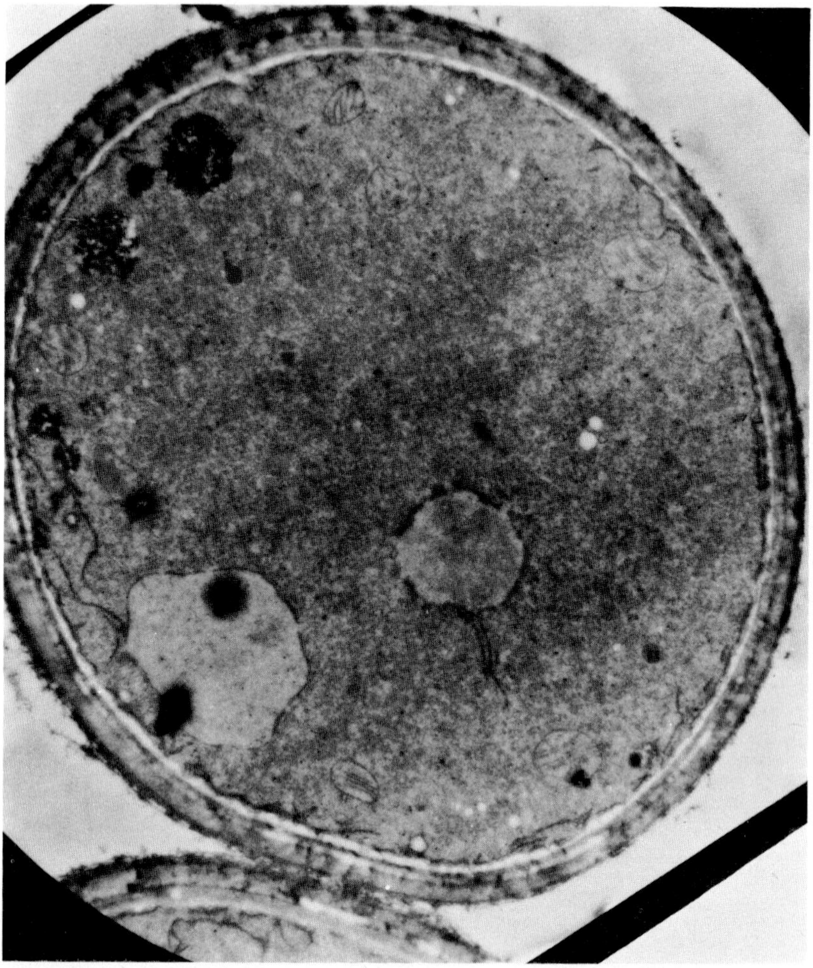

Fig. 3. Yeastlike form from *M. spirifera.* Note the comparatively low number of mitochondria, abundance of peripheral ER-membranes and thick, layered cell wall. (Fixation glutaraldehyde/potassium permanganate; embedding Epon; magnification 12.000 ×.)

observations on other dimorphic fungi. Unfortunately, detailed chemical analyses of the Mycotypha cell walls have not as yet been undertaken. However, preliminary experiments with commercial snail enzyme and β-glucosidase suggest that M-form cell wall contains more β-glucans and poly-β-glucuronic acids than Y-cell wall does. If these structural tendencies are verifiable, they are in agreement with those found in other dimorphic fungi, where Y-form cell wall was

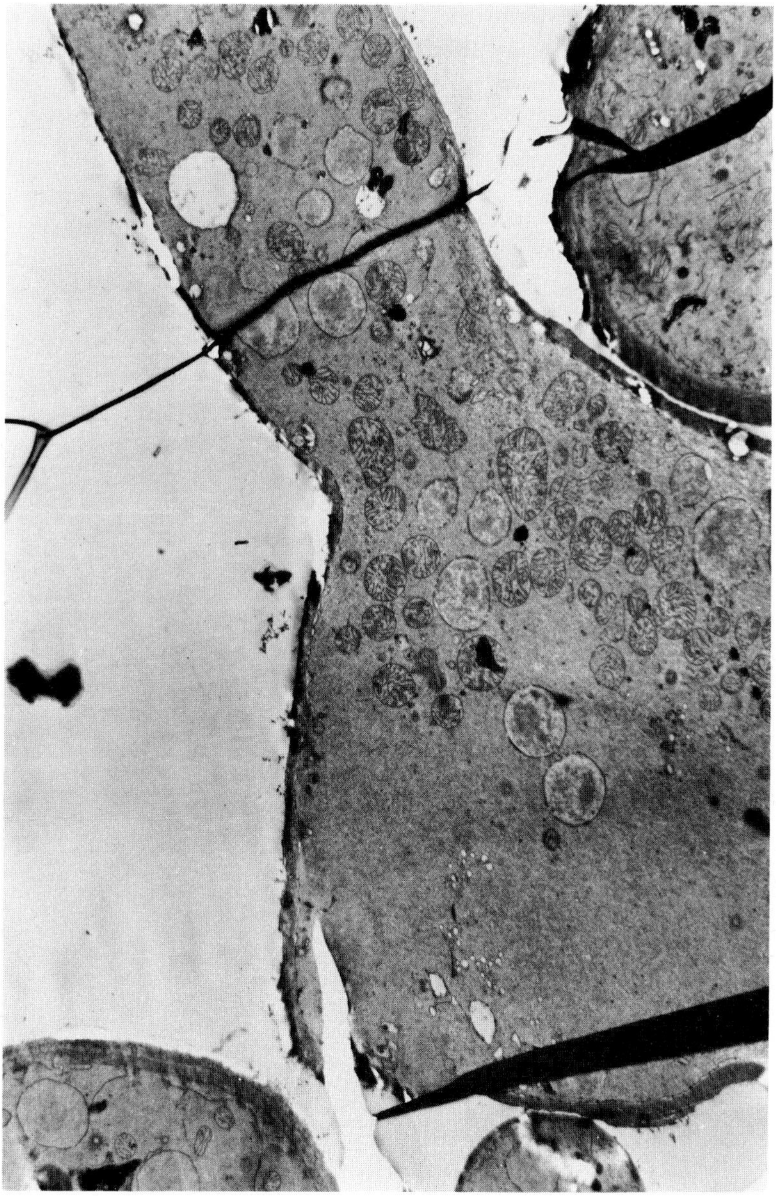

Fig. 4. Dimorph from *M. africana* (fructose-peptone-yeast extract medium, 28° C, shaken 16 h). Note the unequal distribution of organelles. (Fixation glutaraldehyde/potassium permanganate, contrast Reynold's, embedding Epon, magnification 4.500 x.)

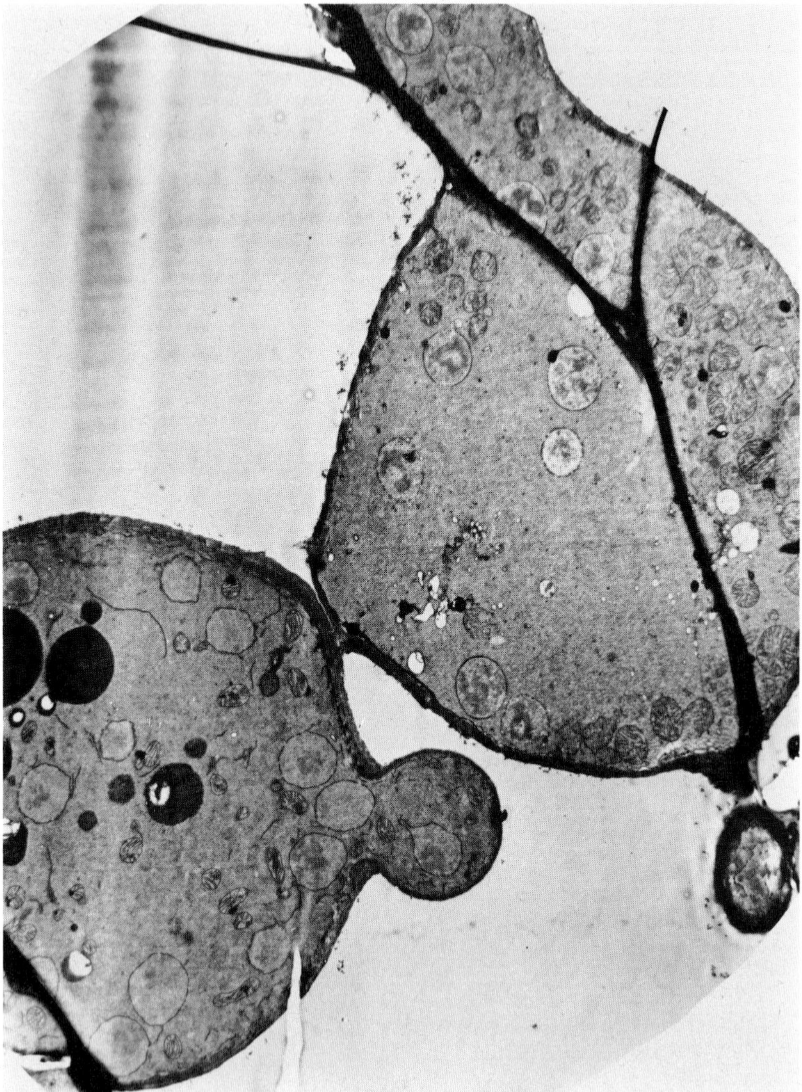

Fig. 5. Dimorph from *M. africana.* Preculture and preparation of cells is the same as Fig. 4. (magnification 4.000 x.)

found to be predominantly α-glucan and M-form cell wall to be predominantly β-glucan [3, 4].

A further verification of the dependence between the metabolic balance of fermentation/respiration and Y/M-dimorphism must result following an inhibition of respiration. Such an inhibition should enhance Y-form. Thus obstruction of the electron transport with antimycin A (10^{-5}M) or KCN (2×10^{-5}M) enhances Y-form. If however, the obstruction is so that electron transport may continue—though at a reduced rate—following for example addition of rotenone (10^{-4}M), amytal (10^{-5}M) or thenoyltrifluoracetone (10^{-5}M), the Y/M-morphology remains unaffected or even M-form is slightly increased (Fig. 6). Inhibition of oxidative phosphorylation by oligomycin (6×10^{-5}M) leads to Y-form, while uncoupling with 2,4-DNP (10^{-4}M) only causes growth inhibition without affecting morphology.

Inhibition of mitochondrial protein synthesis with chloramphenicol (1,5 mg ml^{-1}) induces Y-form while inhibition of cytoplasmic protein synthesis with cycloheximide (20 μg ml^{-1}) only caused growth inhibition without an effect on morphology. It is still

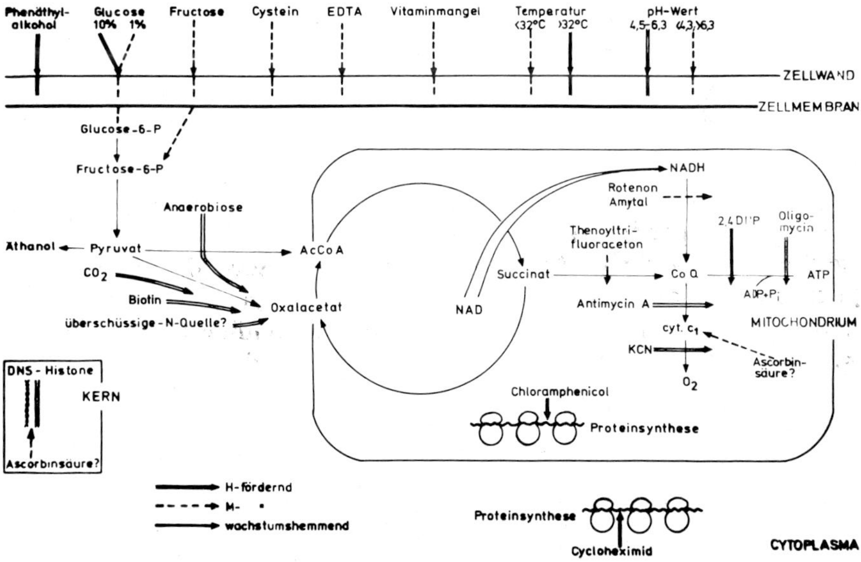

Fig. 6. Schematic summary of agents tested; the position indicates their presumed site of action. double line: Y-form stimulated; dotted line: M-form stimulated; full line: inhibition only.

unclear if the Y-inducing effect of phenethyl alcohol (0.17%, v/v) is caused by a selective inhibition of mitochondrial protein synthesis or by cell membrane damage.

Addition of 2-5% CO_2 to the atmosphere aerobically (air) as well as anaerobically (nitrogen) results in an enhancement of Y-form. Preliminary manometric experiments with *M. spirifera* in a 5% CO_2-atmosphere, indicate that CO_2 inhibits respiration without affecting fermentation rate. It is remarkable in this context that addition of biotine (10^{-6} M), the coenzyme of CO_2-fixation, also promotes Y-morphology. The stimulation of Y-form by increased CO_2-supply is not due to a lowered pH, since pH values lower than 4,5 encourage M-form. In fact, under standard conditions only in the pH range 4,5–6,3 does Y-form predominate. Above and below, growth is primarily M-type.

High glucose concentrations known to repress the respiratory system stimulate Y-growth, while low glucose supply or fructose encourage M-production.

Excessive N-source favors Y-growth, whereby the effectiveness of peptone and yeast extract is most striking.

On the other hand limitations of growth brought about by restricted or exhausted nutrient supply or by vitamin deficiency usually promote mycelial differentiation.

In *M. africana* and *M. spirifera* ascorbic acid (10^{-2} M) encourages a predominance of M-form even within the pH range of optimal Y-growth.

In the same 2 species cysteine (5×10^{-3} M) and EDTA ($2{,}5 \times 10^{-5}$ M) cause an enhancement of M-form.

The data listed above clearly show that dimorphism in *Mycotypha*–particularly in *M. africana* and *M. spirifera*– is very sensitive to a variety of environmental factors. Several of these factors are also effective in other dimorphic fungi, but to our knowledge there is no other single fungus responding to such a broad spectrum of agents. Considering the diversity and complexity of factors affecting morphologic expression in a potentially dimorphic system, one should be cautious with the interpretation. A relation of morphological duality to the balance of fermentation/respiration is strongly suggested. Nevertheless, this could be an oversimplification of the complex metabolic networks which, for example, might contribute to the differences in composition and structural organization of the cell walls produced. Since the change from a spherical cell with apolar budding into hyphae with polar growth–eventually followed by fructification–has to be interpreted as a differentiation process, the complexity of the problem should not be a surprise. Therefore, further attention should be drawn to the possibility that there may be more than one specific metabolic site triggering the Y/M-alternative.

References

1. Bartnicki-Garcia, S. and Nickerson, W. J. (1962). *J. Bact.*, **84**, 841.
2. Scherr, G. W. and Weaver, G. (1953). *Bact. Rev.*, **17**, 51.
3. Kanetsuna, F., Carbonell, L. M., Azuma, I. and Yamapupa, Y. (1972). *J. Bact.*, **110**, 208.
4. Kanetsuna, F. and Carbonell, L. M. (1971). *J. Bact.*, **106**, 946.

PART III

Physiology and Biochemistry of Yeast Protoplasts

Control of Synthesis and Secretion of Exoenzymes by Yeast Protoplasts

J. O. LAMPEN, S.-C. KUO and F. R. CANO

*Institute of Microbiology, Rutgers University,
The State University of New Jersey, New Brunswick,
New Jersey, U.S.A.*

A characteristic and important feature of yeasts is the existence of large quantities of cell-bound enzymes outside of the plasma membrane and accessible to external substrate. By the use of protoplasts, it has been possible to examine both the synthesis and secretion of these external enzymes and to compare this process with the formation of enzymes that occur inside the plasma membrane. This paper will focus on factors that determine the occurrence of synthesis under various conditions or that modify the process, and not with the nature of the biosynthetic pathway. The following paper by Gascón, [1] on biosynthesis of invertase will deal more directly with the latter question.

The yeast enzymes we have studied [2] and their significant properties are summarized in Table I. Two forms of invertase have been detected and purified. The large enzyme, mainly external, is a mannan protein of about 270,000 m wt [3] ; the small internal enzyme lacks mannan and is not identical with the protein moiety of the external enzyme although related to it in many ways [4] . Acid phosphatase is also a large external mannan protein [5] , but in this instance the enzyme retained by the protoplast does not appear to be a smaller form. In contrast to these external enzymes, internal enzymes such as alkaline phosphate and α-glucosidase are not glycoproteins.

A major factor controlling invertase production appears to be catabolite repression by hexoses. α-Glucosidase is also repressible by hexoses and is inducible by maltose [6, 7] . The phosphatases are usually repressible by inorganic phosphate [8] .

The repressibility of invertase production by hexoses led to obvious experimental difficulties in studying invertase secretion, since hexoses are usually the preferred energy source for production

TABLE I

Properties of Yeast Enzymes Studied

Enzyme	Location	Composition	Mol. wt.	Control
Invertase	Internal	Protein	135,000	Hexose-R
	External	Mannan-protein	270,000	Hexose-R
Acid phosphatase	External	Mannan-protein	290,000	P_i-R
Alkaline phosphatase	Internal	Protein	—	P_i-R
Alpha-glucosidase	Internal	Protein	85,000	Maltose-I

R: repressible; I: inducible. [Adapted from Lampen, *et al.* (2)]

of the glycoprotein by protoplasts [9, 10]. To eliminate some of these problems, we chose a *Saccharomyces* strain [303-67] shown by Winge and Roberts [11] to contain only the single gene SUC-2 for invertase production, and by two-stage UV-irradiation produced a series of resistant mutants. Table II summarizes the relevant properties of the three organisms used for most of our work. The original strain 303-67 is highly sensitive to hexoses; mutant FH4C is relatively resistant and has been used for production and characterization of the invertase forms. Mutant 1016 is partially resistant in that it is still repressible by high levels of hexose but produces large quantities of enzyme at lower hexose concentrations. All three strains produce α-glucosidase; in mutant FH4C, production of this enzyme is repression resistant (as for invertase).

TABLE II

Characteristics of Repression-Resistant Mutants

	Strain 303-67	Mutant 1016	Mutant FH4C
Hexose repression			
Invertase	++++	++	±
Alpha-glucosidase	+	+	—
Sporulation	+	+	—
Flocculation	—	++	++++
Cyclic AMP (pmoles/mg cells)			
High glucose	8.4	2.0	10.8
Low glucose	14.3	2.8	12.8

Mutant FH4C proved to be haploid and no longer sporulates; also it grows in clumps which appear to have a continuous cell wall around most of the cells.

In light of recent reports that the concentration of cyclic AMP in yeast cells is higher in derepressed than in repressed conditions [12], it was of interest to compare the levels of cyclic AMP in this series of closely-related mutants. Cultures were grown to mid-exponential phase either in high or in low glucose media and their cyclic AMP content measured. As will be noted from Table II, the level of cyclic AMP is somewhat higher in the less repressive low glucose medium than in high glucose. However, there is no simple relation between the cyclic AMP levels of these strains and their relative sensitivity to repression. FH4C produces 20-50 times as much invertase as 303-67, yet their cyclic AMP levels are similar. Mutant 1016 is intermediate in repressibility, but its cyclic AMP level is much lower than in the other two strains. In other studies with mutant 1016 (not presented in detail) we have shown that it contains an adenyl cyclase, probably membrane bound, and has an extremely high level of phosphodiesterase. It would appear to be this high phosphodiesterase activity that is reflected in the observed low cyclic AMP levels. As others have reported [13], invertase production can be stimulated by added cyclic AMP; however, the results are extremely erratic. In summary, it is clear that cyclic AMP does affect the rate of invertase production by yeast protoplasts, but our results offer no clue to its specific role.

I. Role of Glycosylation Process in Regulating Glycoprotein formation

Following the report by Farkas *et al.* [14] that 2-deoxy-D-glucose (2-dG) inhibited mannan-protein synthesis by yeast protoplasts at concentrations which did not substantially affect total protein synthesis, we examined this phenomenon using conditions optimal for glycoprotein formation by our protoplast system. We were able to confirm the general results of Farkas *et al.*, but observed as well that inhibition occurred only after a 20-30 min period (Fig. 1). Synthesis of the mannan proteins invertase and acid phosphatase was inhibited as well as formation of the intracellular enzyme α-glucosidase, but alkaline phosphatase (also internal) was unaffected. Synthesis of mannan and glucan and the total uptake of fructose and maltose were also inhibited after a similar lag period. Under conditions in which maltose uptake was inhibited at least 80%, there was no leakage of glucose from the protoplasts; thus it is probable that

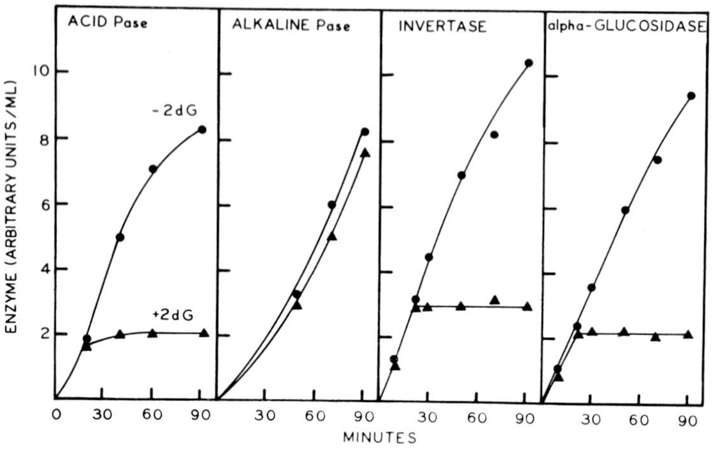

Fig. 1. Effect of 2-deoxyglucose (2-dG) on the formation of enzymes by protoplasts. Protoplasts (preincubated where necessary in the presence of maltose) were suspended at 5 x 10^7 per ml in 0.8 M sorbitol medium containing 100 mM fructose (for phosphatase experiments) or 20 mM maltose (for invertase and α-glucosidase experiments) and incubated at 30°C [15]. (●), no 2-dG added; (▲) 2.5 mM or 0.5 mM 2-dG added at 0 time (ratio 2-dG/fructose or maltose = 1/40).

the actual transport of maltose into the cell (and not the subsequent cleavage and phosphorylation stages) is reduced in the presence of 2-dG.

During the lag period before the inhibitions were evident, 2-deoxyglucose (2-dG-^{3}H was used) was taken up by protoplasts and accumulated primarily as 2-dG-6-P; intracellular levels of 12-15 mM were attained. No 2-dG-^{3}H was detected at this time (40 min) in the nucleotide sugars or in wall polysaccharides.

Yeast phosphomannose isomerase, and especially phosphoglucose isomerase were inhibited *in vitro* by 2-dG-6-P; yet in the protoplasts there was no accumulation of 6-phosphates of mannose, glucose or fructose, in spite of the probable inhibition of hexose phosphate metabolism. Thus the reduced transport of sugars cannot be a consequence of increased levels of these hexose phosphates. Inhibition of glucose transport by a high intracellular level of glucose-6-phosphate has been reported [16, 17]. We propose in turn that the restriction of transport in the presence of 2-dG is a result of feedback inhibition by the accumulated 2-dG-6-P.

Our interpretation of the various effects of 2-dG on glycoprotein formation is presented schematically in Fig. 2. We suggest that the

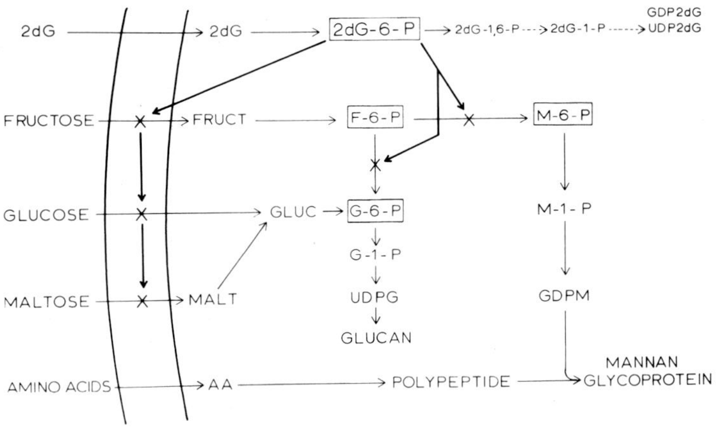

Fig. 2. Schematic drawing of major stages in the transport and metabolism of hexoses and of 2-deoxyglucose by yeast protoplasts. The heavy lines and stages marked X indicate the major points at which inhibition by 2-dG-6-P is suggested to occur.

intracellular 2-dG-6-P is the active agent. It directly inhibits interconversion of the 6-phosphates of fructose, glucose and mannose. At the same time it restricts the transport of the sugars into the cell and thus prevents any significant accumulation of hexose phosphates. The result is an effective blockade of polysaccharide and glycoprotein synthesis, while the limited continuing uptake of the sugars, at the 2-dG level used, still provides sufficient energy for the production of proteins that do not have major carbohydrate moieties.

Further evidence for the critical role of reduced sugar transport is the observation that the sensitivity of glycoprotein synthesis is about the same when mannose or glucose is used as the energy source in place of fructose (S.-C. Kuo, unpublished data). If the phosphohexose isomerases were the only pertinent reactions inhibited by 2-dG-6-P, mannose and glucose could still be converted to their 6-phosphates, thus possibly permitting the synthesis of mannan and invertase from mannose or of glucan from glucose. The uniform sensitivity on the three hexoses is readily explained by the concept presented in Fig. 2.

In concluding our comments on the action of 2-dG, we should re-emphasize that our results support the suggestion by Farkas *et al.* [14] that in the absence of an adequate supply of the appropriate glycosyl units for a glycoprotein enzyme, formation of the active enzyme is prevented. Our data do not offer any indication of the mechanism by which this restriction occurs. We can only say that attempts to demonstrate the accumulation of incompletely glycosylated invertase in the presence of inhibitory levels of 2-dG have not been successful.

II. Studies on Lomofungin

In investigating the mechanisms controlling enzyme production, it is frequently important to know if formation of messenger-RNA (m-RNA) is essential for a particular process or if control is exerted primarily at a translational level. With most bacterial cells, it has been convenient to use actinomycin D and rifampicin to inhibit the process of transcription, and to determine the half-life on m-RNA under various conditions. Unfortunately these antibiotics are relatively ineffective against yeast either because of poor permeability or failure to inhibit the yeast RNA polymerases. One could increase the permeability of the yeast cell or protoplast to the inhibitor by a detergent such as Brij 58 [18] or with a polyene antibiotic, e.g., amphotericin B [19]; however, when one wishes to study the formation of extracellular enzymes, alteration of the plasma membrane with possible direct effects on the release reaction, seems an undesirable complication.

With this in mind, we turned to the antibiotic lomofungin which was reported by Gottlieb and Nicolas [20] to inhibit RNA synthesis by *Saccharomyces cerevisiae*; DNA synthesis was considerably less sensitive, and protein synthesis the least affected. *In vitro* protein synthesis was not inhibited by lomofungin. We have confirmed their findings, and have investigated the effect of lomofungin on the isolated nuclear RNA polymerases of yeast and on synthesis of both external and internal enzymes by yeast protoplasts.

Lomofungin is a phenazine (1-carbomethoxy-5-formyl-4, 6, 8-trihydroxyphenazine [21]), a class of substances that contains many known antimicrobial agents. It is inhibitory for both gram-positive and gram-negative bacteria and for yeasts and other fungi [22].

The growth of *S. cerevisiae* mutant 1016 in a synthetic medium with glucose or maltose as the carbon source is relatively sensitive to lomofungin. At a concentration of 5 μg per ml growth was inhibited almost completely within 1 hour, and even 1 μg per ml reduced the growth rate by about 50%.

The effect of lomofungin on the synthesis of nucleic acid and protein by protoplasts of mutant 1016 is shown in Table III.

TABLE III

Effect of Lomofungin on Nucleic Acid and Protein Synthesis by Protoplasts

Lomofungin µg/ml	RNA	Protein
	% of control value	
0	100 (44070)*	100 (54402)*
5	83	99
10	35	95
20	22	92
40	16	88

Protoplasts from cells grown in 0.05 M maltose were suspended at 5×10^7 per ml in 0.8 M sorbitol medium containing 50 mM fructose and incubated at 30°C [15]. After 80 min incubation 2.5 ml samples of the suspension received lomofungin as indicated and 10 µCi uridine-5-^{3}H (specific activity, 20 Ci/mM) or 0.62 µCi protein hydrolysate-^{14}C. Samples were removed after 5 min incubation and analysed for the amount of radioactivity incorporated into RNA and protein. * cpm/ml.

Incorporation of ^{3}H-uridine into RNA and of ^{14}C-amino acids into protein during a 5-minute period was measured. As Gottlieb and Nicolas [20] had found, RNA synthesis was sensitive and protein synthesis inhibited only marginally. At 40 µg per ml, lomofungin produced an 85% inhibition of RNA synthesis. During the same 5 min period, synthesis of protein had decreased only 12%. The time course of inhibition of RNA and protein synthesis by lomofungin is presented in Fig. 3. RNA synthesis by protoplasts was inhibited within 5 min and by 10 min had essentially ceased. Protein synthesis was not significantly affected until much later—in most experiments only after 30 min of incubation.

A. *Effect of lomofungin on isolated RNA polymerases*

Since these observations indicated that RNA synthesis was the primary target of lomofungin in yeast protoplasts, we tested the effects of lomofungin on isolated RNA polymerases from *Escherichia coli* and yeast. The DNA-dependent RNA polymerase of *E. coli* was studied first, since this material was available in purified form. Under

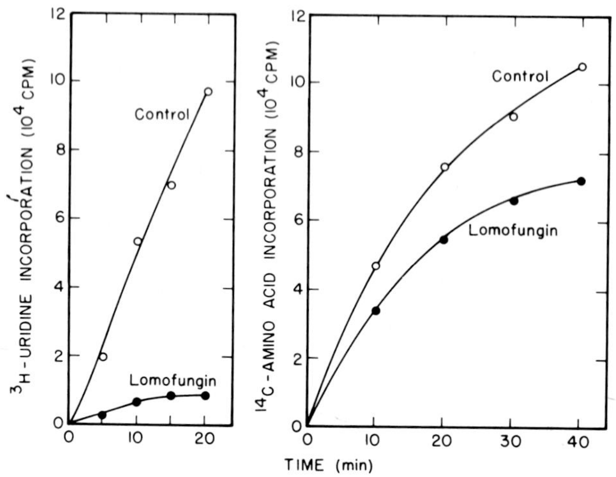

Fig. 3. Time course of lomofungin inhibition of RNA and protein synthesis. The conditions were the same as in Table III. The protoplasts were preincubated for 80 min, then lomofungin (40 μg/ml) was added (0 min). All values are expressed as cpm/0.5 ml samples.

the general conditions of Burgess [23] the enzyme was inhibited by 5-40 μg/ml of lomofungin with the amount required dependent on the concentration of the enzyme. In order to determine whether or not lomofungin primarily prevented the initiation of new RNA chains, as rifampicin does [24], or was similar to streptolydigin in halting the chain elongation process [25]. the kinetics of inhibition of the *E. coli* polymerase were examined. As shown in Fig. 4, polynucleotide formation was completely prevented when rifampicin or lomofungin was added at the same time as the DNA template. If addition of the antibiotics was delayed for 5 min, during which time a number of nucleotide chains had been initiated, lomofungin still halted ^{3}H-UTP incorporation immediately, but rifampicin allowed elongation to continue. Thus lomofungin definitely inhibits chain elongation, whereas rifampicin inhibits initiation, but not elongation.

Finally, we determined whether rifampicin inhibits *E. coli* RNA polymerase by reacting with the enzyme protein or with the DNA template. Incubating the DNA and lomofungin together for 5 min before addition of the enzyme did not affect the subsequent responses to suboptimum levels of DNA. In contrast, if lomofungin and increasing levels of the enzyme were incubated together before the DNA was added, no polymerase activity was subsequently detected at low enzyme levels although higher amounts showed

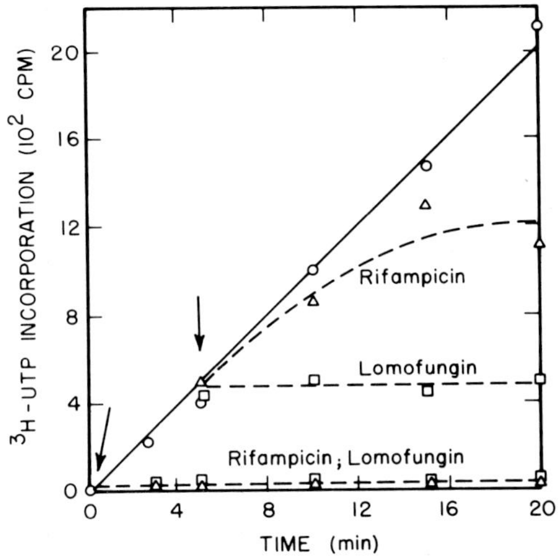

Fig. 4. Kinetics of antibiotic inhibition of *E. coli* RNA polymerase. The assay employed native calf thymus DNA as template and measured the incorporation of ^{3}H-UTP into TCA-insoluble polynucleotides as described in Table III. Lomofungin and rifampicin concentrations, when added (↓), were 20 μg/ml. Control (○); rifampicin (△), and lomofungin (□).

activity. Lomofungin at 20 μg per ml appeared to inhibit completely about 0.7 μg of RNA polymerase with no effect on additional amounts of the enzyme. We conclude from this that lomofungin inhibits by binding with the RNA polymerase itself and not with the DNA template (as would be the case with actinomycin D).

The nuclear RNA polymerases from mutant 1016 were then examined. These were extracted and separated by the procedure of Ponta, *et al.* [26] and the three peaks of polymerase activity obtained are shown in Fig. 5. The active fractions from each peak were pooled and their sensitivity to lomofungin, rifampicin and to α-amanitin tested (Table IV). Lomofungin at 20 μg per ml inhibited all three of the nuclear polymerases, rifampicin only peak A and α-amanitin primarily peak C. The results with α-amanitin on peak A were extremely variable for reasons that are obscure, and we cannot present any information on the sensitivity of this fraction to α-amanitin. Since our prime interest was to know if the yeast polymerases are sensitive to lomofungin, we have not investigated the mechanism of their inhibition and can only assume that it is similar to that for *E. coli* enzyme.

From these results and our previous studies with protoplasts, we

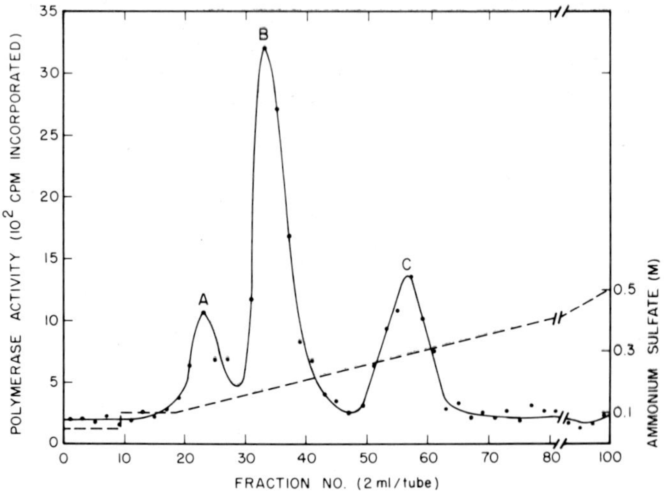

Fig. 5. Separation of yeast nuclear RNA polymerases on DEAE-cellulose. Cells of mutant 1016 were disrupted in a Braun homogenizer and the extract chromatographed on DEAE- cellulose according to Roeder and Rutter [27]. Fractions (25 μl) were assayed (26) with denatured calf thymus DNA as template and ^{3}H-UTP as the labelled trinucleotide. Both Mn^{2+} and Mg^{2+} were present.

TABLE IV

Effect of Antibiotics on Yeast Nuclear RNA Polymerases

Antibiotic	μg/ml	RNA Polymerases* A	B % Inhibition	C
None	—	0	0	0
α-Amanitin	20	—	31	80
Rifampicin	20	99	0	0
Lomofungin	20	100	96	92

* The polymerases represent peaks A (fractions 22-24), B (fractions 32-34) and C (fractions 56-58). The separation and assay conditions are described in Fig. 5. Control values in cpm are 333 for A, 995 for B, and 846 for C.

conclude that lomofungin is an effective inhibitor of yeast RNA polymerases *in vivo* as well as *in vitro*. Thus it may offer a relatively specific means of manipulating the rate of m-RNA synthesis in yeast and of evaluating the relative importance of synthesis of m-RNA and of its translation in controlling the formation of exoenzymes by yeast protoplasts.

B. *Effect of lomofungin on enzyme synthesis by protoplasts*

The synthesis of three enzymes was examined under conditions in which the synthesis of RNA but not of protein was promptly inhibited by lomofungin (Fig. 3). As seen in Fig. 6, production of acid phosphatase, an external glycoprotein enzyme, did not decrease until 30-40 min after addition of lomofungin (40 μg/ml) to derepressed protoplasts which were actively synthesizing this enzyme. A similar 30-40 min lag was obtained with invertase, another external glycoprotein enzyme.

The induction and synthesis of α-glucosidase was studied next since this is an internal carbohydrate-free enzyme and thus valuable for comparison with the secreted glycoproteins. Lomofungin at 40 μg/ml prevented almost completely the induction of α-glucosidase

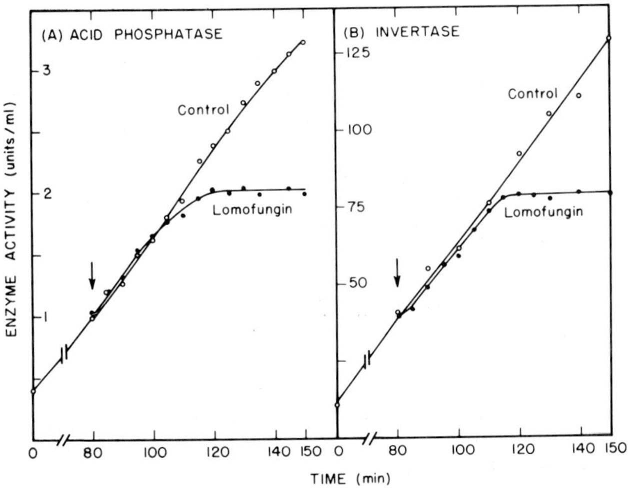

Fig. 6. Effect of lomofungin on the synthesis of acid phosphatase and invertase by protoplasts. Protoplasts were suspended at 5 x 10^7/ml in 0.8 M sorbitol medium containing 100 mM (acid phosphatase experiment) or 50 mM fructose (invertase experiment) and were incubated at 30°C. Lomofungin (40 μg/ml) was added at 80 min(↓).

(Fig. 7; addition at 0 min), but once the protoplasts were induced (40 min), continued synthesis of the enzyme was not severely impaired for another 40 min after addition of the antibiotic.

Since production of acid phosphatase, invertase, and α-glucosidase by protoplasts did not decrease until 20-30 min after RNA synthesis had ceased (cf. Fig. 3), it is probable that continued synthesis of m-RNA is not essential for formation of these enzymes. From the available data, one cannot decide whether synthesis of the enzymes eventually halted because the supply of m-RNA had been depleted or if there was a generalized breakdown of protein formation, since the drop in total protein synthesis which occurred at about the same time might also reflect a general lack of m-RNA molecules.

It should be pointed out that our findings are consistent with the existence of relatively long-lived m-RNA's for the three enzymes examined; but we cannot completely rule out the possibility that the

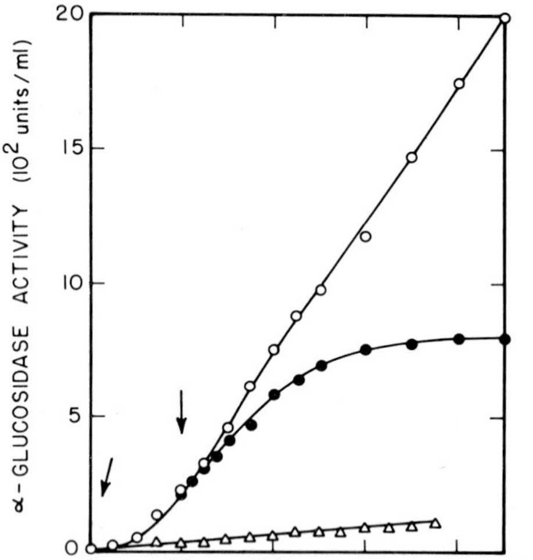

Fig. 7. Inhibition by lomofungin of the induction and synthesis of α-glucosidase. Protoplasts from cells grown in 0.2 M glucose medium were suspended at 5×10^7 per ml in 20 mM citrate-phosphate buffer containing 0.8 M sorbitol, 20 mM fructose and 0.5% Casamino acids. After preincubation at 30°C for 90 min the protoplasts were suspended in fresh mediun to which 20 mM maltose (inducer) was added (0 time). Lomofungin (40 μg/ml) was added (↓) either at zero time (△) or after 40 min (●) in the presence of maltose. At various time intervals samples were removed, dialysed against phosphate buffer (1 mM; pH 6.8) at 4°C for overnight and assayed for α-glucosidase activity.

protoplasts contain large pools of these m-RNA's which have a short half life but do not fall to a rate-limiting level until 30-40 min after addition of lomofungin. Answers to these important problems will be obtained only by a great deal of additional study utilizing a variety of experimental approaches.

Acknowledgments

This research was supported in part by U.S. Public Health Service Grant AI-04572 from the National Institute of Allergy and Infectious Diseases.

References

1. Gascón, S. (1973). This volume. p. Academic Press, New York, London.
2. Lampen, J. O., Kuo, S.-C., Cano, F. R. and Tkacz, J. S. (1972). *In:* Fermentation Technology Today. p. 819 (ed. Terui, G.). Society of Fermentation Technology, Japan.
3. Neumann, N. P. and Lampen, J. O. (1967). *Biochemistry,* **6**, 468.
4. Gascón, S. Neumann, N. P. and Lampen, J. O. (1968). *J. biol. Chem.,* **243**, 1573.
5. Van Rijn, H. J. M., Boer, P. and Steyn-Parvé, E. P. (1972). *Biochim. biophys. Acta,* **268**, 431.
6. Van Wijk, R. (1968). *Proc. Kon. Ned. Akad. Wetensch. Ser. C. Biol. Med. Sci.,* **71**, 60.
7. Gorts, C. P. M. (1969). *Biochim. biophys. Acta,* **184**, 299.
8. McLellan, W. L. and Lampen, J. O. (1963). *Biochim. biophys. Acta,* **67**, 324.
9. Islam, M. F. and Lampen, J. O. (1962). *Biochim. biophys. Acta,* **58**, 294.
10. Kuo, S.-C. and Lampen, J. O. (1971). *J. Bact.,* **106**, 183.
11. Winge, O. and Roberts, C. (1957). *C. R. Trav. Lab. Carlsberg, Ser. Physiol.,* **25**, 419.
12. Van Wijk, R. and Konijn, T. M. (1971). *FEBS Letters,* **13**, 184.
13 Megnet, R., Schlanderer, G. and Dellweg, H., *In:* Fourth International Fermentation Symposium, Abstracts, Kyoto, Japan, March 1972, p. 224.
14. Farkas, V., Svoboda, A. and Bauer, S. (1970). *Biochem. J.,* **118**, 755.
15. Kuo, S.-C. and Lampen, J. O. (1972). *J. Bact.,* **111**, 419.
16. Sols, A. (1967). *In:* Aspects of Yeast Metabolism. p. 47. (Mills, A. K. and Krebs, Sir. H., eds0 Blackwell Scientific Publishers, Oxford.
17. Azam, F. and Kotyk, A. (1969). *FEBS Letters,* **2**, 333.
18. Hereford, L. M. and Hartwell, L. H. (1971). *Nature New Biol.,* **234**, 171.
19. Medoff, G., Kobayashi, G. S., Kwan, C. N., Schlesinger, D. and Venkov, P. (1972). *Proc. natn. Acad. Sci. U.S.A.,* **69**, 196.
20. Gottlieb, D. and Nicolas, G. (1969). *Appl. Microbiol.,* **18**, 35.
21. Tipton, C. D. and Rinehart, K. L. Jr. (1970). *J. Am. Chem. Soc.,* **92**, 1425.
22. Johnson, L. E. and Dietz, A. (1969). *Appl. Microbiol.,* **17**, 755.

23. Burgess, R. R. (1969). *J. biol. Chem.*, **244**, 6160.
24. Sippel, A. and Hartmann, G. (1968). *Biochim. biophys. Acta*, **157**, 218.
25. Schleif, R. (1969). *Nature, Lond.*, **223**, 1068.
26. Ponta, H., Ponta, U. and Wintersberger, E., (1971). *FEBS Letters*, **18**, 204.
27. Roeder, R. G. and Rutter, W. J. (1969). *Nature, Lond.*, **224**, 234.

Secretion of Invertase and α-Galactosidase by Yeast

S. GASCÓN, P. S. LAZO, F. MORENO and A. G. OCHOA

Departamento de Microbiología, Instituto de Biología Celular, Facultad de Ciencias, Universidad de Salamanca, Spain

The biosynthesis and secretion of glycoproteins and especially exocellular enzymes by yeast cells and protoplasts is an interesting system on which one can study the secretion process in great detail. A considerable amount of information regarding some of these enzymes, in particular yeast invertase (β-D-fructofuranoside fructohydrolase, EC 3.2.1.26), has accumulated.

After a short review of facts known about invertase secretion, this paper is concerned with some of our latest results on invertase and is followed by a presentation of our preliminary findings on an almost forgotten enzyme, α-galactosidase or melibiase (α-D-galactoside galactohydrolase EC 3.2.2.22).

More than a decade ago it was known through the work of de la Fuente and Sols [1] that invertase is located in contact with the external environment of the cell in such a way that the metabolism of sucrose by the yeast is initiated by the splitting of sucrose outside the cells, followed by the uptake of monosaccharides, and their intracellular fermentation. Direct confirmation of invertase outside the cytoplasmic membrane was obtained by Friis and Ottolenghi [2] and by Sutton and Lampen [3] who observed that the enzyme is released into the medium during protoplasts formation. Protoplasts prepared from depressed cells are capable of secreting invertase into the surrounding medium while those from repressed cells cannot do so.

Invertase content in yeast cells is dependent on growth conditions and especially on glucose concentration which is an important factor controlling the synthesis of the enzyme [4-8]. When yeast grows in low glucose media the cells are derepressed for the synthesis of invertase and most of it is released during protoplasting, although a portion of the enzyme, located within the protoplast, is not affected by the process. In cells grown in high levels of glucose, that is in fully repressed cells, most of the enzyme is intracellular.

There exists a correlation between the enzyme localization on the cell and the molecular form of the enzyme. The presence of at least two molecular forms of invertase can be demonstrated by various means, although they were first investigated by gel filtration of cell free extracts on Sephadex G-200 columns [9]. Extracts from derepressed cells show an enzyme distribution after gel filtration as shown in Fig. 1. Most of the enzyme appears close to the elution volume and has been called large invertase. The enzyme which appears in an elution volume, void volume ratio of 1.7 is called small invertase. Figure 2 shows the elution profile obtained using extracts from repressed cells. In this instance, the predominant form is the small one, although it should be noted that the amount of small invertase per cell does not change significantly with growth conditions. Changes in enzyme concentration that take place as a result of glucose concentration of the culture medium are due to the large enzyme [9]. Protoplast formation removes the large molecular form of the enzyme. Under appropriate conditions protoplasts are capable of synthesizing the cell wall components including extracellular enzymes [10-12].

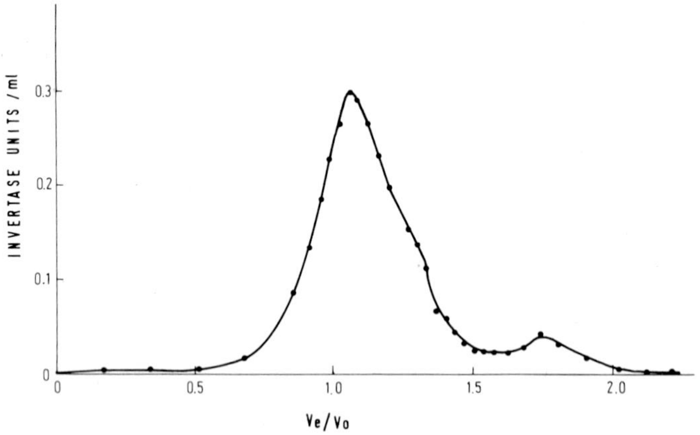

Fig. 1. Gel filtration in Sephadex G-200 of cell free extracts prepared from derepressed cells. *Saccharomyces* strain 303-67 was grown in a rotatory shaker at 28°C in a medium containing 1% yeast extract and 1% glucose. After 16 h the glucose had disappeared. The quantitative anlysis of heavy and light invertases was performed as described previously [9]. The cells were broken in a Braun MSK homogenizer and after centrifugation, the cell free extract containing 20 invertase units was applied to a Sephadex G-200 column (2.5 x 90 cm) equilibrated with 0.05 M Tris-HCl buffer pH 7.5, using reversed flow. The invertase was assayed as described previously [19].

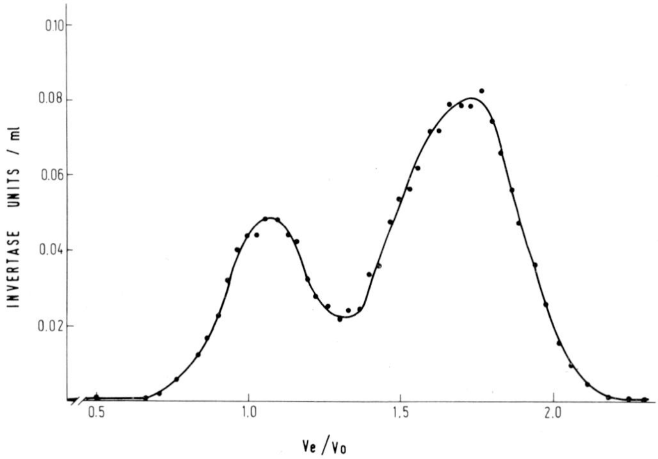

Fig. 2. Gel filtration in Sephadex G-200 of cell free extracts prepared from repressed cells. Cells were grown in a medium containing 1% yeast extract and 3% glucose and were collected after 16 h. At this time there was more than 1% glucose left in the supernatant. Ten invertase units were applied to the column.

It has been demonstrated that the enzyme actively secreted by the protoplasts as well as the enzyme found in the culture supernatant is all in the large form [9, 13].

The conclusion from this kind of experiment is that invertase outside the cytoplasmic membrane is on the large form, that the small form is always inside the protoplast and that a small proportion of the large enzyme might also be present within the protoplast.

The large form of the enzyme has been purified in several laboratories and in the strains we use for our work on invertase, *Saccharomyces* 303-67 [14] and FH4C [15], the enzyme is a glycoprotein containing 50% mannose and 3% glucosamine [16]. In view of the fact that most mammalian extracellular proteins contain carbohydrate [17] and that carbohydrate addition might be needed for protein secretion [18] it was necessary to isolate the small enzyme to check if it contained carbohydrate. The small invertase was purified and found to have a molecular weight of approximately 135,000, similar to the protein component of the heavy enzyme, and the specific activity was 2900 units per mg of protein [19]. A comparative study of the properties of the purified forms of both enzymes showed broad similarities but also two essential differences [20]. The small invertase is devoid of carbohydrate, and disparities were found in the amino acid composition. The absence of carbohydrate in the small enzyme concurred with our hypothesis, but no satisfactory explanation for the disparities in amino acid composition has been found.

Four years ago working at Dr. J. R. Villanueva's Department in Salamanca and knowing that the small form of invertase is devoid of carbohydrate, Dr. P. Beteta started to look for the localization of invertase inside the protoplasts, using the method of isolation of vacuoles or vesicles in a Ficoll gradient as described by Matile and Wiemken [21]. We think that our results show that both light and heavy invertase are present on those vesicles, probably derived from the endoplasmic reticulum. A working hypothesis that is being used by us for the design of new experiments on the study of exocellular enzyme synthesis and secretion by yeast cells and protoplasts is presented in Fig. 3 [22]. There can obviously be no claims on the purity of the vesicles present in one of the three fractions obtained after breaking the yeast protoplast, and in this Symposium Dr. Cabib [23] has presented yet another example of a biochemical function ascribable to particulated material present in this fraction. Unpublished results obtained by A. G. Ochoa and F. Moreno in our Department lead us to believe that not only invertase but melibiase and acid phosphatase might also be located in this fraction although much more work needs to be done in that direction.

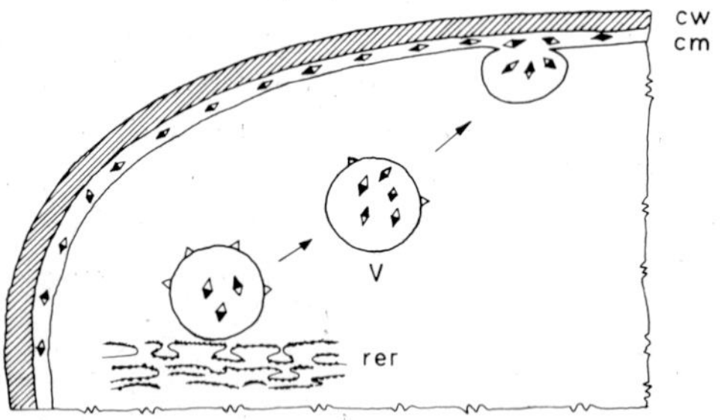

Fig. 3. Proposed scheme for the formation and secretion of invertase in yeast. RER = rough endoplasmic reticulum. V = vesicle, CM = cytoplasmic membrane; CW = cell wall. △ light invertase (no carbohydrate); ◆ heavy invertase (50% mannan and 3% glucosamine).

In some stages of growth it is difficult to isolate protoplasts from yeasts, especially if one is looking for high yields [10]. To overcome this problem we have tried with some success to isolate the vesicles by starting with a cell free extract obtained by mechanical breakage of intact cells by shaking with glass beads in a homogenizer (Braun, A. G. Melsungen, Germany).

These results at least show that the intracellular localization of invertase and for that matter of exocellular enzymes, is still an open question, as is the case for the pathways of biosynthesis and secretion of these enzymes. The fact that more than 90% of these enzymes are found outside the cytoplasmic membrane makes things even more difficult than they usually are when one tries to find the intracellular localization of an enzyme.

In an attempt to further characterize the system Dr. Liras in our Department studied the effect of cycloheximide and 2-deoxyglucose on the synthesis and secretion of invertase by yeast cells and protoplasts [24]. The results are in general similar to those reviewed by Drs Lampen [25] and Sentandreu [26] in this Symposium. For this reason we shall not enter into the details of our results. Suffice to say that using protoplasts and in all the experimental conditions studied, the relative amount of both forms remained constant which indicates that although cycloheximide and 2-deoxyglucose exert their action at two different points in the biosynthesis of glycoproteins, either of them can stop the process of synthesis and secretion of yeast invertase.

Invertase synthesis is so sensitive to the concentration of glucose on the medium that we thought, as did a number of people in other laboratories, that it would be interesting to check if externally added cyclic AMP would be capable of overcoming catabolic repression. First Paloma Liras and then most of the people in our group have tried to prove this, but as Dr. Lampen pointed out earlier, the results are extremely erratic.

Until now we have probably given the impression that there are two and only two forms of invertase. This would imply that the 50% content of mannan of the heavy enzyme is added simultaneously to the small invertase. However, this sounds improbable and we have been looking for the presence of invertase forms of intermediate molecular weight that might reveal the existence of a sequential addition of the carbohydrate moiety. While doing gel-filtration in a 90 × 2.5 cm column of Sephadex G-200 for other experimental purposes we analysed invertase in the fractions and the results are expressed in Figs 4 and 5.

The existence of at least two heavy forms of the enzyme has already been reported [9] and although the resolution obtained now is higher than it was before, we do not know exactly which growth

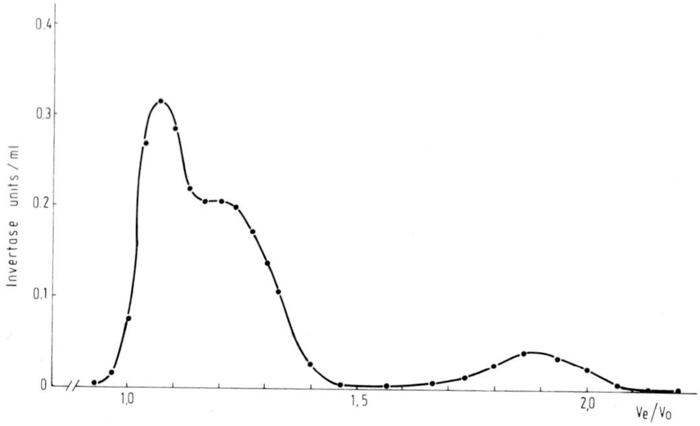

Fig. 4. Gel filtration in Sephadex G- 200 of cell free extracts prepared from cells grown in 1% yeast extract and 1% glucose. No glucose was left after 16 h of growth when the cells were harvested. Fifteen invertase units were applied to the column.

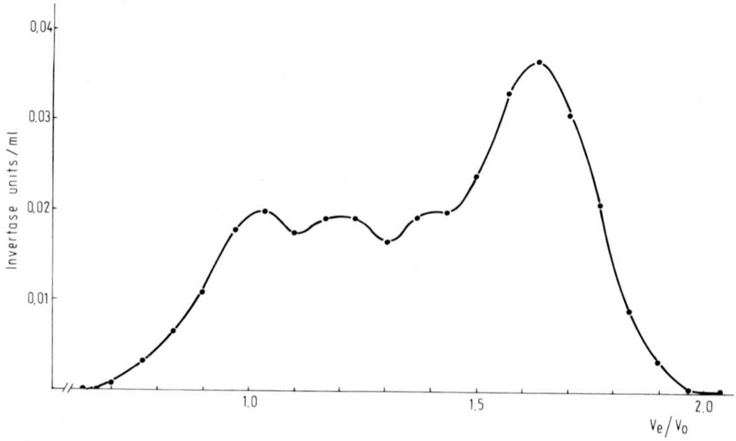

Fig. 5 Gel filtration in Sephadex G-200 of cell free extracts prepared from cells grown in 1% yeast extract and 6% glucose. More than 3% glucose remained after 16 h of growth when the cells were harvested. Five invertase units were applied to the column.

conditions determine the appearance of several forms. As we have already mentioned, the enzyme in the culture supernatant is in the heavy form and it elutes from Sephadex G-200 columns as a symmetrical peak.

Unpublished results by A. G. Ochoa and F. Moreno in our Department show that crude extracts applied to a DEAE-Sephadex column show an elution pattern with three or four peaks but it is difficult to interpret these experiments because changes in phosphate content could affect the affinity of invertase to the ion exchanger.

There is a great deal of controversy at this moment over the interpretation of data on intracellular localization and molecular forms of invertase. Professors Lampen and Matile and their colleagues favour the idea that the only functional form of invertase is the heavy enzyme. According to their hypothesis intermediate and light forms will be formed from material diverted from the main pathway either before or during the glycosilation process. Our working hypothesis is that described previously and would imply the biosynthesis of the small form first and the sequential addition of carbohydrate to it. This would take place in the rough endoplasmic reticulum and in vesicles derived from it, in a similar way to the much better understood process by which animal glycoproteins are synthesized and secreted [27]. We are working actively in this area and planning new experiments in order to clarify certain details of this scheme.

Finally, we will present some of our preliminary findings on α-galactosidase or melibiase from yeast.

The work of de la Fuente and Sols [1] and of Friis and Ottolenghi [2] on invertase had a reply by the same authors on melibiase with practically identical results [1, 28]. Melibiose is hydrolysed outside the cytoplastic membrane and the constituent hexoses are transported into the cell where they are metabolized. In induced cells most of the enzyme is outside the membrane and is released, during protoplasting. Protoplasts from induced cells are capable of synthesizing the enzyme and readily secrete it.

There are however, two considerable differences between these two enzymes. The first one might be advantageous to some experimental approaches, because melibiase is an inducible enzyme. The second difference is disadvantageous, since, as far as we know, very little work has been done in connection with the enzyme during the last few years.

In our work we have used *Saccharomyces* strain 303-49 of Winge and Roberts [14] which is sucrose negative and melibiose positive. It grows slowly on galactose as can be seen in Fig. 6, which also shows the utilization of galactose and the appearance of melibiase in the culture supernatant and in the washed cells. The activity first appears in the cells and later part of it goes to the supernatant. The total activity of the culture is proportional to its growth and therefore the specific activity is constant while the galactose is present in the

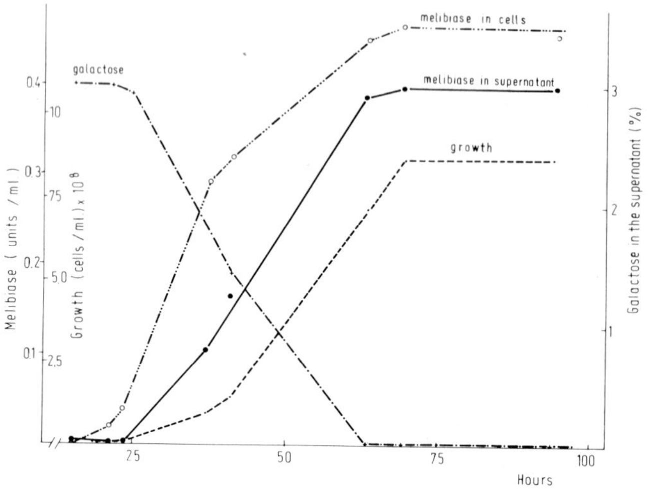

Fig. 6. Synthesis of melibiase by *Saccharomyces* strain 303-49. Cells were grown in a medium containing 1% yeast extract and 3% galactose. Galactose in the supernatant was estimated by the galactose-oxidase method. Melibiase was assayed in a two steps method similar to that used for invertase, estimating the glucose liberated from melibiase by the glucose-oxidase method. One unit of melibiase is defined as the amount of enzyme which hydrolyses 1 μmol of melibiose in 1 min at 30°C in 0.05 M sodium acetate buffer, pH 5.0 containing 0.03 M melibiose.

medium. If glucose is added at any time the synthesis of melibiase ceases. Likewise, cells grown in a mixture of glucose and galactose, synthesize melibiase only after the glucose has disappeared completely from the medium.

The melibiase present in the culture supernatant is found after gel-filtration on Sephadex G-200 in a peak which appears close to the elution volume (Fig. 7). This would imply a molecular weight greater than 250,000 daltons. From the results obtained after two experiments with gel-filtration of a cell free extract, the enzyme in the cells seems similar to that found in the supernatant and no form equivalent to the small invertase was detected, although this should not be regarded as proof of the absence of this hypothetical small melibiase.

In order to understand better the process of secretion and the characteristics of melibiase we have started the purification of the enzyme. Most melibiase in the culture supernatant as well as that from cell free extracts obtained from induced cells, is soluble in ammonium sulphate at 70% saturation. It would be interesting to

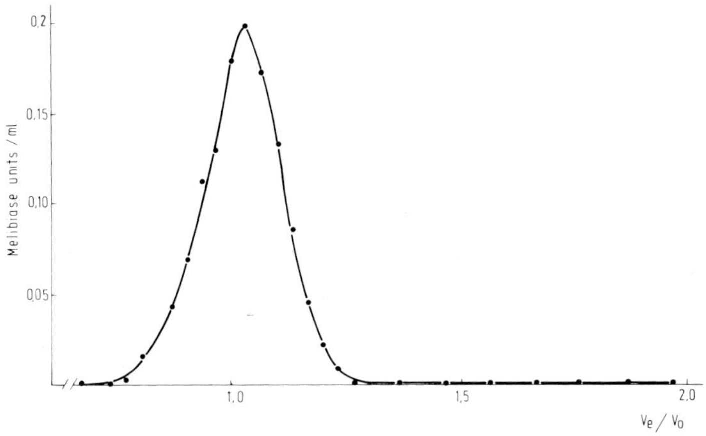

Fig. 7. Gel filtration in Sephadex G-200 of culture supernatant from induced *Saccharomyces* 303-49. The cells were grown in conditions identical to those described in Fig. 6 and were harvested after 75 h. Ten melibiase units were applied to the column.

know if the extracellular melibiase is a glycoprotein and if a small carbohydrate free of the enzyme is present inside the protoplast.

Acknowledgements

The authors gratefully acknowledge the valuable suggestions made by Professor Julio R. Villanuva and his continued interest in these investigations. We are also indebted to Pilar Herrero for her help in some experiments on α-galactosidase.

This work was supported in part by grants from the Ministerio de Educación y Ciencia and from the Fundación Juan March.

References

1. De la Fuente, G. and Sols, A. (1962). *Biochim. biophys. Acta,* **56**, 49.
2. Friis, J. and Ottolenghi, P. (1959). *C. R. Trav. Lab. Carlsberg,* 31, 259.
3. Sutton, O. D. and Lampen, J. O. (1962). *Biochim. biophys. Acta,* **56**, 303.
4. Davies, R. (1953). *Biochem. J.,* **55**, 484.
5. Davies, A. (1956). *J. Gen. Microbiol.,* **14**, 109.
6. Suomalainen, H. and Oura, E. (1957). *Arch. Biochem. Biophys.,* **68**, 425.
7. Dodyk, F. and Rothstein, A. (1964). *Arch. Biochem. Biophys.,* **104**, 478.
8. Gascón, S. and Ottolenghi, P. (1972). *C. R. Trav. Lab. Carlsberg,* 39, 85.
9. Gascón, S. and Ottolenghi, P. (1967). *C. R. Trav. Lab. Carlsberg,* 36, 85.

10. Villanueva, J. R. (1966). *In:* The Fungi Vol. II, p. 3 (Ainsworth, G. C. and Sussman, A. S., eds). Academic Press, London and New York.
11. Villanueva, J. R. and García Acha, I. (1971). *In:* Methods in Microbiology, Vol. 4, p. 665 (Booth, C., ed.). Academic Press, London and New York.
12. Villanueva, J. R. and Garcia Acha, I. (1972). *Rev. lat-amer. Microbiol.*, **14**, 49.
13. Lampen, J. O., Neumann, N. P., Gascón, S. and Montenecourt, B. S. (1967). *In:* Organizational Biosynthesis, p. 363 (Vogel, H. J., Lampen, J. O. and Bryson, V., eds). Academic Press, New York.
14. Winge, O. and Roberts, C. (1957). *C. R. Trav. Lab. Carlsberg, Ser. Physiol.*, **25**, 419.
15. Symington, E. B. and Lampen, J. O. (1966). *In:* Proceedings of the Ninth International Congress of Microbiology, p. 147 (Slakhanova, V. M., ed.). Moscow, July 1966.
16. Neumann, N. P. and Lampen, J. O. (1967). *Biochemistry*, **6**, 468.
17. Eylar, E. H. (1965). *J. Theor. Biol.*, **10**, 89.
18. Soodak, M. (1966). *In:* Current Aspects of Biochemical Energetics, p. 459 (Kaplan, H. O. and Kennedy, E. P., eds). Academic Press, New York and London.
19. Gascón, S. and Lampen, J. O. (1968). *J. biol. Chem.*, **243**, 1567.
20. Gascón, S., Neumann, N. P. and Lampen, J. O. (1968). *J. biol. Chem.*, **243**, 1573.
21. Matile, Ph. and Wiemken, A. (1967). *Arch. Mikrobiol.*, **56**, 148.
22. Beteta, P. and Gascón, S. (1971). *FEBS Letters*, **13**, 297.
23. Cabib, E., Farkas, V., Ulane, R. E. and Bowers, B. (1973). This volume. Academic Press, London and New York.
24. Liras, P. and Gascón, S. (1971). *Eur. J. Biochem.*, **23**, 160.
25. Lampen, J. O., Kuo, S. C. and Cano, F. (1973). This volume. Academic Press, London and New York.
26. Sentandreu, R. and Elorza, M. V. (1973). This volume. Academic Press, London and New York.
27. Spiro, R. G. (1970). *Ann. Rev. Biochem.*, **39**, 599.
28. Friis, J. and Ottolenghi, P. (1959). *C. R. Trav. Lab. Carlsberg*, **31**, 272.

Structure and Function of the Yeast Cell Envelope

HEIKKI SUOMALAINEN and TIMO NURMINEN

Research Laboratories of the State Alcohol Monopoly (Alko), Helsinki, Finland

The parallel developments that have occurred in techniques for electron-microscopic studies of the fine structure of the yeast cell, and of cell rupture and the isolation of individual organelles have enabled biochemists and microbiologists to relate many metabolic functions to ultrastructure and chemical composition.

Most of the yeast enzymes are located within the protoplast. Figure 1 illustrates some typical examples of the location of intracellular enzymes. In the aerobic yeast cell, succinate dehydrogenase and the terminal electron-transport activities are present only in the mitochondrial particles [1], which also contain oligomycin-sensitive ATPase [2], and are capable of the complete oxidation of pyruvate via the tricarboxylic acid cycle [3]. However, clear evidence has been put forward that most enzymes of the tricarboxylic acid cycle are located in both the mitochondrial and soluble cytosol fractions [1, 4]. The cytosol further contains the enzymes of the glycolytic and oxidative pentose phosphate pathways [5], ethanol dehydrogenase [5], and the key enzymes of the glyoxylate cycle [1, 4]. The vacuole represents the lysosome of the yeast cell, and has a content of various hydrolytic enzymes [6]. The microsomal fraction obtained by the application of differential centrifugation is rather heterogeneous. Structures derived from the endoplasmic reticulum [5], fragments of the plasma membrane [7] and large protein particles containing fatty acid synthase activity [8] are to be found in this fraction. A prominent function commonly attributed to the rough endoplasmic reticulum is the synthesis of protein; similarly, the NADPH-cytochrome *c* reductase [9] of the microsomal membrane fraction, and its ability to incorporate acetate into a variety of lipids [10] can be so attributed.

Naturally enough, the subcellular distribution of enzymes inside the yeast cell is more complicated than this schematic diagram would suggest. For instance, as a rule the site of synthesis of an enzyme

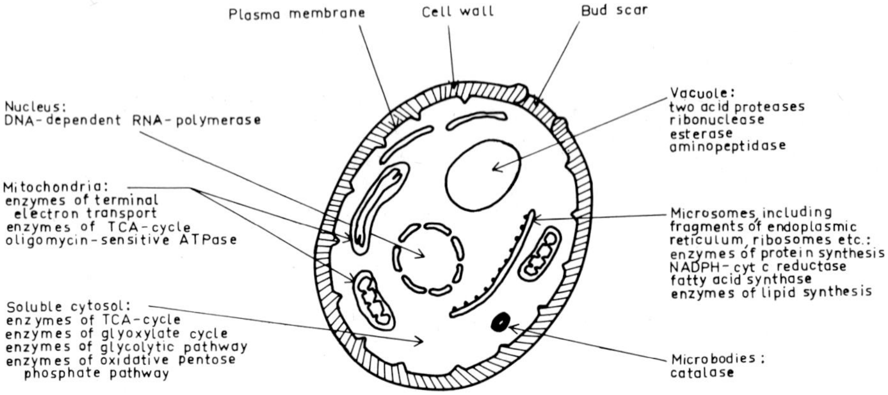

Fig. 1. Schematic diagram illustrating the localization of the intracellular enzymes of an aerobic yeast cell (*Saccharomyces cerevisiae*).

differs from its final location. Even less is known about the enzymic and chemical composition of the cell envelope of the yeast cell, particularly in regard to the plasma membrane. In this connection, the term "cell wall" is restricted to material outside the plasma membrane. The cell wall and the plasma membrane together comprise the cell envelope. In what follows, a presentation is given of some results obtained in relation to the enzymic and chemical composition of these envelope fractions of yeast, mainly baker's yeast (*Saccharomyces cerevisiae*).

I. Isolation of Plasma Membrane Preparations from Cell Envelopes

A method applied in our laboratories for study of the composition of the cell wall and plasma membrane was based upon the observation that carefully isolated cell envelopes of yeast contain not only the wall but also fragments of the plasma membrane [11, 12]. The principle of isolation of the plasma membrane fragments from the cell envelopes is illustrated in Fig. 2. Enzymic digestion of the isolated cell envelopes was effected by means of purified snail-gut enzymes. The resulting digested preparation was collected and washed by centrifugation, and further fractionated for removal of the wall remnants. The bulk of the carbohydrates of the cell envelopes could be removed by enzymic digestion. For the main part, the purified plasma membrane preparation contained proteins and lipids, although a significant amount of carbohydrate was still present (Table I). An electron-micrograph of the plasma membrane

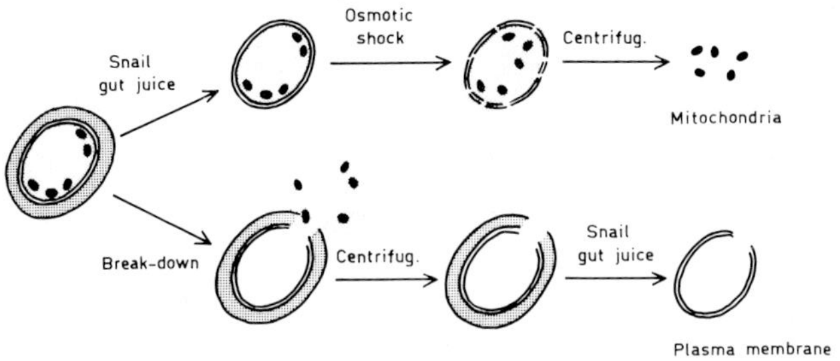

Fig. 2. Scheme of the preparation of mitochondria and plasma membrane from the yeast cell using snail-gut enzyme [13].

TABLE I
Overall chemical composition of isolated cell envelopes and plasma membrane preparations obtained from them, expressed as percentages of dry weight

Preparations	Carbo-hydrates	Proteins	Lipids
Cell envelopes	83	10	5.1–6.4
Plasma membranes	26–28.5	32–38	33.3–42

preparation obtained from the isolated yeast cell envelopes is reproduced in Fig. 3, with the preparation shadowed by an alloy of platinum and palladium.

II. Enzymic Composition of the Envelope Fractions

A study of the distribution of various enzymes among subcellular fractions [14] indicated the absence of known intracellular enzymes from the isolated cell envelopes, indicating that these were not contaminated by intracellular material (Table II). It also follows that these enzymes are not present in the plasma membrane of yeast.

Considerable amounts of saccharase, acid phosphatase, inorganic pyrophosphatase, ATPase activity at pH 3.5, and Mg^{2+}-dependent ATPase with an optimum activity at pH 6.8, were discernible in the isolated cell envelopes (Table III). During enzymic digestion, most of the enzymes apparent in the cell envelopes were released into the surrounding medium, but the Mg^{2+}-dependent ATPase remained in the plasma membrane preparation [14, 15]. The release of saccharase and acid phosphatase corresponded to that noted in the

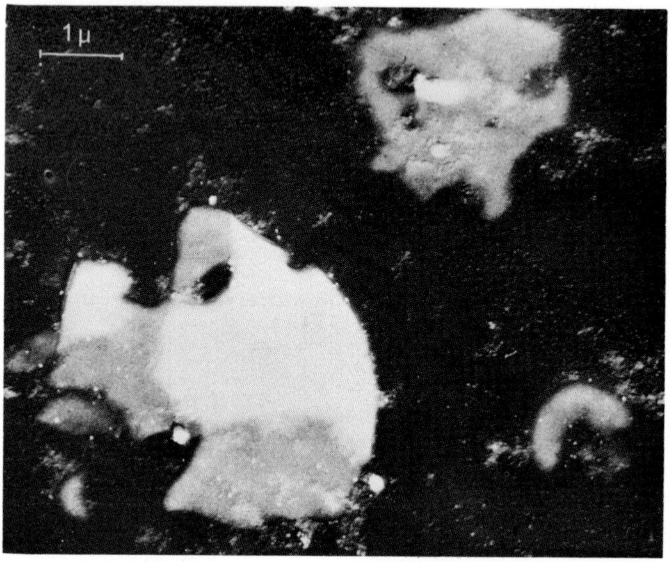

Fig. 3. Electron-micrograph of a plasma membrane preparation obtained from isolated yeast cell envelopes. The preparation has been shadowed with an alloy of platinum and palladium [11].

TABLE II
Total activities of some enzymes in the cell envelope and in cell interior [14]

Enzyme	Substrate	Cell envelope	Cell interior
Hexokinase	D-Glucose	0	100
Alkaline phosphatase	*p*-Nitrophenyl phosphate	1	99
Esterase	Phenyl propionate	2	98
	Triacetin	0	100
$NADH_2$ oxidase system	$NADH_2$	2	98
Deamidase	L-Asparagine	2	98
	Nicotinamide	3	97

preparation of yeast protoplasts, and thus indicated that the released enzymes were mainly located in the cell wall outside the plasma membrane. As yet, it is not known with any certainty whether the small activities of the mainly solubilized enzymes residual to the plasma membrane preparations represent remnants of the wall, or intrinsic characteristics of the plasma membrane. Nevertheless, it

TABLE III
Behaviour of saccharase, acid phosphatase, inorganic pyrophosphatase and ATPases of isolated cell envelopes during enzymic digestion and subsequent fractionation

Preparation	Saccharase	Acid phosphatase	Inorganic pyro-phosphatase	ATPase at pH 3.5	Oligomycin-insensitive Mg^{2+}-ATPase at pH 6.8
Cell envelopes	100	100	100	100	100
Released into medium	85	78	82	105	0
Purified plasma membranes	4	7	4	12	102

The total activities of the cell envelopes are denoted as 100.

may for example be suggested that the saccharase in the plasma membrane is in transit to the wall from its site of synthesis [16].

The Mg^{2+}-dependent ATPase of the yeast plasma membrane differs from the mitochondrial ATPase of yeast in its insensitivity to oligomycin. Moreover, it is the only enzyme with a location definitely confirmed as being in the plasma membrane by all three methods known for the preparation of plasma membranes from yeast [7, 14, 15, 17]. In contrast to ATPase activity at pH 3.5, the Mg^{2+}-dependent ATPase is detectable only in the isolated cell envelopes, but not with intact cells (Fig. 4).

Clearly all the yeasts synthesize the intracellular enzymes involved in the turnover of membrane lipids. However, little work has been done on the lipolytic enzymes of the envelope fractions of yeast. The lipase and phospholipase activities that we found [18] in various preparations of aerobic baker's yeast are shown in Table IV. These activities are mainly located in the cell interior, although considerable amounts are also demonstrable in the isolated cell envelopes. On the other hand, it appears that resting cells of baker's yeast do not contain noteworthy amounts of extracellular or water-extractable lipolytic enzymes.

After partial enzymic digestion and subsequent fractionation, the bulk of the lipolytic activities present in the cell envelopes was recovered in the plasma membrane preparations (Table V). Exposure of the cell envelopes to successive washings resulted in removal of the fragments of the plasma membrane, as indicated by release of the Mg^{2+}-dependent ATPase, and the simultaneous loss in the bulk of lipid components. The resulting purified walls were completely devoid of lipase activity, whereas the phospholipase was more firmly bound to the preparation. These results suggest that the lipase

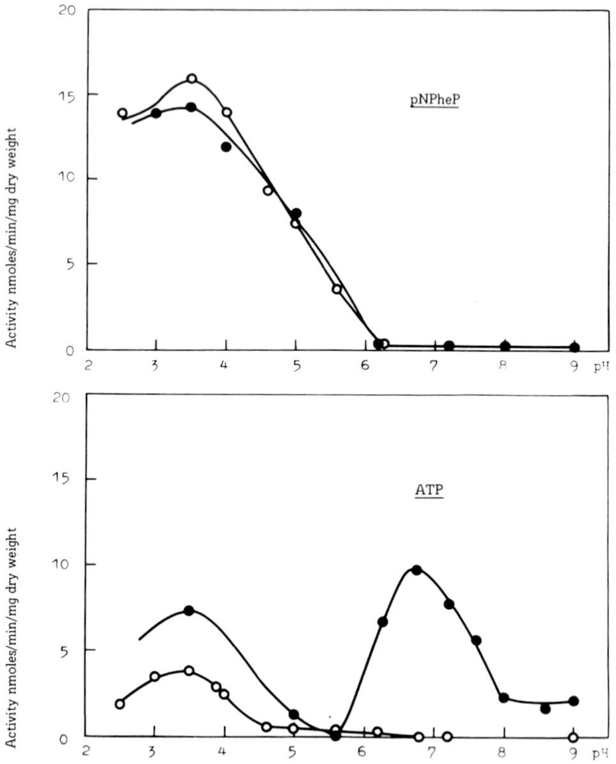

Fig. 4. The pH-activity curves of the hydrolysis of externally added *p*-nitrophenyl phosphate (pNPheP) and of ATP by intact cells (○—○) and by isolated cell envelopes (●—●) [14].

TABLE IV

Total lipolytic activities of preparations of baker's yeast, as μmol/min per preparation obtained from 1 g fresh weight of cells [18]

Preparation	Lipase	Phospholipase: Lysophosphatidylcholine degraded	Phospholipase: Glycerylphosphorylcholine formed
Intact cells	0.163	14.6	3.8
Water extract of intact cells	0.009	0.9	0.5
Disintegrated cells	0.366	105.5	66.7
Cell envelopes	0.095	4.6	2.6

TABLE V
Total lipolytic activities in cell envelopes and preparations obtained from the isolated cell envelopes of baker's yeast

Preparation	Lipase	Phospholipase	
		Lyso-phosphatidyl-choline degraded	Glyceryl-phosphoryl-choline formed
Cell envelopes	100	100	100
Crude plasma membranes	96	74	100
Purified plasma membranes	38	37	40
Purified walls	0	40	57

The total activities of the cell envelopes are denoted as 100.

activity of the cell envelope is associated with the plasma membrane alone, whereas the phospholipase is located both in the plasma membrane and in the cell wall.

III. Lipid Composition of the Envelope Fractions

Relatively few investigations have been concerned with the lipid composition of the yeast cell wall. On the other hand, the lipid composition of the plasma membrane is interesting, since the permeability of the membrane depends upon a number of lipid parameters [19].

Neutral lipids, including free and esterified sterols, amounted to about 50% of the total lipids in the cell envelope of baker's yeast (Table VI). The total lipid content of the walls, on a dry weight basis, was no more than about a third of that of the whole envelope, but proportionately more (up to 75%) of the total lipids were neutral. The principal lipid groups among the neutral lipids are triglycerides, sterol esters, free sterols and lower glycerides. The content of free sterols is obviously higher in whole envelopes than in the walls. This is in agreement with observations made with animal plasma membranes, in which the sterols exist unesterified. However, it can be concluded that a large proportion of the sterols in the yeast plasma membrane are esterified. The observation that the wall also seems to contain a certain amount of sterols is rather surprising. The main sterol in the envelope fractions of aerobic baker's yeast has been found by us to be ergosterol, comprising about two-thirds of

TABLE VI
Neutral lipid composition of whole envelopes and cell walls of baker's yeast, expressed as mg/g dry weight of the preparation

	Whole envelope	Cell wall
Total lipids	51.1	16.2
Neutral lipids		
Total	25.3	12.0
Squalene	0.1	0.2
Sterol esters	7.0	3.4
Triglycerides	9.2	4.4
Free sterols	3.2	0.9
Di- and monoglycerides and free fatty acids	2.2	2.1

the total sterol content. Another main sterol component was found to be zymosterol. The envelope lipids of aerobic baker's yeast contained only a small amount of squalene. These results are in contrast with those obtained for the envelope fractions of anaerobic brewer's yeast; on investigation, this proved to contain about ten times as much squalene and one-tenth as much ergosterol. Under anaerobic conditions, squalene is accumulated by the yeast; on exposure of the cells to oxygen, this hydrocarbon is converted into sterols.

It was found that phospholipids amounted to about 25% of the envelope lipids of aerobic baker's yeast, and the phospholipid compositions of whole cells, cell envelopes and plasma membrane preparations were qualitatively similar (Table VII). However, the whole cells contained phosphatidylcholine in a clearly larger

TABLE VII
Phospholipid compositions of whole cells, cell envelopes and plasma membranes obtained from the isolated cell envelopes [20]

Phospholipid	% of total P in phospholipids		
	Whole cells	Cell envelopes	Plasma membranes
Phosphatidylcholine	55	45	20
Phosphatidylethanolamine	20	15	20
Phosphatidylinositol + phosphatidylserine	15	30	35
Phosphatidic acid	<5	5	5
Diphosphatidylglycerol	<5	<5	5
Unidentified components	5	5	10

proportion than did the cell envelopes, which were in turn clearly richer in phosphatidylinositol [20]. Combined figures are given here for phosphatidylinositol and phosphatidylserine, but we have established subsequently that phosphatidylserine is no more than a minor envelope phospholipid. In contrast to a part of the neutral lipids, it is probable that the phospholipids are not components of the wall, but are at least for the most part, constituents of the plasma membrane. The fact that they are recoverable in the remaining plasma membrane preparation after enzymic digestion of the isolated cell envelopes lends support to this view.

Knowledge is still lacking on the nature of a significant proportion of the lipids in the cell wall and in the plasma membrane. A complex sphingolipid, containing inositol, phosphorus and mannose, has been isolated by us from whole envelopes [21] and appears to be the predominant glycolipid present.

The major fatty acids of the whole envelopes and the walls are palmitoleic and oleic acids [20]. Whole envelopes contain proportionately more C_{18} acids than do the walls (Table VIII), which is in

TABLE VIII

Fatty acid compositions of whole envelopes of baker's yeast, and of cell walls, a complex sphingolipid and other lipid fractions derived from the whole envelopes [21, 22]

Fatty acid	Whole envelope	Cell wall	Sphingo-lipid	Neutral lipids	Neutral glyco-lipids	Glycero-phospha-tides	Acidic glyco-lipids
	% of total fatty acids						
16 : 1 + 16 : 0	28.6	30.8	7.1	34.4	16.6	39.2	27.4
18 : 1 + 18 : 0	38.6	30.2	10.3	34.6	14.1	35.7	31.4
Fatty acids with more than 18 carbon atoms	6.3	1.9	82.6	3.2	30.1	12.1	14.6
Per cent distribution of the main fatty acids with more than 18 carbon atoms (Components below 10% omitted)							
19 : 0				14.7			
20 : 0		11.2		21.6	11.0		
22 : 1						10.7	
22 : 0					21.0	19.9	17.6
24 : 0						25.9	
26 : 0	12.9	17.7	21.1	16.0			
26 : h	42.6	21.3	68.3				31.5

agreement with the observation [20] that oleic acid is predominant in the plasma membrane fraction. The whole envelopes contained fatty acids with more than 18 carbon atoms to a higher proportion than was the case with the walls, but in both instances the principal, very long-chain fatty acids were 2-hydroxy-$C_{26:0}$ acid and $C_{26:0}$ acid [21]. Since these two fatty acids were also the major fatty acids of the complex sphingolipid isolated, it may be inferred that their presence chiefly denotes the distribution of this lipid, which is consequently present both in the wall and plasma membrane of baker's yeast. Several other glycolipids also appear to be minor components of the cell envelope. Thus, the neutral glycolipid fraction mentioned in Table VIII consists of three kinds of glycolipids [22]. There are two sterol glycosides, two cerebrosides and two alkali-labile glyceroglycolipids. The acidic glycolipids detected are a galactose-containing sulphatide and the inositol-, phosphorus- and mannose-containing spingolipid mentioned above. No knowledge is possessed as yet in regard to the distribution of the minor glycolipids between the cell wall and the plasma membrane, or their exact structure.

IV. Isolation of Plasma Membrane Preparations by Zonal Centrifugation

With a view to acquiring larger amounts of yeast plasma membranes in less time, a study was made of the fractionation of subcellular organelles by zonal centrifugation [23]. Homogenates of anaerobically-grown cells, containing neither detectable cytochrome *c* oxidase nor mitochondria, were used in the present experiments. It could accordingly be expected that the greater part of the membrane material was derived from the plasma membrane.

Figure 5 illustrates the distribution of protein and some enzymes after zonal centrifugation in continuous density gradients of sucrose or iso-osmotic Ficoll. In both cases, the bulk of soluble proteins and solubilized enzymes was found within the lowest range of density. Mg^{2+} -dependent ATPase is coincident with the second protein peak at a median density of 1.10 in both gradients [22]. In sucrose, two other peaks of this enzyme were apparent at higher densities, indicating the presence of different kinds of membrane fragments.

Figure 6 presents a distribution diagram of phospholipid, sterol and nucleic acids in sucrose and Ficoll gradients. To some extent these compounds were non-sedimentable, but for the main part occurred in the same particulate range as most of the Mg^{2+} -dependent ATPase. Plasma membrane vesicles contaminated with

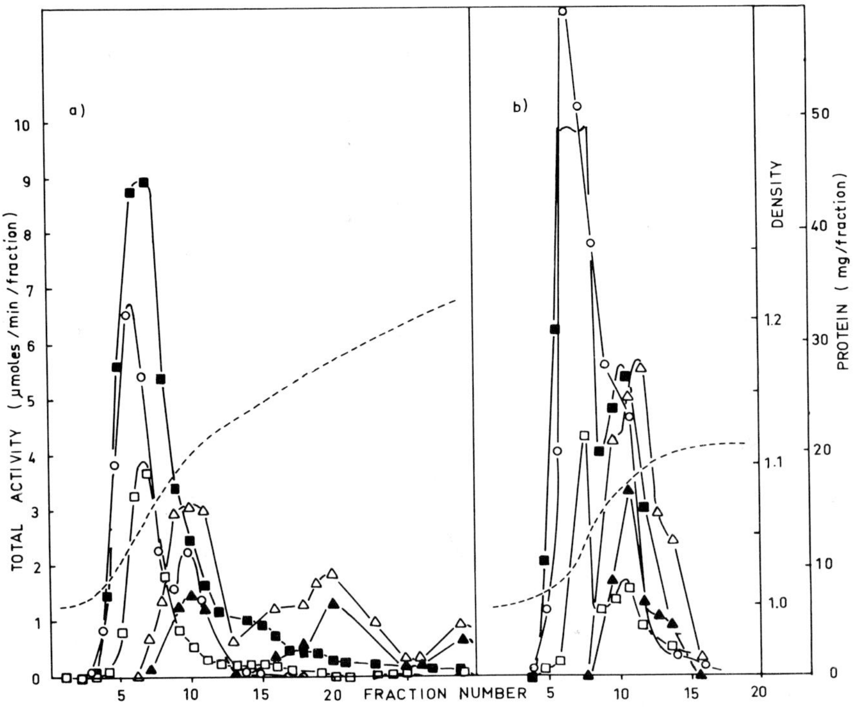

Fig. 5. Fractionation of a homogenate of anaerobically-grown baker's yeast by zonal centrifugation in density gradients of (a) sucrose and (b) Ficoll. Centrifugation was 5 x 10^6 $g \cdot$ min at the sample zone. Density (- - - -) and the distributions of protein (○), acid phosphatase (□), alkaline phosphatase (■), Mg^{2+}-dependent ATPase (△) and oligomycin-insensitive Mg^{2+}-dependent ATPase (▲) have been shown.

cytoplasmic material, presumably derived mostly from the rough endoplasmic reticulum, were obviously present in this particulate range, as indicated by electron-microscopy. However, they were found at somewhat higher densities in a purer state; in sucrose, at a shoulder of the phospholipid and first Mg^{2+} -dependent ATPase peak, or in Ficoll at a shoulder of the peak of the Mg^{2+} -dependent ATPase, which here contained the second phospholipid peak. An electron-micrograph (Fig. 7) of a thin section of a pellet, fixed with glutaraldehyde and osmium tetroxide and stained with lead citrate, reveals the presence of empty membrane vesicles as the predominant organelles in this fraction. In the higher magnifications (Figs 8 and 9), the triple-layered unit membrane structure is clearly visible.

The heavier membranes (Fig. 10) were characteristic organelles,

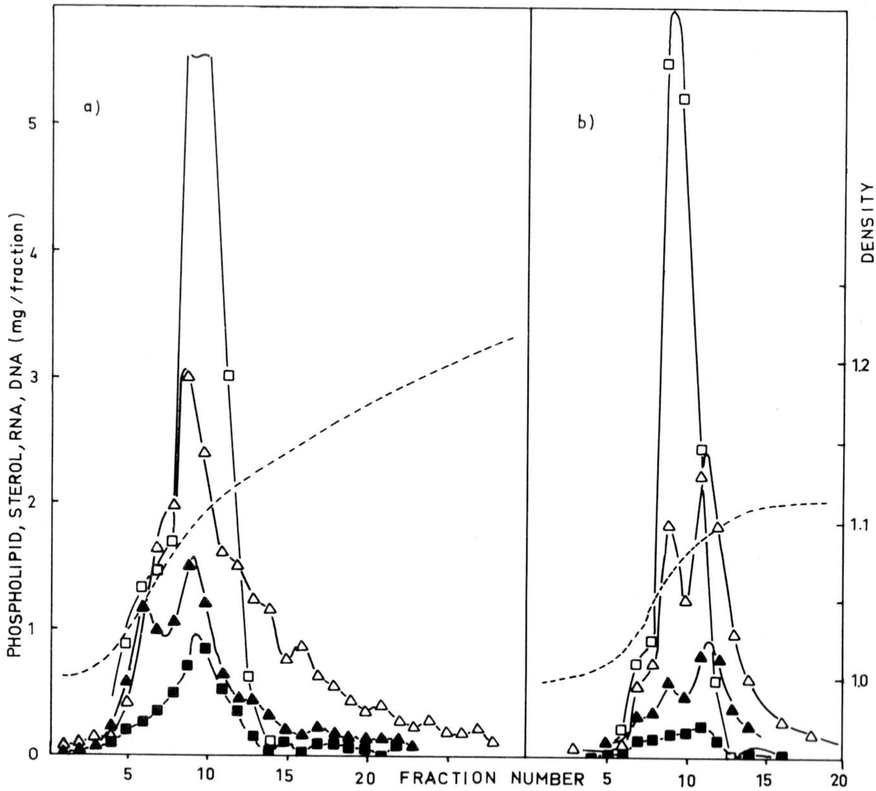

Fig. 6. Fractionation of a homogenate of anaerobically-grown baker's yeast by zonal centrifugation in density gradients of (a) sucrose and (b) Ficoll. Centrifugation was 5 x 10^6 *g* · min at the sample zone. Density (- - - -), phospholipid (△), sterol (▲), RNA (□) and DNA (■) have been assayed.

found in sucrose at higher densities corresponding to the second and third peaks of the Mg^{2+}- dependent ATPase. Occasionally, a third peak of protein, and most of the sterol and phospholipid, were also observable here, whereas the nucleic acid content was always low. Furthermore, no mitochondrial or microsomal oxydoreductases were present. It may be assumed from this that these heavy membranes are also fragments of the plasma membrane. Conceivably, more carbohydrate is retained in these heavy membranes than in the

---

Fig. 8. Electron-micrograph of a thin section of a zonal fraction containing light vesicles of yeast plasma membrane. The preparation has been fixed with glutaraldehyde and osmium tetroxide, and stained with lead citrate. Note the triple-layered membrane structure.

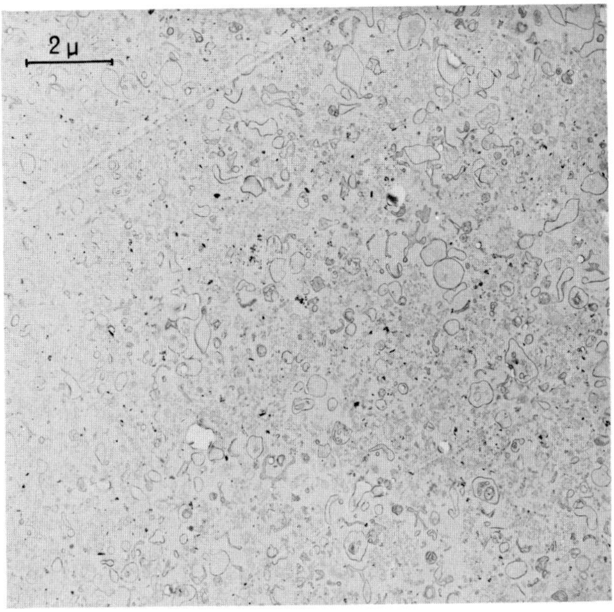

Fig. 7. Electron-micrograph of a thin section of a zonal fraction containing light vesicles of yeast plasma membrane. The preparation has been fixed with glutaraldehyde and osmium tetroxide, and stained with lead citrate.

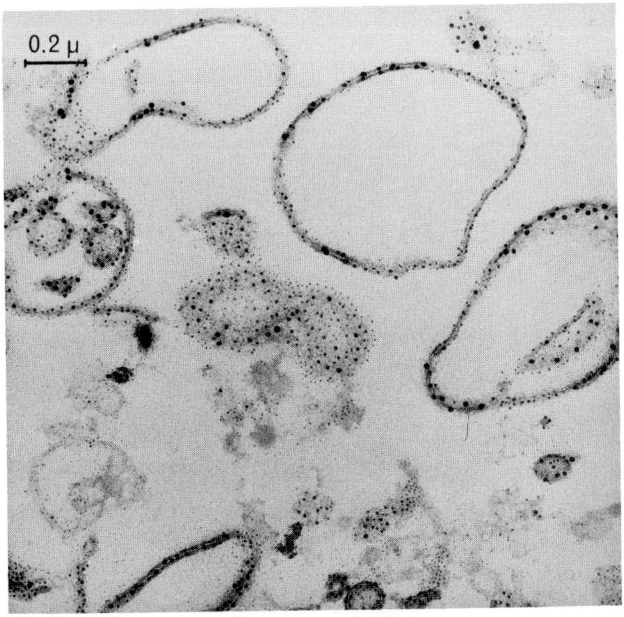

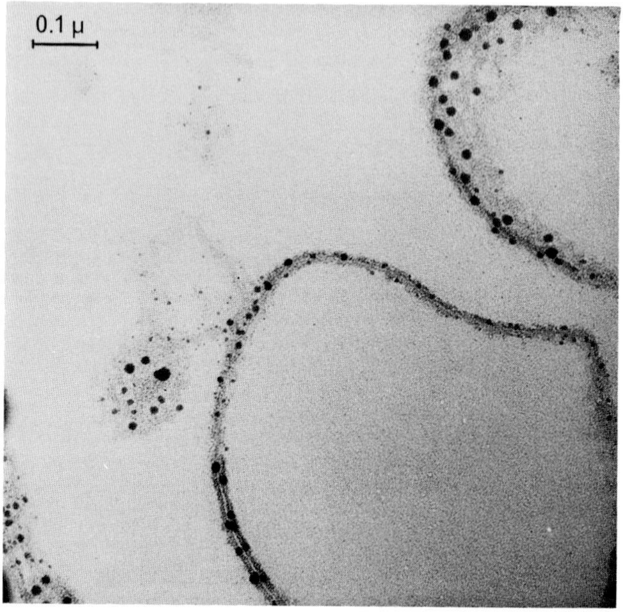

Fig. 9. Electron-micrograph of a thin section of a zonal fraction containing light vesicles of yeast plasma membrane. The preparation has been fixed with glutaraldehyde and osmium tetroxide, and stained with lead citrate. Note the triple-layered membrane structure.

plasma membrane vesicles found at the lower density ranges (cf. refs 7, 24).

Although further information is needed about the distribution patterns of subcellular organelles of yeast after zonal centrifugation in suitable gradients [25], it is evident that zonal centrifugation offers promise of larger-scale isolation of the plasma membrane preparations from yeast.

V. Functions of the Envelope Fractions

The specific shape of the yeast cell arises from the rigid cell wall [26]. The stability of this shape is attributable to the existence of a large number of bonds between the individual wall constituents [16]. Disulphide bridges (Fig. 11), and also the inner layer of the wall, composed mainly of glucan, are thought to be responsible for the structural integrity of the cell wall. Another feature of the yeast cell wall is its elasticity, which becomes apparent when the cells shrink in the presence of high concentrations of osmotically-active

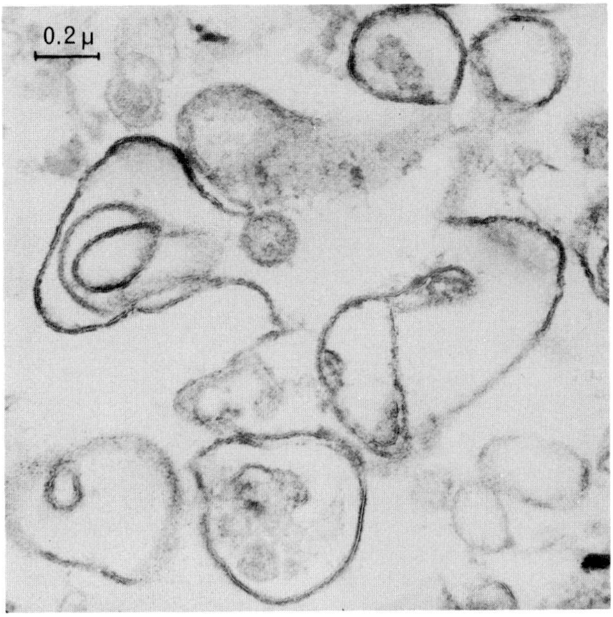

Fig. 10. Electron-micrograph of a thin section of a zonal fraction containing heavy fragments of yeast plasma membrane. The preparation has been fixed with glutaraldehyde and osmium tetroxide, and stained with lead citrate.

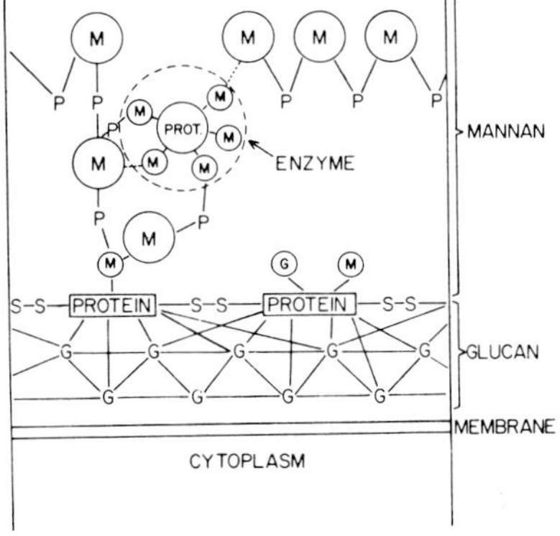

Fig. 11. Schematic structure of yeast wall according to Lampen [16]. M–mannan; G–glucan; Enzyme–saccharase.

substances. The cell wall also acts as a filter, which allows only molecules of moderate size to diffuse through it. Since it is evident that mannan-protein forms the outer surface of the wall, this plays a key role in yeast flocculation. Similarly, it is probable that the wall mannan determines the antigenic properties [27].

Figure 12 and Table IX summarize the distribution of some enzymes in the envelope fractions of *Saccharomyces* yeast. Some observations in regard to this location have already been mentioned, but present knowledge is, of course, based upon many studies made by very many researchers.

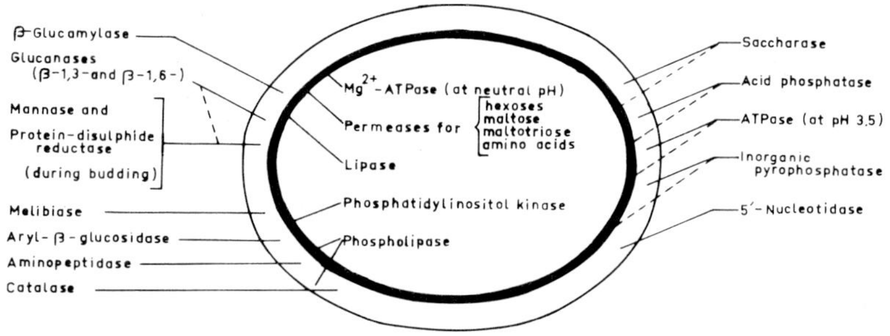

Fig. 12. Distribution of enzymes in the envelope fractions of *Saccharomyces*-yeast.

Most of the external enzymes known to be present in the wall are hydrolases. Many of these enable the yeast to utilize substrates which cannot permeate the plasma membrane For instance, saccharase, melibiase and glucamylase enable the yeast to utilize the respective carbon sources. A similar observation can be made about sugar phosphates and other phosphate esters, which are hydrolysed by the phosphatases in the wall, but are incapable of passing through in an unhydrolysed state. Participation of the external enzymes in the growth and alteration of the cell wall is indicated by the evidence that beta-glucanases [36, 51], mannanase [40] and protein-disulphide reductase [26] function in the budding process. A scanning electron-micrograph of dividing cells of baker's yeast showing several bud scars is shown in Fig. 13.

The yeast plasma membrane has several functions: it forms an expansible cover and protective barrier to the protoplast, and it acts as an organelle which controls the entry and exit of solutes into the cell; moreover, it possibly participates in the synthesis of cell-wall components.

TABLE IX

Distribution of enzymes in the envelope fractions of *Saccharomyces*-yeast

Enzyme	Cell wall	References	Plasma membrane	References
Saccharase	+	[28, 29]	±	[14, 24]
Acid phosphatase	+	[30-32]	±	[14]
ATPase (at pH 3.5)	+	[14, 33]	±	[14]
5′-Nucleotidase	+	[34]		
Inorganic pyrophosphatase	+	[14, 34]	±	[14]
Melibiase	+	[35]		
Glucanases (β-1,3- and β-1,6-)	+	[36]		
Aryl β-glucosidase	+	[37]		
β-Glucamylase	+	[38]		
Aminopeptidase	+	[39]		
Mannanase } during budding	+	[40]		
Protein-disulphide reductase } during budding	+	[26, 41]		
Catalase	+	[42]		
Phospholipase	+	[18, 43-45]	+	[18]
Phosphatidylinositol kinase			+	[46]
Lipase			+	[18]
Mg^{2+}-ATPase (at neutral pH)			+	[7, 14, 15, 17]
Permeases for hexoses			+	[47, 48]
Permeases for maltose			+	[47, 48]
Permeases for maltotriose			+	[49]
Permeases for amino acids			+	[50]

Several of the mechanisms by which the plasma membrane mediates transport have been studied in detail. The presence of transporting systems containing binding sites specific for certain substrates or groups of structurally-related compounds has been recognized. The activity of the permeases, or carriers transporting molecules through the plasma membrane, has features characteristic of enzymes. Thus, enzyme activities which assist the passage of hexoses [47, 48], maltose [47, 48], maltotriose [49] and amino acids [50] are known to be present in the yeast plasma membrane, even though it has proved impracticable to isolate and identify them.

In the presence of Mg^{2+}, K^+ or Na^+ cannot increase the activity of Mg^{2+}-dependent ATPase [14]. In this respect, the ATPase of the yeast plasma membrane clearly differs from the enzyme system, which in many animal tissues is involved in the active transport of Na^+ outward, and K^+ inward, across the plasma membrane.

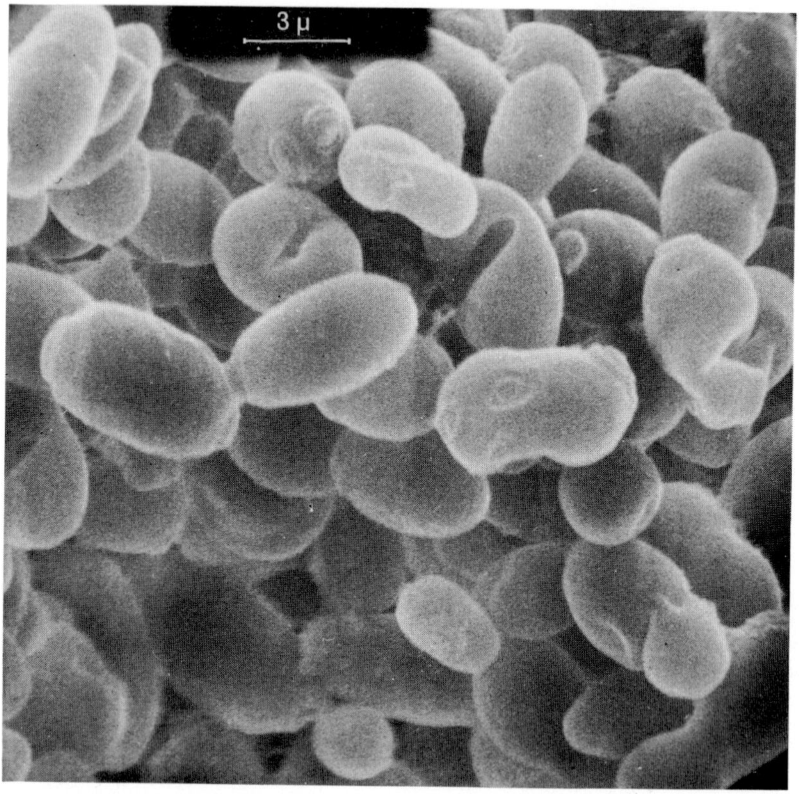

Fig. 13. Scanning electron-micrograph of the dividing cells of *Saccharomyces cerevisiae* showing several bud scars.

Consequently, it is probable that the Mg^{2+}-dependent ATPase of the yeast plasma membrane is involved in the active transport of other substances.

The activity of the lipolytic enzymes present in the cell envelope must be controlled [52]. Thus, the permeability of the plasma membrane may be modified, for example by the activation of the phospholipase, with subsequent degradation of membrane phospholipids. An alternative role may also be suggested. The plasma membrane phospholipids of rat liver participate in the uptake of non-esterified fatty acids, a process that involves lysophosphatidylcholine-phosphatidylcholine cycles [53]. However, it is not known as yet whether similar processes operate in the yeast plasma membrane.

References

1. Perlman, P. S. and Mahler, H. R. (1970). *Arch. Biochem. Biophys.*, **136**, 245.
2. Schatz, G. (1965). *Biochim. biophys. Acta*, **96**, 342.
3. Duell, E. A., Inoue, S. and Utter, M. F. (1964). *J. Bact.*, **88**, 1762.
4. Duntze, W., Neumann, D., Gancedo, J. M., Atzpodien, W. and Holzer, H. (1969). *Eur. J. Biochem.*, **10**, 83.
5. Matile, P., Moor, H. and Robinow, C. F. (1969). *In:* The Yeasts, Vol. 1, p. 219 (Rose, A. H. and Harrison, J. S., eds). Academic Press, London and New York.
6. Matile, P. and Wiemken, A. (1967). *Arch. Mikrobiol.*, **56**, 148.
7. Matile, P., Moor, H. and Mühlethaler, K. (1967). *Arch. Mikrobiol.*, **58**, 201.
8. Klein, H. P., Volkmann, C. M. and Chao, F. C. (1967). *J. Bact.*, **93**, 1966.
9. Schatz, G. and Klima, J. (1964). *Biochim. biophys. Acta*, **81**, 448.
10. Klein, H. P., Volkmann, C. M. and Leaffer, M. A. (1967). *J. Bact.*, **94**, 61.
11. Suomalainen, H., Nurminen, T. and Oura, E. (1967). *Fed. Eur. Biochem. Soc. Meet.*, 4th, Abstr. Commun. p. 111.
12. Suomalainen, H., Nurminen, T. and Oura, E. (1967). *Suom. Kemistilehti*, **B40**, 323.
13. Suomalainen, H. (1969). Antonie van Leeuwenhoek, *J. Microbiol. Serol.*, **35**, 83.
14. Nurminen, T., Oura, E. and Suomalainen, H. (1970). *Biochem. J.*, **116**, 61.
15. Nurminen, T., Oura, E. and Suomalainen, H. (1968). *Fed. Eur. Biochem. Soc. Meet.*, 5th, Abstr. Commun. p. 115.
16. Lampen, J. O. (1968). Antonie van Leeuwenhoek, *J. Microbiol. Serol.*, **34**, 1.
17. Boulton, A. A. (1965). *Expl Cell Res.*, **37**, 343.
18. Nurminen, T. and Suomalainen, H. (1970). *Biochem. J.*, **118**, 759.
19. Van Deenen, L. L. M. (1972). *Chem. Phys. Lipids*, **8**, 366.
20. Suomalainen, H. and Nurminen, T. (1970). *Chem. Phys. Lipids*, **4**, 247.
21. Nurminen, T. and Suomalainen, H. (1971). *Biochem. J.*, **125**, 963.
22. Suomalainen, H. and Nurminen, T. (1972). *Proc. 4th Int. Ferment. Symp.*, Kyoto 1972, p. 825.
23. Anderson, N. G., Harris, W. W., Barber, A. A., Rankin, C. T., Jr. and Candler, E. L. (1966). *Nat. Cancer Inst., Monogr.*, **21**, 253.
24. Matile, P. (1970). *FEBS Symp.*, **20**, 39.
25. Cartledge, T. G. and Lloyd, D. (1972). *Biochem. J.*, **126**, 381.
26. Nickerson, W. J. (1963). *Bact. Rev.*, **27**, 305.
27. Cambell, I. and Gilmour, R. H. (1969). *J. Gen. Microbiol.*, **59**, 193.
28. Burger, M., Bacon, E. E. and Bacon, J. S. D. (1958). *Nature, Lond.*, **182**, 1508.
29. Friis, J. and Ottolenghi, P. (1959). *C. R. Trav. Lab. Carlsberg*, **31**, 259.
30. Suomalainen, H., Linko, M. and Oura, E. (1960). *Biochim. biophys. Acta*, **37**, 482.
31. McLellan, W. L., Jr. and Lampen, J. O. (1963). *Biochim. biophys. Acta*, **67**, 324.
32. Tonino, G. J. M. and Steyn-Parvé, E. P. (1963). *Biochim. biophys. Acta*, **67**, 453.
33. Rothstein, A. and Meier, R. (1948). *J. Cell. Comp. Physiol.*, **32**, 77.
34. Schwenke, J., Farias, G. and Rojas, M. (1971). *Eur. J. Biochem.*, **21**, 137.

35. Friis, J. and Ottolenghi, P. (1959). *C. R. Trav. Lab. Carlsberg,* **31**, 272.
36. Phaff, H. J. (1966). *Int. Congr. Microbiol.,* 9th, Moscow, Abstr., p. 93.
37. Kaplan, J. G. (1965). *J. Gen. Physiol.,* **48**, 873.
38. Hopkins, R. H. (1958). *Wallerstein Lab. Commun.,* **21**, 309.
39. Matile, P. (1969). *In:* Proc. 2nd. Symp. Yeasts., p. 503. Publishing House of the Slovak Academy of Sciences, Bratislava.
40. Maddox, I. S. and Hough, J. S. (1971). *J. Inst. Brew.,* **77**, 44.
41. Nickerson, W. J. and Falcone, G. (1956). *Science, N.Y.,* **124**, 318.
42. Kaplan, J. G. (1965). *Nature, Lond.,* **205**, 76.
43. Kokke, R., Hooghwinkel, G. J. M., Booij, H. L., Van den Bosch, H., Zelles, L., Mulder, E. and Van Deenen, L. L. M. (1963). *Biochim. biophys. Acta,* **70**, 651.
44. Van den Bosch, H., Bonte, H. A. and Van Deenen, L. L. M. (1965). *Biochim. biophys. Acta,* **98**, 648.
45. Kokke, R. (1966). Doctoral Thesis, University of Leiden.
46. Wheeler, G. E., Michell, R. H. and Rose, A. H. (1972). *Biochem. J.,* **127**, 64P.
47. Burger, M., Hejmová, L. and Kleinzeller, A. (1959). *Biochem. J.,* **71**, 233.
48. de la Fuente, G. and Sols, A. (1962). *Biochim. biophys. Acta,* **56**, 49.
49. Harris, G. and Thompson, C. C. (1961). *Biochim. biophys. Acta,* **52**, 176.
50. Grenson, M., Mousset, M., Wiame, J. M. and Béchet, J. (1966). *Biochim. biophys. Acta,* **127**, 325.
51. Abd-El Al, A. and Phaff, H. J. (1967). *Bact. Proc.,* p. 32.
52. Harrison, J. S. and Trevelyan, W. E. (1963). *Nature, Lond.,* **200**, 1189.
53. Wright, J. D. and Green, C. (1971). *Biochem J.,* **123**, 837.

The Biosynthetic Pathway of Yeast Mannan Glycoproteins

R. SENTANDREU and M. V. ELORZA

Departamento de Microbiología, Instituto de Biología Celular CSIC, Facultad de Ciencias, Universidad de Salamanca, Spain

I. Introduction

The cell wall of *Saccharomyces cerevisiae,* like the wall of many organisms, consists of a two-phase system: One of them is formed by an insoluble β-linked glucan which constitutes a framework of crossed macromolecules of considerable strength. This glucan is the structural polymer responsible for the cell shape and its protective properties are obvious. The structural polymer is embedded by the second phase which is built up by a gelatinous matrix of polysaccharides and/or glycoproteins, lipids and probably other substances that interpenetrate each other.

The major component of the yeast matrix is a glycoprotein(s) whose carbohydrate portion is known as "yeast mannan". The chemical structure of this glycoprotein is shown in Fig. 1. The peptide moiety is rich in hydroxyl amino acids, serine and threonine. One of the linkages between carbohydrate and amino acids in this macromolecule is an *O*-mannosyl bond from mannose and mannose oligosaccharides to serine and threonine. The other linkage connects the mannan, highly branched and of high molecular weight to the peptide, and is probably a nitrogen glycosyl bond between *N*-acetyl glucosamine and aspartamide [1,2].

Several exoenzymes are associated with the wall structure in some unknown manner; they are also glycoproteins with "mannan" as the carbohydrate moiety [3,4]. Though the relationship between the catalytic and non-catalytic glycoproteins is not clear, it seems that in general, the matrix materials play an essential role in maintaining the cell in its particular environment. The presence of melibiase or invertase, for instance, represents the needs of the cell in a particular set of growth conditions.

The present paper records progress on biosynthesis of the glycoprotein "yeast mannan". For other aspects of the yeast wall,

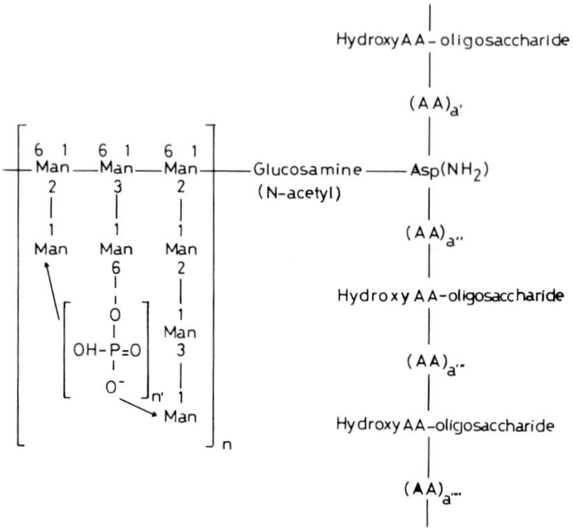

Fig. 1. Chemical structure of "yeast mannan" glycoprotein.

such as cytology, chemical structure, morphogenesis, etc. the reader is referred to recent reviews [5-8].

II. Initial Studies

Initial studies on the metabolism of yeast mannan were carried out with the assumption that carbohydrates were the only constituents. Bretthauer, Wilken and Hansen [9] demonstrated in *Hansenula holstii* that the formation of an extracellular phosphorylated mannan proceeded from glucose as well as mannose with all the carbons remaining intact. They also showed that the isomerization of glucose to mannose was not as nucleoside diphospho-dexose but as hexose-phosphate.

The demonstration by Leloir and collaborators that a large number of sugar interconversions and polysaccharide syntheses involve esters of monosaccharides and nucleoside diphosphates, and the belief that the synthesis of wall polymers might be associated with the cell plasmalemma moved Algranati, Carminatti and Cabib [10] to study a particulate enzymic preparation obtained from lysed protoplasts. They showed that the mannosyl moiety of guanosine diphosphate-mannose (GDP-mannose) was incorporated into a product with characteristics of yeast mannan. Behrens and Cabib [11] found that the enzymic system obtained from yeast protoplasts was highly specific for GDP-mannose, with a pH range of activity from 5.5 to 7.2. Mn^{2+} was shown to be a powerful activator of the transfer reaction. The radioactive mannan formed was degraded by acetolysis [12] and it was found that the particulate preparation had synthetized a branched polymer that was similar to that formed by growing cells [11]. This result is remarkable, because in other systems, like that obtained from *Micrococcus lysodeikticus* [13] the transfer of mannose from GDP-mannose to the membrane mannan takes place preferentially at a single, non-reducing terminal position.

III. The Effect of Cycloheximide on the Synthesis of "mannan" Glycoproteins

The demonstration that yeast mannan exists as a mannoprotein rich in hydroxylaminoacids [1, 14, 15] permits the study of its synthesis by following the incorporation of radioactivity from glucose-^{14}C or threonine-^{14}C. Glucose-^{14}C is incorporated into the cell wall in the structural glucan whilst isomerized mannose-^{14}C is incorporated into the mannoproteins. Threonine-^{14}C is also incorporated into mannoproteins [16].

Further progress in the knowledge of the biosynthetic pathway of the mannoprotein has been brought about by the combined studies with both cells and protoplasts. These studies have been carried out using different antimetabolites, mainly 2-deoxyglucose and cycloheximide. These compounds permit the dissection of the overall sequence of a pathway into a number of component steps because they interfere with the process at specific positions.

Nečas [17-19] studied the formation of the wall by yeast protoplasts and found that in media without an adequate physical support, given by an adequate concentration of agar or gelatin, protoplasts synthetized the "fibrillar" component of the wall whereas the formation of the matrix was reduced or absent. Later, it was confirmed that protoplasts are able to synthetize all or almost all wall constituents in a solid or liquid media, but in the second condition the major part of the polymers of the matrix diffuse away, at least initially, because they do not find the adequate chemical structures to anchor them to the fibrillar network.

In a systematic study on inhibitors of wall formation in protoplasts, Nečas, Svoboda and Kopecká [20] found that cycloheximide prevents the synthesis of the wall matrix, but formation of the "fibrillar" network is unaltered.

Cycloheximide is a glutarimide antibiotic that stops protein synthesis in some eukaryotic cells by inhibiting the movement of the aminoacyl-tRNA or its entry into the peptidyl site of the ribosomes [21].

Studies on the incorporation of threonine and glucose into the wall by growing cells in the presence and absence of cycloheximide showed that the drug halted the incorporation of the peptide material into the cell wall and this inhibition was parallel to that of the mannan synthesis. Glucan formation was little prevented in these conditions [16]. This result was interpreted as due either to a primary inhibition at peptide level of the glycoprotein formation or to the failure of resynthesis of the enzymes which participate in the mannan synthesis. It has been shown that the inhibition of RNA by cycloheximide [22], is a secondary effect resulting from the halt of protein synthesis [23].

Inhibition of the peptide may be the result of, at least, a blockage of the uptake of amino acids or to their polymerization at the ribosomes. Amino acid uptake was not inhibited but, on the contrary, an increase in the labelled amino acids into the pool was detected after treatment of the cells with cycloheximide [61].

The effect of cycloheximide on the amino acid polymerization at the ribosomes (protein synthesis) and its relationship with mannan formation was determined using several concentrations of antibiotic. The incorporation of threonine into the total cell protein and the formation of radioactive mannan was parallel; different levels of inhibition of protein synthesis resulted in a proportional inhibition on mannan formation. Glucan was insensitive to protein synthesis inhibition. The formation of mannan was also followed in *Saccharomyces fragilis,* a species whose growth is insensitive to cycloheximide and no inhibition on either protein or mannan synthesis was found [24].

These results clearly indicate that mannan formation is dependent on the continuous synthesis of protein. The need for protein synthesis could be related either to the necessity of the formation of acceptor molecule(s) or to the decay of enzymes involved in the transfer of mannosyl residues to the growing polymer. If the first hypothesis is right, a sharp decrease in mannan formation must take place as soon as the acceptor molecules present in the moment of protein synthesis inhibition are glycosilated. If the second hypothesis turns out to be true, a slow decrease in mannan formation should be detected. This decrease in the rate of mannan synthesis must be proportional to time of antibiotic treatment.

The small inhibition on glucan formation in the presence of the drug suggests that the enzymes involved in the glycosylation of polysaccharides are quite stable and show a low rate of decay. To get additional information about this point, we have studied the turnover of the mannosyl transferase activities and the nature of the mannose acceptors present in the cells after inhibition of protein synthesis. This was done using a particulate membrane preparation obtained from cells preincubated with the drug. The decrease on mannosyltransferase activity detected as the mannose incorporated from GDP-mannose followed a triphasic behaviour. During a few minutes after the addition of the drug, the particulate preparation showed the same activity as the preparation obtained from untreated cells. This was followed by a sharp decrease in activity and finally a further, slower decrease which was proportional to time of preincubation [25].

These results, together with the known cycloheximide effect on protein synthesis, were interpreted as an indication that the addition of the drug to the cells produces an inhibition of the synthesis of mannosyl acceptors at the ribosomes, but glycosylation of those already formed goes on till depletion. The moment in which the cells are depleted of mannosyl acceptors corresponds to the sharp decrease in mannosyl-transferase activity. This interpretation suggests that the inhibition on "mannan" formation by cycloheximide is not related to the turnover of the enzymes involved, and is in agreement with the fact that the decrease on mannose incorporated *in vitro* is not proportional to time of treatment. The decay of the enzymes is probably represented by the last part of the graph; that is to say, extrapolation of the last part of the graph till the ordinate would give the real rate of enzyme decay.

The nature of the mannosyl acceptors present in the particulate preparations obtained from cells pretreated with cycloheximide was found to be almost all mannolipids and no glycoproteins [25] indicating that the previous interpretation was correct. Treatment of the yeast with cycloheximide results in the inhibition of wall mannoproteins because the drug inhibits the synthesis of mannose acceptors at the ribosomes.

We have also considered the possibility that the inhibition of "mannan" formation was the result of an interference of cycloheximide with the pathway initiated with the uptake of carbohydrates and passing through the interconversion of hexoses, leading to the formation of GDP-mannose, the sugar nucleotide donor for mannan synthesis. This possibility was ruled out because the addition of the drug to a cell culture results in the accumulation of GDP-mannose in large amounts [24].

We have also considered the chance that cycloheximide might inhibit the mannosyl transferase activities in some way. This last possibility was discarded because the transfer of mannose carried out by the particulate membrane preparation is insensitive to additions of cycloheximide in concentrations up to 24 fold of that needed to halt protein synthesis in intact cells [24, 26].

The described results show that protein and mannan synthesis take place in a parallel form, that both processes are sensitive to protein inhibitors, such as cycloheximide, in antimetabolite sensitive yeast, and insensitive in resistant species. In the presence of the drug the cells accumulate GDP-mannose, the sugar nucleotide donor for mannan synthesis, and in the particulate membrane system obtained from them, glycosylation of lipids is not inhibited. In a particulate preparation of untreated cells the presence of cycloheximide produces no decrease on the glycosilation of endogenous acceptors [24, 26]. All these results provided us with information for the proposal of the following pathway of the mannoproteins of the wall (Fig. 2).

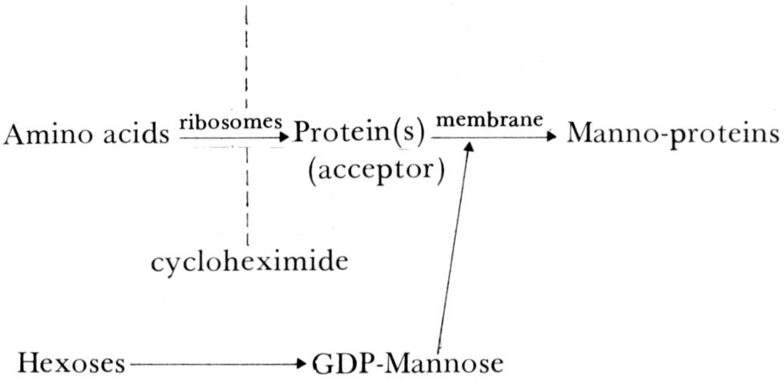

Fig. 2. Yeast mannan glycoproteins: Biosynthetic pathway.

IV. Participation of a Lipid Intermediate on the synthesis of "mannan" Glycoproteins

Investigations on polysaccharide biosynthesis have suggested the presence of two mechanisms which take part to form the total molecule. First, the glycosydic moiety of a nucleoside diphosphate-sugar is transferred to an acceptor molecule. The transfer of several glycosydic residues (it may be always the same, or different ones) may produce an oligosaccharide or polysaccharide. The second step is the transglycosylation of the sugar or an oligosaccharide formed in the acceptor to another part of the same molecule, forming a branched structure (glycogen), or to a polysaccharide or oligosaccharide previously formed, with the result of an increase in the chain length of the receptor (*M. lysodeikticus* mannan, capsular polysaccharides, etc., Fig. 3).

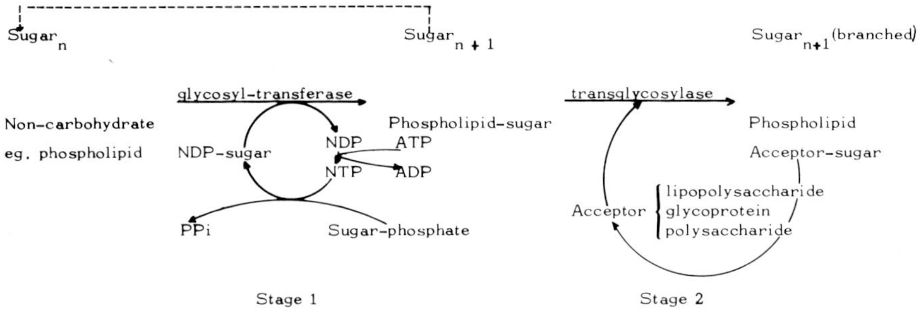

Fig. 3. Steps in the biosynthesis of polysaccharides. The product of Stage 2 may act as further acceptor material in this glycosylation step (broken lines). (Modified from Northcote, [62]).

The biosynthesis of some capsular polysaccharides or lipopolysaccharides takes place by the sequential transfer of several glycosyl residues of a repeating unit to a phospholipid identified as C_{50} to C_{60} prenylpyrophosphate. The resulting carrier serves as a donor of the repeating unit in the formation of the polysaccharide [27]. A similar procedure and lipid have also been shown to participate in the synthesis of the peptidoglycan of bacterial cell walls [28], mannan of *M. lysodeikticus* [29].

Eucaryotic cells contain dolichols which are C_{70} to C_{100} cis-trans polyprenols with a saturated unit at the hydroxyl end of the molecule. Behrens, Parodi and Leloir [30] have shown that in a liver microsomal preparation, dolichols monophosphate may be glycosylated and then serve as sugar donor to an oligosaccharide bound to a lipid through a phosphate or pyrophosphate bridge.

The presence of a lipophilic intermediate which might be the immediate precursor of yeast mannan has also been suggested and the linkage between the mannosyl residue and the lipid might be a phosphodiester [31]. Phosphorylated dolichols from yeast or liver can serve as mannosyl acceptors in a particulate system from yeast [32].

We have investigated the biosynthetic pathway of the mannoproteins of the yeast wall matrix with a special emphasis on the characterization of a possible lipid intermediate.

We have employed a particulate preparation obtained by fractionation of broken cells that has made the mannosyltransferase complex conveniently available in large amounts [33]. The particulate preparation (P_{40}) in the presence of GDP-mannose-^{14}C transfers mannose to several endogenous acceptors; one of them shows the characteristics of mannan glycoproteins, the others of mannolipids.

Chromatographic analyses combined with "chase" experiments revealed the presence of at least three mannolipids and the radioactivity of one of those lipids could be "chased" away by "cold" GDP-mannose [33, 34].

We have studied the stability of these lipids to alkaline ethanolysis [35] and found that two of them are deacylated and the third one whose radioactivity could be "chased" is stable in conditions in which a phosphodiester linkage is affected little, if at all. In *M. lysodeikticus* the lipid intermediate which participates in mannan synthesis is also stable to the alkaline treatment [29].

The chemical nature of the lipids deacylated during alkaline treatment can be inferred by their elution from silicic acid columns by chloroform-methanol (1 : 1) but not by acetone [33]. This behaviour is typical of phosphatides and rules out the presence of a lipid, similar to galactosyl diacyl glycerol, detected by Baraud and colleagues [36]. Preliminary experiments indicate that inositol and/or perhaps galactose may form part of their molecules.

When the kinetics of the formation of the alkali-stable lipid were followed independently of the other lipids, we found that its synthesis took place before the incorporation of radioactive mannose into the mannoproteins. The mannose-^{14}C associated with this lipid was constant when the substrate was present in excess; in other words, there was a small pool that was saturated almost instantaneously and did not change in amount during the experiment. Addition of an excess of non-radioactive GDP-mannose resulted in the disappearance of the label from this fraction and a concomitant increase of the radioactivity associated with the mannoproteins [33]. These are the expected characteristics of a lipid participating as intermediate in the synthesis of wall mannoproteins.

Analyses of the lipid intermediate were carried out with the material purified by combination of silicic acid, DEAE-cellulose and Sephadex LH-20 columns with alkaline ethanolysis [34]. The purified lipid intermediate gives a single radioactive spot on thin layer chromatography using several solvents. This spot could be detected chemically either by *p*-anisaldehyde [37] or ammonium sulphatesulphuric acid reagent [38]. The first one is semi-specific for isoprenoid alcohols and gives with them an olive-green colour on gentle heating.

Total hydrolysis (2N HCl, 100° C 2 h) released a single radioactive compound that chromatographed on paper as authentic mannose. Chemical analysis of the purified lipid intermediate gave a ratio of mannose to phosphorus of 1.27 : 1.0. The lipid intermediate, as mentioned above, is resistant to mild alkaline treatment, but is unstable to mild acid hydrolysis (O, 1N HCl, 100° C 20 min). This treatment converts all the radioactive mannose to a water soluble form. This acid lability together with the increased rate of mannose incorporation carried out by the membrane preparation in the presence of bacitracin [61] suggests that the hydrophilic part of the lipid intermediate is a mannosyl-l-phosphate. This and the relative stability of the radioactive lipid to alkaline treatment rules out a pyrophosphoryl linkage.

Chromatography of the lipidic material after mild acid hydrolysis showed a non-radioactive spot which gave an olive-green colour when reacted with *p*-anisaldehyde-sulphuric acid reagent.

The chromatographic analysis of both the lipid intermediate and the lipophilic part of it after acid hydrolysis, together with the ratio of mannose to phosphorous found and the effect of bacitracin, suggest that the lipid intermediate that participates in the synthesis of mannoproteins of the yeast cell wall is a phosphodiester of mannose and a polyprenoid alcohol. Some of the characteristics of this lipid are summarized in Table I.

The nature of this polyprenoid alcohol is unknown at the present time. Tanner, Jung and Behrens [32] have used an indirect experimental approach to ascertain the chemistry of this lipid. This method, employed previously by Behrens and Leloir, [39] is used to

TABLE I
Properties of the lipid intermediate

(a)	Kinetics of formation 1. Instantaneous labelling with GDP-mannose-^{14}C 2. Chase of radioactivity with cold GDP-mannose
(b)	Single olive-green spot when TLC is developed with *p*-anisaldehyde-sulphuric acid reagent
(c)	Acid-unstable
(d)	Alkali-stable
(e)	Ratio Phosphorus: mannose = 1.0 : 1.27

avoid the need of isolation of sufficient amounts of mannolipid in a pure state to enable it to be characterized unequivocally. Tanner and his collaborators showed that a particulate membrane preparation obtained from yeast was able to use exogenous dolichol phosphates (Fig. 4) as acceptors of the mannosyl moiety of GDP-mannose. Nevertheless, Jankowski and Chojnacki, [40] have found that dolichol- and ficaprenol-phosphates (a mixture of C_{50}, C_{55} and C_{60} polyprenol produced by *Ficus elasticus,* Fig. 4) can serve as acceptors of the glycosylic part of UDP-glucose by rat liver and brain transferases. These results indicate that the shortening of the chain length of the polyisoprenol and the presence of a terminal unsaturated unit are not critical for the enzymic system present in preparations which transfer glycosyl residues to lipid carriers.

Another interesting point is that dolichol phosphates or pyrophosphates may also serve as the final acceptors of glycosyl residues as has been suggested by Behrens, Parodi and Leloir [30].

$$CH_3-\overset{\overset{CH_3}{|}}{C}=CH-CH_2-\left[CH_2-\overset{\overset{CH_3}{|}}{C}=CH-CH_2\right]_n-CH_2-\overset{\overset{CH_3}{|}}{C}=CH-CH_2-OH$$

Ficaprenols $_n$ = 8 to 10

$$CH_3\overset{\overset{CH_3}{|}}{C}=CH-CH_2-\left[CH_2-\overset{\overset{CH_3}{|}}{C}=CH-CH_2\right]_n-CH_2-\overset{\overset{CH_3}{|}}{CH}-CH-CH_2-OH$$

Dolichols yeast $_n$ = 12 to 16 liver $_n$ = 15 to 20

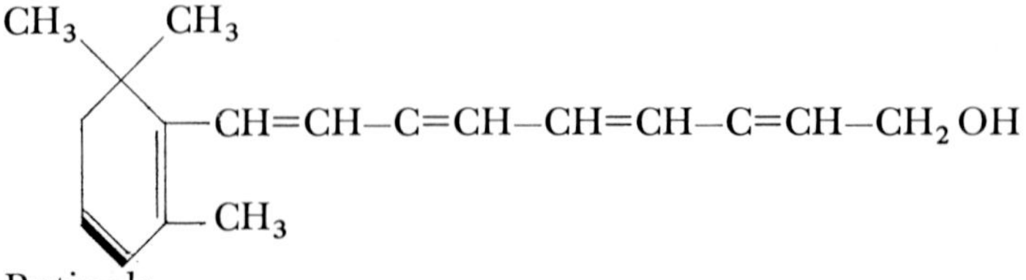

Retinols

Fig. 4. Glycosyl phospho- or pyrophosphorylated lipid carriers. Lipophilic moieties.

Ficaprenol-pyrophosphate and ficaprenol phosphate-type prenoids have been shown to be the lipophilic moiety of the lipids intermediate that participate in the synthesis of bacterial peptidoglycan [28] and salmonella O-amtigen [27] and the membrane mannan found in *M. lysodeikticus* [29]. Small amounts of ficaprenol-type compounds associated with yeast dolichols may have gone unnoticed due to the limitation of the analytical methods used.

Retinal metabolites have also been reported to act as mannosyl carriers. De Luca, Rosso and Wolf [41, 42] have shown that in an intestinal membrane fraction, a mannolipid derived from retinal can be used as the carbohydrate donor to a glycoprotein fraction which seems to be affected by vitamin A.

It cannot be decided at the moment which lipid among the three groups of isoprenoids (dolichol-, ficaprenol- and retinol-type) indicated above is the participating one in the yeast "mannan" synthesis. The method of purification described by Sentandreu and Lampen [34] may afford enough material to characterize it by mass spectrometry and other methods as has been done in some cases [27-29, 43].

Figure 5 shows a modification of the metabolic pathway depicted in Fig. 2 so as to include the participating lipid intermediate.

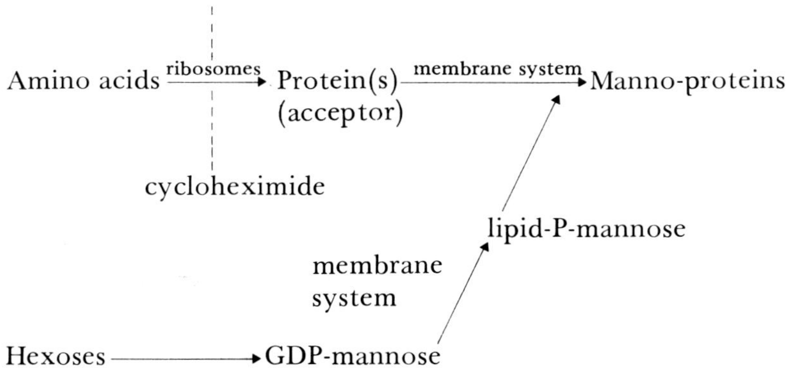

Fig. 5. Yeast mannan glycoproteins: biosynthetic pathway.

Cellular Sites of Synthesis of "mannan" Glycoproteins

The "mannan" synthetase described in this paper is isolated as a particulate bound enzyme(s) that is not rendered soluble by either washing with low ionic solutions or by extracting the lipids with acetone. Treatment with Triton X-100 inactivates the transferase activity of the preparation (results to be published elsewhere). This suggests that the enzyme(s) is firmly attached to the membrane.

Investigations on dividing cells of *Sacch. cerevisiae* have revealed that bud formation is initiated by the accumulation of vesicles. These vesicles, derived probably by proliferation of the endoplasmic reticulum [44], travel toward the periphery of the cell and fuse with the plasmalemma. The extension of the cell wall at the point of bud formation may occur using the materials that are carried by these vesicles and passed across the plasmalemma by reverse pinocytosis [45]. It has also been postulated that these are rich in enzymes that weaken and dissolve the wall during bud formation [44], and Matile and colleagues [46], in an attempt to solve the dilemma, isolated by isopycnic centrifugation a particulate preparation, called "glucanase particles", which contained exo- and endoglucanases, mannan and protein. These investigators claimed that the particles isolated were vesicles derived from the endoplasmic reticulum.

The identity between "mannan synthetases", "glucanase particles" and the vesicles observed by electron microscopy has not been confirmed, but it has been shown by autoradiography that in higher plants some of the matrix material of the cell wall is isomerized in the cisternae of the Golgi bodies and progressively transferred to the wall [47].

VI. Regulation of the Biosynthetic Pathway of "mannan" Glycoproteins

Understanding regulatory phenomena in biological systems is one of the fundamental objectives of molecular biology.

Although we have some information about the chemical structure and enzymatic mechanisms involved in the assembly of mannoproteins, we know little about the control of the biosynthetic pathways.

In yeast, as in many organisms, there is a correlation between cell cycle and the concentration of several enzymes. This phenomenon occurs in the absence of changes imposed by the external environment. Halvorson and his colleagues [48] have interpreted this finding as being the result of the "linear reading" of the structural genes involved. Each gene is transcribed at a limited period of time during the cell cycle. This periodical synthesis of enzymes affords a method of controlling the concentration of some enzyme products. Yeast invertase, one of the catalytic mannoproteins found in the yeast wall, is synthetized in a periodic manner [49].

We have investigated the possible fluctuation during the cell cycle of the two main components of the yeast wall, namely glucan and mannoproteins, and found that both are incorporated into the wall in a progressive form [50].

This finding indicates that the genes involved are always available for transcription or that the mRNAs or the glucan and mannan synthetases have a low rate of decay. As Lampen has suggested [51], it is possible that the mRNAs of the acid phosphatase and invertase, two of the catalytic mannoproteins of the wall, have a relatively long life, and as we have demonstrated [25] the glucan- and mannan-synthetases present a low decay. The continuous synthesis of glucan and mannan during the whole cell cycle may well be the result of the combination of long life in mRNAs and glucan- and mannan synthetases.

One possible site for regulation of the synthesis of these glycoproteins is at the point of carbohydrate attachment, which is thought not to be under direct genetic control. The concentration and specificity of the enzymes and factors related to substrate, donor and the environment may well be of some importance.

The presence of two types of interconnecting linkages between the protein and carbohydrate, in which they carry mannose units at different degrees of polymerization, might represent a gradual assembly of the high molecular mannan molecule. The first steps would be the transfer of mannose from the donor lipid intermediate to serine or threonine in the peptide to form low molecular chains of mannosyl units linked by $1 \rightarrow 2$ and $1 \rightarrow 3$. In a second step the transfer of these subunits by a transglycosylation process occurs to give the mannan with $1 \rightarrow 6$ bounds [1].

In a very elegant work, Spencer, Gorin and Rank [52] have demonstrated that the presence or absence of α-($1 \rightarrow 3$)-linked mannopyranosyl end units of the side chains of yeast mannan is under the control of a single mendelian gene. These results suggest the existence of specific enzymes for each type of linkage present in these mannoproteins.

Another point of regulation is suggested by the fact that certain immunological properties, like blood-group specificity are shown both by glycoproteins as well as by glycolipids. This is attributable to similar chemical structures which are the result of a series of transferases common to both groups [53]. Inhibition of any of these enzymes may lead either to failure of glycoprotein biosynthesis or to products with deranged or incomplete structure. In this context a similar situation might be possible during the synthesis of mannoproteins. Inositol deficiency during the growth of several yeasts leads to altered cell wall composition and morphology. The particulate preparation described in this paper transfers mannose for GDP-mannose to the lipid intermediate and also to some other lipids. Preliminary studies have shown that they may be inositol containing mannolipids, and other mannolipids may be related to the sphingolipid II (mykoglycolipid) described by Wagner and Zofcsik [54]. The complex inositol-containing lipids might induce the synthesis of a non-specific long-chain lipid mannosyl transferase that could transfer mannose to both the inositol-containing lipids and the polyprenol phosphate that participate as an intermediate in the synthesis of mannoproteins. An alternative hypothesis suggests a control of the polyprenol phosphate mannosyl transferase by inositol-containing lipids in a way similar to that described for several microsomal enzymes. In these cases, the enzymatic activities are regulated either because the lipids are co-factors or a specific lipid environment is needed to produce an active enzyme.

This could be the reason why in *Saccharomyces carlsbergensis* the bud fails to separate at suboptimal levels of inositol, which it requires as a growth factor and shows abnormal wall composition with 2-3 times more glucan than mannan. Control cells show about equal amounts of the two polymers. In *Schizosaccharomyces pombe*, the hypovitaminosis of inositol also is reflected in the amount of mannan synthetized, which is reduced to 50% of its normal amount, and the formation of transverse walls is prevented [55].

VII. The Biosynthetic Pathway of "yeast mannan" Glycoproteins: A Working Hypothesis

We suggest that the wall mannoproteins are synthetized following a pathway similar to other glycoproteins of mammalian cells. The assembly of the polypeptide backbone on ribosomes is under RNA control according to the established pathway of protein synthesis. Glycosylation of these proteins takes place in a vesicular system during the transport of the wall materials to the deposition site in the wall (Fig. 6). A similar mechanism has been suggested in the transport of glycoproteins in mammalian cells [57-59], but in this case the transport is, as in the polysaccharides of the matrix of higher plants [56], mediated by the Golgi apparatus. The orientated transport of the vesicles (accumulation at the point of bud initiation, etc.) might be controlled by arrangements of microtubules which can act to channel the vesicles to particular sites of the cell surface. The mechanisms governing the reverse pinocytosis are unknown, but it

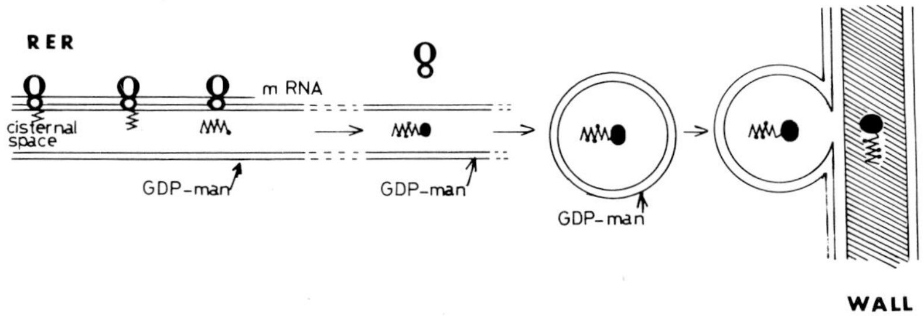

Fig. 6. The biosynthetic pathway of "yeast mannan" glycoproteins. Mannose (●) is added to the polypeptide chain in the rough endoplasmic reticulum (RER) and in a vesicular system during the transport to the wall.

can be suggested that various aspects of membrane specificity are likely to be involved.

The secreted material now has to find the adequate structure to be retained within the wall. The final morphogenetic aspects of wall assembly are primarily dependent on a process of assembly of the relatively simple glucan subunits to form the integrated lattice structure. Once this has been assembled, other wall components, primarily mannan glycoproteins, can then be accreted on to the inner and outer surfaces of the lattice to give added strength, rigidity and the specific enzymatic activities. The relationship between the structural, highly insoluble glucan component and the mannoproteins of the matrix is unknown at the present time. It may be that other matrix components, although present in relatively small amounts, form bridges between the glucan and the mannoproteins of the matrix. The absence of adequate amounts of glucan and/or the bridging material may be the reason for the liberation of mannoproteins into the medium in which protoplasts are growing [3, 60].

Finally, before we can understand the mechanism of biosynthesis and the vectorial assembly of mannoproteins, glucan and the other wall components, we will probably require the combined efforts of genetics, biochemists, microbiologists and other specialists.

Acknowledgements

The major part of the work discussed in this paper was carried out at Dr. J. O. Lampen's laboratory at the Institute of Microbiology, Rutgers University. Dr. Lampen made it possible with his interest and support. We are grateful to Dr. D. H. Northcote F. R. S. for his stimulating discussions and to E. Tacoronte and J. M. Sierra who worked with us in some experiments. Finally, we are greatly indebted to Prof. J. R. Villanueva, with whom we have collaborated for many years.

References

1. Sentandreu, R. and Northcote, D. H. (1968). *Biochem. J.*, **109**, 419.
2. Sentandreu, R. and Northcote, D. H. (1969). *Carbohyd. Res.*, **10**, 584.
3. Lampen, J. O. (1968). *Antonie van Leeuvenhoek*, **34**, 1.
4. Gascón, S. Lazo, P. S., Moreno, F. and Ochoa, A. G. (1972). *In:* Yeast, Mould and Plant Protoplasts p. 157 (Villanueva, J. R., Acha, I. G., Gascón, S. and Uruburu, F. eds). Academic Press, London and New York.
5. Matile, Ph., Moor, H. and Robinow, C. F. (1969). *In:* The Yeast, Vol. 1, p. 219. (Rose, A. H. and Harrison, J. S., eds). Academic Press, London and New York.
6. Phaff, H. J. (1971). *In:* The Yeast, Vol. 2, p. 135. (Rose, A. H. and Harrison, J. S., eds). Academic Press, London and New York.
7. Bartnicki-Garcia, S. and McMurrough, I. (1971). *In:* The Yeast, Vol. 2, p. 441. (Rose, A. H. and Harrison, J. S., eds). Academic Press, London and New York.
8. Hunter, K. and Rose, A. H. (1971). *In:* The Yeast, Vol. 2, p. 211. (Rose, A. H. and Harrison, J. S., eds). Academic Press, London and New York.
9. Bretthauer, R. K., Wilken, D. R. and Hansen, R. G. (1963). *Biochim. biophys. Acta*, **78**, 420.
10. Algranati, I. D., Carminatti, H. and Cabib, E. (1963). *Biochem biophys. Res. Commun.*, **12**, 504.
11. Behrens, N. H. and Cabib, E. (1968). *J. biol. Chem.*, **243**, 502.
12. Lee, Y. C. and Ballou, C. E. (1965). *Biochemistry*, **4**, 257.
13. Scher, M. and Lennarz, W. J. (1969). *J. biol. Chem.*, **244**, 2777.
14. Cawley, T. N. and Letters, R. (1969). *Biochem. J.*, **115**, 9p.
15. Thieme, T. R. and Ballou, C. E. (1972). *Biochemistry*, **11**, 1115.
16. Elorza, M. V. and Sentandreu, R. (1969). *Biochem. biophys. Res. Commun.*, **36**, 741.
17. Nečas, O. (1961). *Nature, Lond.*, **192**, 580.
18. Nečas, O. (1962). *Folia. Biol. (Praha)*, **8**, 256.
19. Nečas, O. (1965). *Folia. Biol. (Praha)*, **11**, 97.
20. Nečas, O., Svoboda, A. and Kopecká, M. (1968). *Exp. Cell. Res.*, **53**, 29.
21. Mckechan, W. and Hardesty, B. (1969). *Biochem. biophys. Res. Commun.*, **36**, 625.
22. Kerridge, D. (1958). *J. gen. Microbiol.*, **19**, 497.
23. Hartwell, L. H. and Hutchison, H. T. (1970). *Molec. Gen. Genet.*, **106**, 347.
24. Sentandreu, R. and Lampen, J. O. (1970). *FEBS Letters*, **11**, 95.
25. Elorza, M. V. and Sentandreu, R. (1972). *In:* Yeast, Mould and Plant Protoplasts p. 205 (Villanueva, J. R., Acha, I. G., Gascón, S. and Uruburu, F. eds). Academic Press, London and New York.
26. Farkas, V., Svoboda, A. and Bauer, S. (1970). *Biochem. J.*, **118**, 1755.

27. Wright, A., Dankert, M., Fennessey, P. and Robbins, P. W. (1967). *Proc. natn. Acad. Sci. U.S.A.*, **57** 1798.
28. Higashi, Y., Strominger, J. L. and Sweeley, C. C. (1967). *Proc. natn. Acad. Sci. U.S.A.*, **57**, 1878.
29. Scher, M., Lennarz, W. J. and Sweeley, C. C. (1968). *Proc. natn. Acad. Sci. U.S.A.*, **59**, 1313.
30. Behrens, N. H., Parodi, A. J. and Leloir, L. F. (1971). *Proc. natn. Acad. Sci. U.S.A.*, **68**, 2857.
31. Tanner, W. (1969). *Biochem. biophys. Res. Commun.*, **35**, 144.
32. Tanner, W., Jung, P. and Behrens, N. H. (1971). *FEBS Letters*, **16**, 245.
33. Sentandreu, R. and Lampen, J. O. (1971). *FEBS Letters*, **14**, 109.
34. Sentandreu, R. and Lampen, J. O. (1972). *FEBS Letters*, **27**, 331.
35. Dawson, R. M. C. (1967). *In:* Lipid Chromatographic Analysis, Vol. 1, p. 272 (Marinetti, G. V., ed.). New York.
36. Baraud, J., Maurice, A. and Napias, C. (1970). *Bull. Soc. Chim. Biol.*, **52**, 421.
37. McSweeney, G. P. (1963). *J. Chromatog.*, **17**, 183.
38. Ziminiski, T. and Borowski, E. (1966). *J. Chromatog.*, **23**, 480.
39. Behrens, N. H. and Leloir, L. F. (1970). *Proc. natn. Acad. Sci. U.S.A.*, **66**, 153.
40. Jankowski, W. and Chojnacki, T. (1972). *Biochim. biophys. Acta.*, **260**, 93.
41. Deluca, L., Rosso, G. and Wolf, G. (1970). *Biochem. biophys. Res. Commun.*, **41**, 615.
42. Deluca, L., Rosso, G. and Wolf, G. (1971). *Fedn. Am. Socs. exp. Biol.*, **30**, 1134 (Abstr.)
43. Anderson, R. G., Hussey, H. and Baddiley, J. (1972). *Biochem. J.*, **127**, 11.
44. Moor, H. (1967). *Archiv für Mikrobiol.*, **57**, 135.
45. Sentandreu, R. and Northcote, D. H. (1969). *J. gen. Microbiol.*, **55**, 393.
46. Matile, Ph., Cortat, M., Wiemken, A. and Frey-Wyssling, A. (1971). *Proc. natn. Acad. Sci. U.S.A.*, **68**, 636.
47. Northcote, D. H. and Pickett-Heaps, J. D. (1966). *Biochem. J.*, **98**, 159.
48. Halvorson, H. O., Carter, B. L. A. and Tauro, P. (1971). *In:* Advances in Microbiology and Physiology. Vol. 6, p. 47. (Rose, A. H. and Wilkinson, J. F., eds). Academic Press, New York and London.
49. Gorman, J., Tauro, P., Laberge, M. and Halvorson, H. O. (1964). *Biochem. biophys. Res. Commun.*, **15**, 43.
50. Sierra, J. M. and Sentandreu, R. (1972). *FEBS Letters*,
51. Lampen, J. O., Kuo, S. C. and Cano, F. (1972). *In:* Yeast, Mould and Plant Protoplasts p. 143 (Villanueva, J. R., Acha, I. G., Gascón, S. and Uruburu, F. eds). Academic Press, London and New York.
52. Spencer, J. F. T., Gorin, P. A. J. and Rank, G. H. (1971). *Can. J. Microbiol.* **17**, 1451.
53. Watkins, W. M. (1970). *In:* Blood and Tissue Antigens, p. 441 (Aminoff, D., ed.). Academic Press, London and New York.
54. Wagner, H. and Zofcsik, W. (1966). *Biochem. Z.*, **344**, 314.
55. Posternak, Th. (1966). *Proceed 2d Meeting FEBS, Vienna*, Vol. 2, p. 35 (Kimdl, H., ed.). Pergamon Press.
56. Northcote, D. H. (1972). *Ann. Rev. Plant. Physiol.*, **23**, 113.
57. Spiro, R. G. (1970). *Ann. Rev. Biochem.*, **39**, 889.
58. Heath, E. C. (1971). *Ann. Rev. Biochem.*, **40**, 29.

59. Marshall, R. D. (1972). *Ann. Rev. Biochem.*, **41**, 673.
60. Sentandreu, R. and Northcote, D. H. (1969). *Biochem. J.*, **115**, 231.
61. Sentandreu, R., Elorza, M. V. and Lampen, J. O. (1972). Molecular Mechanisms of Antibiotic Action on Protein Synthesis and Membranes. Proceedings of a Symposium. p. 432. (Muñoz, E., Garcia-Ferrand, Z. and Vazquez D. eds). Elsevier, Amsterdam.
62. Northcote, D. H. (1969). *In:* Essays in Biochemistry, Vol. 5, p. 89 (Campbell, P. N. and Greville, G. D., eds). Academic Press, London and New York.

The Effect of Cycloheximide on Mannosyl Transferase Activity of a Membrane Preparation from *Saccharomyces cerevisiae*

M. V. ELORZA and R. SENTANDREU

Departamento de Microbiología, Instituto de Biología Celular, CSIC, Facultad de Ciencias, Universidad de Salamanca, Spain

The metabolism of glycoproteins is being increasingly studied at all levels of cell complexity because these macromolecules participate in basic phenomena such as cell agglutination, surface antigenicity [1], transient modification of the antigenic nature of cell membranes [2], cell wall plasticity [3], etc.

We have chosen *Saccharomyces cerevisiae* as a test system to investigate the general process of synthesis and control of glycoproteins by eucaryotic cells.

The major component of the *Sacch. cerevisiae* wall matrix is a series of glycoproteins [4-6] whose carbohydrate moeity is known as "yeast mannan".

We have studied the formation of these mannoproteins in *Sacch. cerevisiae* growing in the presence and absence of cycloheximide, an inhibitor of protein synthesis in this yeast, and shown that their formation is dependent on the continous synthesis of protein [7].

The effect of the drug upon incorporation of glucose-^{14}C into the wall *in vivo* is shown in Fig. 1. The presence of cycloheximide in a cell culture produces no effect during a short period; after about five minutes the radioactivity incorporated decreases, and finally further incorporation of mannose is prevented. Synthesis of glucan, on the other hand, is little affected.

The decrease in mannan synthesis *in vivo* by cycloheximide was interpreted as either the result of the inhibition of polypeptide acceptor formation or the fact that protein synthesis might be required to maintain adequate levels of the enzymes involved in glycosylation of acceptor molecules.

If the first hypothesis is right, a sharp decrease in mannan formation must take place as soon as the acceptor molecules present

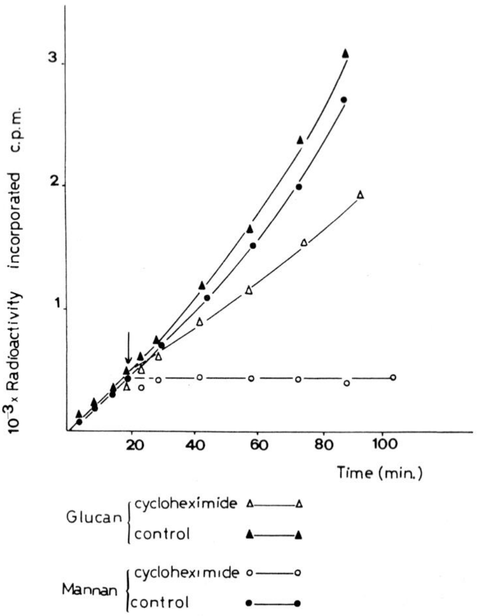

Fig. 1. Incorporation of radioactivity from Glucose-^{14}C into the yeast cell wall. The cycloheximide was added at 20 min from the start of the incubation.

at the moment that protein synthesis is inhibited are glycosylated. If the second hypothesis is correct, a slow decrease in mannan formation should be detected. This decrease in the rate of mannan synthesis must be proportional to the time of antibiotic treatment.

The effect on mannosyl transferase activity was determined in particulate membrane preparations obtained from cells preincubated with the drug by following the amount of mannose-^{14}C transferred into endogenous acceptors. These membrane preparations were obtained as previously described [8].

The amount of radioactive mannose incorporated by the particulate membrane preparation was studied using two levels of GDP-mannose, the sugar donor for the mannan synthesis. We detected no differences in mannose incorporated from GDP-mannose-^{14}C by the preparations obtained from untreated cells or from those preincubated with cycloheximide for two minutes (Table I). After long preincubation (10 minutes and 30 minutes) the transferred mannose decreased by 60 to 70%.

These results are summarized in Fig. 2; this graph shows that the decrease in transferase activity determined by the radioactive mannose incorporated by the preparation follows a triphasic

TABLE I

Incorporation of radioactivity from GDP-mannose-^{14}C into endogenous acceptor(s) by the particulate preparation obtained from *S. cerevisiae* treated with cycloheximide

Treatment (min)	*GDP-mannose* (n moles)	(n Ci)	Mannose incorporated (n moles)	Relative activity
0	20.14	2	0.910	100
2	20.14	2	0.890	98
10	20.14	2	0.380	42
30	20.14	2	0.250	28
0	80.14	2	1.560	100
2	80.14	2	1.440	93
10	80.14	2	0.560	36
30	80.14	2	0.520	33

Sacch. cerevisiae LK2G12 was incubated with cycloheximide (4μg/ml) and at the times indicated, the cells were collected and broken. Particulate preparations (100 μg protein) in 500 μl Tris-maleate buffer pH 6.8-Mn Cl_2-mercaptoethanol [7] were incubated at 30° C for 5 minutes. GDP-mannose-^{14}C (200 μl) was then added and the mixtures incubated at 30° C for 30 min. The reaction was stopped by precipitation with 5% trichloroacetic acid at 0° C for 30 min and the precipitate collected on glass fibre discs and counted in a liquid scintillation spectrometer [7].

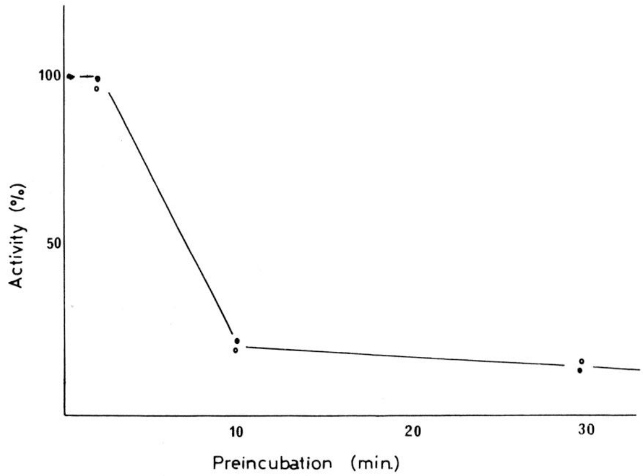

Fig. 2. Cycloheximide: effect of preincubation on mannosyl transferase activities.

behaviour: during the first 2 minutes of treatment we found no decrease in the mannosyl transferase activity; this phase was followed by a sharp drop in mannosyl transferase activity; finally, we detected a decrease in activity proportional to time of incubation.

These results together with the known effect of cycloheximide on protein synthesis, suggest that addition of the drug to the cells produces an inhibition of mannose acceptor synthesis at the ribosomes, but during the first two minutes of treatment, glycosylation of acceptors previously formed is not affected. The sharp drop in the mannose incorporated, detected later on, is probably not related to the turnover of the mannosyl transferases involved because the decrease of mannose incorporated *in vitro* is not proportional to time of treatment. We think that this drop in mannose incorporated corresponds to the moment the cells have been depleted of polypeptide acceptor. The last part of the graph probably corresponds to the decay of enzymes.

To check the above hypothesis, we have determined the nature of the residual mannosyl acceptors present in the cells after treatment with the antibiotic.

The experiment described (Table I) was repeated (only control and cells treated for 30 minutes) and the radioactivity incorporated by the particulate preparation fractionated into mannolipids and mannoproteins.

The mannose incorporated by the preparation obtained from antibiotic-treated cells was only 40% of that incorporated by the control preparation. 75% of this radioactivity was incorporated into mannolipids (Table II).

The radioactivity found in the mannoproteins was less than 19% of that incorporated by the control preparation. This amount is probably a maximum figure, because chloroform-methanol [2 : 1], being one of the best solvents for lipid extraction, fails to dissolve some phosphoinositides and probably other lipids.

In summary, the cells, after treatment with cycloheximide do have mannosyl transferase activity and although we have not checked the specific enzyme involved in mannan glycosylation, the results support the view that the turnover of the mannosyl transfer enzymes is very slow. Treatment of the yeast with cycloheximide results in the inhibition of wall mannoproteins because the drug inhibits the synthesis of mannose acceptors at the ribosomes. Kanfer and Richards [9] have suggested that the impairment of protein synthesis interferes with ganglioside synthesis by inhibiting the synthesis of a protein or peptide which binds ganglioside precursors.

TABLE II

Transfer of radioactivity from GDP-mannose-^{14}C into mannolipids and mannan glycoprotein acceptors by the particulate membrane preparations obtained from *Sacch. cerevisiae* treated with cycloheximide

Treatment (min)	*GDP-mannose* (n moles)	(n Ci)	*Mannose incorporated* total (n moles)	lipids (n moles)	glycoproteins (n moles)
0	80.14	20	0.80	0.37	0.43
30	80.14	20	0.32	0.24	0.08

Sacch. cerevisiae LK2G12 was incubated with cycloheximide (4 μg/ml) and at the times indicated, the cells were collected and broken. Two sets of each particulate preparation (240 μg protein) in 800 μl Tris-maleate buffer pH-6.8-$MnCl_2$-mercaptoethanol (as Fig. 1) were incubated at 30° C for 5 min. GDP-mannose-^{14}C (200 μl) was then added and the mixtures incubated at 30° C for 30 min. In one set of tubes, the reaction was stopped by precipitation with 5% trichloroacetic acid. The precipitate formed after 30 min at 0° C was collected on glass filter disc and counted in a liquid scintillation spectrometer. To the other set of tubes, 20 vol chloroform-methanol [2 : 1] was added and after 3 h at room temperature with occasional shaking the suspension was filtered, washed with water (0.2 vol) taken to dryness and counted.

A secondary effect of cycloheximide on yeast cells corresponds to the decay of enzymes involved in the glycosylation of at least the mannolipids (Table II). The glycosylation is proportional to time and the mannolipids are not resynthetized in the presence of the drug. This effect is, of course, not restricted to mannolipid synthesis and is similar to that found by Morris [10] on the synthesis of the wall polysaccharides in *Chlorella* and by Shah and Peterson [11] on the transfer of galactose to the hydroxyl fatty acid ceramides in the rat brain.

References

1. Winzler, R. T. (1970). *Int. Rev. Cytol.*, 29, 77.
2. Peters, J. H. and Hansen, P. (1971). *Eur. J. Biochem.*, **19**, 502.
3. Lamport, D. T. A. (1969). *Biochemistry*, **8**, 1155.
4. Sentandreu, R. and Northcote, D. H. (1968). *Biochem. J.*, **109**, 419.
5. Lampen, J. O. (1968). *Antonie van Leeuvenhoek*, **34**, 1.
6. Gascon, S., Lazo, P. S., Moreno, F. and Ochoa, A. G. (1972). *In:* Yeast, Mould and Plant Protoplasts p. 157 (Villanueva, J. R., Acha, I. G., Gascón, S. and Uruburu, F. eds). Academic Press, London and New York.
7. Sentandreu, R. and Lampen, J. O. (1970). *FEBS Letters*, **11**, 95.
8. Sentandreu, R. and Lampen, J. O. (1971). *FEBS Letters*, **14**, 109.
9. Kanfer, J. N. and Richards, R. L. (1967). *J. Neurochem.*, **14**, 513.
10. Morris, I. (1967). *J. Expl Bot.*, **18**, 54.
11. Shah, S. N. and Peterson, N. A. (1971). *Biochim. biophys. Acta*, **239**, 126.

Environmentally Induced Changes in the Lipid Composition of Cells and Membranes of *Saccharomyces cerevisiae*

J. A. HOSSACK, G. E. WHEELER AND A. H. ROSE

Microbiology Laboratories, School of Biological Sciences, University of Bath, Bath, England

I. INTRODUCTION

During the first half of this Century, the biochemist was conspicuously successful in charting the major metabolic pathways that occur in living cells. With this knowledge available, it seemed inevitable that cell biologists would attempt to discover the intracellular location of individual metabolic processes. More recently, cell biologists have added a vectorial component to the overall picture provided by metabolic maps, and this has led to a renaissance in cell physiology. This development was accompanied by an interest in relationships between composition and function of cell membranes which form a permeability barrier between a cell and its environment and, by delimiting subcellular organelles, give rise to several discrete metabolic compartments within the cell.

To date, research has inevitably concentrated on biological membranes prepared from those tissues and organisms that are amenable to laboratory manipulation. The erythrocyte membrane and the Schwan cell membrane, which forms the myelin sheath, have been favoured by many workers. But, although these membranes can readily be obtained in bulk quantity, they are specialized membranes and so may not be ideal for studying basic membrane phenomena [1]. Other workers have turned to micro-organisms as convenient sources of biological membranes, largely because microbial membrane composition can be manipulated by altering the composition of the growth medium or other environmental conditions used for growing the organism. Attempts have been made to study the composition and function of several bacterial membranes, including (a) *Escherichia coli* (although because of the chemically complex cell wall in this bacterium it is not an ideal organism [2]); (b) strains of

Bacillus which can easily be converted into protoplasts; and (c) the mycoplasmas which do not form walls and so are attractive model organisms [3]. However, a major drawback to such studies on prokaryotic organisms is that the membrane composition is radically different from that of membranes derived from eukaryotic micro-organisms, and animal and plant cells.

An obvious solution to the problem of selecting an ideal model organism for composition-function studies on biological membranes is to choose a eukaryotic micro-organism, the membrane composition of which can be changed in a specific and predictable fashion. If the organisms selected can be studied and manipulated genetically so much the better. Strains of *Saccharomyces cerevisiae*, particularly those which are susceptible to conversion into sphaeroplasts, go a long way towards being ideal model organisms. This explains the considerable increase in research on yeast membranes in recent years.

The present paper shows how the lipid and membrane composition of *Sacch. cerevisiae* can be altered in a specific and predictable manner, providing a very useful method for studying the role of individual membrane components (e.g. phospholipids, sterols, fatty-acid residues) in membrane function.

II. Lipid Composition of *Saccharomyces cerevisiae*

For a long time, knowledge of the lipid composition of *Sacch. cerevisiae* was fragmentary compared with that available for other yeast-cell components. This situation simply reflected the fact that, for many years, studies on lipids lagged behind those on other classes of biological molecule, mainly because of the unattractive physical properties of lipids. Indeed, it is only since accurate and sensitive methods for analysis of lipids have become available that reliable data on the lipid composition of cells have been reported.

Several analyses of the lipid composition of *Sacch. cerevisiae* have been published [4]. From these analyses, it appears that the lipid composition of *Sacch. cerevisiae* resembles that of other eukaryotic organisms (Table I). There are, however, important differences between the lipid composition of *Sacch. cerevisiae* and that of prokaryotic micro-organisms. Firstly, lipid extracts of *Sacch. cerevisiae* contain appreciable amounts of sterols (free and esterified) and of triacylglycerols, two classes of lipid that are virtually absent from prokaryotic micro-organisms [5]. Secondly, glycolipids are major components of Gram-positive bacteria and have also been detected in some Gram-negative bacteria [6, 7]. However, they are

TABLE I
Lipid content and composition of *Saccharomyces cerevisiae* NCYC 366

Lipid content (% dry weight cell)	Lipid composition (mg/100 mg dry weight cells)					
	Phospholipids	Triacylglycerols	Diacylglycerols	Free fatty acids	Free sterols	Sterol esters
12.5	3.76	1.22	0.09	0.38	0.23	1.32

Cells were grown batchwise at 30°C. Data from Hunter and Rose [15].

present in only very small amounts in *Sacch. cerevisiae* [8, 9]. Finally, this basic distinction between the lipid compositions of bacteria and *Sacch. cerevisiae* extends to the fatty-acid composition. Lipids from *Sacch. cerevisiae* contain predominantly residues of straight-chain saturated and unsaturated fatty acids, with chain lengths of 16 or 18 carbon atoms [10]. Bacterial lipids, on the other hand, contain fatty-acid residues with a greater variation in chain length, and also branched chain and cyclopropane fatty-acid residues [5].

Since the review by Hunter and Rose [4] was prepared, the main emphasis of research on yeast lipids has been on the quantitatively minor components. Reference has already been made to the recently published data on yeast glycolipids [8, 9]. In addition, Lester and his colleagues have made an intensive study of the inositol-containing phospholipids in *Sacch. cerevisiae.* At least seven of these have been detected [11]. The predominant representative is phosphatidylinositol, but there are also small amounts of diphosphatidylinositol and triphosphatidylinositol [12]. Four inositol-containing sphingolipids have been detected in *Sacch. cerevisiae* [11], the main component having the composition: ceramide-(P-inositol)$_2$-mannose [13].

III. Methods for Effecting Variations in the Lipid Composition of *Saccharomyces cerevisiae*

The great advantage of using micro-organisms in any physiological study is that their composition and metabolic activity can be manipulated, with some degree of precision, by varying the composition of the growth medium or other environmental conditions (such as growth temperature). There is a wealth of data available on ways in which environmental factors, such as medium composition, pH value, dissolved oxygen content and growth temperature, influence the lipid composition of bacteria [5] and yeasts [4].

When the aim is to use environmentally induced changes in the lipid composition of microbial membranes in studies on inter-relationships between composition and function it is highly desirable, and for definitive studies essential, that any change effected be stoicheiometric, i.e. that one lipid component be replaced by an equivalent amount of a closely related lipid. Although this ideal situation is difficult to achieve, only in this way can the role of individual components in membrane function be demonstrated.

The purpose of this contribution is to show how it is possible to achieve changes in the lipid composition of *Sacch. cerevisiae* approaching this ideal.

A. *Variations in Growth Temperature*

One of the best documented environmentally induced changes in the lipid composition of micro-organisms is the increase in proportion of unsaturated fatty-acid residues when the temperature at which a micro-organism is grown is lowered below the optimum for growth [14]. In this Laboratory we have examined the effects of variations in growth temperature as possible means of enriching the lipids in *Sacch. cerevisiae* NCYC 366 with unsaturated fatty-acid residues, as a prelude to a study of the role of lipid unsaturation in the functioning of the yeast plasma membrane. A detailed examination of the effect of growth temperature on the lipid composition of this yeast by Hunter and Rose [15] revealed that the changes induced by variations in growth temperature over the range 30°C to 15°C, either in batch-grown or continuously grown cells, are confined not only to variations in the degree of unsaturation of the fatty-acid residues in the lipids but also include changes in the amount of phospholipids synthesized. Indeed, with this strain of *Sacch. cerevisiae,* the proportion of unsaturation in cell lipids is much less affected, over this temperature range, than with many other yeasts including *Candida* strains [16]. Sphaeroplasts from cells of *Sacch. cerevisiae* NCYC 366 grown batchwise at 30°C or 15°C differ in their susceptibility to osmotic shock [17]. But the complex nature of the changes in lipid composition induced by lowering the growth temperature makes it impossible to state precisely the molecular basis of this difference in susceptibility.

B. *Provision of Exogenous Choline and Ethanolamine*

Saccharomyces cerevisiae is known to employ the methylation pathway for synthesis of phosphatidylcholine [4, 18]. The initial step on this pathway is a reaction between CDP-diacylglycerol and serine to form phosphatidylserine, which is then decarboxylated to yield phosphatidylethanolamine. Phosphatidylcholine is then formed by N-methylation of phosphatidylethanolamine. A second pathway, the cytidine nucleotide pathway, operates in many although not all mammalian cells; its operation in eukaryotic micro-organisms has been very little studied.

However, there is evidence that the cytidine nucleotide pathway can be used in *Sacch. cerevisiae.* The pathway involves, initially, the formation of CDP-choline or CDP-ethanolamine from the free bases. These cytidine nucleotides then react with 1,2-diacylglycerols to produce, respectively, phosphatidylcholine and phosphatidylethanolamine. The nucleotides CDP-choline and CDP-ethanolamine have been detected in *Sacch. cerevisiae* by Kennedy and Weiss [19]. Subsequently, Lester and his colleagues [20, 21] reported incorporation of choline into phosphatidylcholine by *Sacch. cerevisiae,* and reported that the presence of choline in the growth medium represses synthesis of enzymes that catalyse synthesis of phosphatidylcholine by methylation of phosphatidylethanolamine.

Waechter and Lester [21] found that incorporation of choline in the growth medium leads to an increased synthesis of phosphatidylcholine by the yeast. This report has recently been confirmed in this Laboratory [22], and we have shown, in addition, that exogenous ethanolamine is incorporated into phosphatidylethanolamine by *Sacch. cerevisiae* NCYC 366 under conditions that also cause synthesis of additional amounts of phosphatidylethanolamine (Table II).

Growth of *Sacch. cerevisiae* in medium supplemented with choline or ethanolamine provides a useful means of obtaining cells that are enriched in phosphatidylcholine or phosphatidylethanolamine although, as yet, they have not been used to study the role of phosphatidylcholine and phosphatidylethanolamine in yeast membranes. A necessary prerequisite to these studies is the need to ascertain the distribution of the additional phospholipids in yeast membranes.

C. *Anaerobic Induction of Lipid Requirements*

Current work in our Laboratory employs a method for varying the lipid composition of *Sacch. cerevisiae* that approaches more closely the ideal situation described earlier. When strains of *Sacch. cerevisiae* are grown under strictly anaerobic conditions, they become auxotrophic for a sterol and an unsaturated fatty acid [23]. The

TABLE II

Effect of exogenous choline and ethanolamine on the phospholipid composition of *Saccharomyces cerevisiae* NCYC 366

Phospholipid	Content (μ mol phosphorus/g dry weight cells) in lipids from cells grown in			
	Unsupplemented medium	Medium supplemented with choline	Medium supplemented with ethanolamine	Medium supplemented with choline + ethanolamine
Phosphatidylcholine	10.7	39.2	15.7	24.4
Phosphatidylethanolamine	14.2	18.5	26.3	13.9
Phosphatidylinositol	13.9	11.1	16.2	14.4
Phosphatidylserine	1.0	3.4	1.9	2.9
Cardiolipin	2.9	2.2	1.5	0.8
Phosphatidic acid	2.7	1.8	5.9	1.0

Cells were grown batchwise at 30°C in basal medium [15] supplemented with choline (1 mM) and/or ethanolamine (10 mM). Data from Ratcliffe *et al.* [22].

specificities for these requirements have been examined, and both have been shown to be fairly broad. Thus, providing that the nutrient (sterol or fatty acid) supplied in the medium is incorporated unchanged into the yeast cell lipids, and this incorporation does not affect the regulation of synthesis of other lipids, the technique could prove valuable as a means of effecting specific changes in the composition of membranes in *Sacch. cerevisiae.*

Anaerobically Induced Fatty Acid Requirements. Bloch and his colleagues [24] found that anaerobically cultured *Sacch. cerevisiae* will not grow on stearic acid ($C_{18:0}$) but would when supplied with a C_{18} fatty acid that has a *cis* double bond between C-9 and C-10 or between C-10 and C-11, or a triple bond between C-9 and C-10. The 12-keto and 11, 12 cyclopropane derivatives supported growth less efficiently. Other acids, including the 10-methyl, 9, 10 epoxy, 9, 10 dihydroxy and 9, 10 *trans* olefinic derivatives, were inactive.

Professor Flavio Alterthum in our Laboratory, using these findings, has grown *Sacch. cerevisiae* NCYC 366 anaerobically in medium containing ergosterol, and oleic acid (18 : 1 *cis* 9 :10), linoleic acid (18 : 2 *cis* 9 : 10 12 : 13), or γ-linolenic acid (18 : 3 *cis* 6 : 7 9 : 10 12 : 13) [25]. Incorporation of any one of these acids in the medium does not affect the rate of growth of the yeast, but does change the duration of the lag phase of growth; this is greatest in medium supplemented with linolenic acid, and shortest in oleic acid-supplemented medium. The duration of the lag phase of growth was extended by about four hours as each acid was replaced by one with a greater degree of unsaturation. Cells grown in medium containing any one of these acids do not differ detectably in size or morphology. Moreover, the total lipid and total phospholipid contents of cells remain unaltered when the fatty-acid composition of the medium is varied. Analyses of the lipids from cells grown in media supplemented with oleic, linoleic, or linolenic acid show that fatty-acid supplement accounts for 65, 57 and 54%, respectively, of the total fatty-acid content (Table III).

Saccharomyces cerevisiae NCYC 366 possesses the considerable advantage of being readily convertible into sphaeroplasts by digesting the cell wall with snail gut juice or an exo-β (1→3) glucanase from Basidiomycete QM 806 [26]. Alterthum and Rose [25] went on to prepare sphaeroplasts from cells grown anaerobically in media supplemented with each of the different fatty acids. The susceptibility to sphaeroplast formation differs among cells grown in these three different media. Those grown in oleic acid-containing medium are converted into sphaeroplasts at a rate comparable with

TABLE III
Fatty-acid composition of lipids of *Saccharomyces cerevisiae* NCYC 366 grown anaerobically on different fatty acids

Fatty acid	Cells grown in medium containing Oleic acid	Linoleic acid	Linolenic acid
12 : 0	2.5	1.7	2.3
13 : 0	1.8	1.1	1.2
14 : 0	3.6	4.8	8.5
14 : 1	tr	1.5	—
16 : 0	18.5	24.2	22.8
16 : 1	4.0	2.1	2.4
18 : 0	2.3	2.7	5.7
18 : 1	65.1	5.2	1.3
18 : 2	1.6	56.7	1.9
18 : 3	—	—	53.9

Values indicate the percentage molar composition. Data from Alterthum and Rose [25].

that for cells grown aerobically. Cells grown in linoleic acid-containing medium are however more refractive and, in order to convert a population of these cells completely into sphaeroplasts in 45-60 min, it is necessary to double the concentration of glucanase protein in the digestion mixture. This finding indicates that the wall composition in these cells differs from that in oleic acid-grown cells. However, in spite of making further increases in the concentration of glucanase in the digestion mixture, it was not possible to convert cells grown in linolenic acid-containing medium into sphaeroplasts. Microscopically, sphaeroplasts can be seen to be formed from these cells, but they are extremely fragile and quickly lyse after extrusion. However, when spermine (10 mM) is included in the digestion mixture [27], it is possible to prepare stable sphaeroplasts of linolenic-grown cells. A plausible explanation of this finding is that spermine protects the extremely fluid plasma membrane of sphaeroplasts emerging from these cells.

This technique for obtaining sphaeroplasts with membranes that are enriched with oleic acid, linoleic or linolenic acid residues allowed Alterthum and Rose [25] to examine the effect of fatty-acid unsaturation on the susceptibility of the sphaeroplasts to osmotic lysis. Not surprisingly, perhaps, in view of the effect on the packing characteristics of phospholipids which contain fatty acyl chains with additional unsaturation, sphaeroplasts from cells grown in the presence of linolenic acid are more susceptible to osmotic lysis in hypotonic solutions of sorbitol than those from cells grown in linoleic acid-containing medium, which in turn are more susceptible to lysis than sphaeroplasts from cells grown in oleic acid-supplemented medium.

The data of Alterthum and Rose [25] are the first to show how anaerobically-induced nutritional requirements of *Sacch. cerevisiae* can be exploited in studies on composition-function relationships in cell membranes. The technique offers numerous possibilities for further exploitation. Different series of fatty acids, including cyclopropane acids, could be examined in view of the broad specificity of the fatty-acid requirement. Other plasma-membrane functions, such as solute transport, might be examined. But probably the most attractive feature of the technique is that it permits a simultaneous examination of variations in the fatty-acid and the sterol composition of yeast membranes. This could be a considerable advantage in view of the suggestion [28] that sterols interact with unsaturated fatty acid acyl residues in membrane phospholipids.

Anaerobically Induced Sterol Requirements. The specificity of the sterol requirement of anaerobically grown *Sacch. cerevisiae* is easier to formulate than that for unsaturated fatty acids. It was examined by Linnane and his colleagues [29], who found, briefly, that the sterol must have a β-hydroxyl group at C-3, the molecule must be planar, and it must possess a long alkyl side chain at C-17.

One of us (J. A. Hossack, unpublished observations) exploited this broad specificity, and examined the growth and composition of *Sacch. cerevisiae* NCYC 366 in media supplemented with Tween 80 (polyoxyethylene sorbitan mono-oleate) to supply the unsaturated fatty acid, and ergosterol, cholesterol, or 7-dehydrocholesterol. These sterols were selected for a preliminary study, since ergosterol is a major yeast sterol, cholesterol is a sterol commonly found in mammalian tissues, and 7-dehydrocholesterol has an intermediate structure, with the steroid nucleus of ergosterol and a side chain identical with that of cholesterol.

There was no significant difference in the duration of the lag phase of growth or the rate of exponential growth of the yeast in media supplemented with the different sterols. Moreover, cells grown in each of the media do not differ either morphologically or in their contents of total lipids or total phospholipids. However, preliminary data show a difference in the amounts of free sterol and sterol ester in the lipids extracted from cells grown in each of the sterol-supplemented media. Ergosterol and cholesterol would seem to be incorporated largely unmodified and account for, respectively, 75 and 80% of the total free sterol in the cells. However, cells grown in media containing 7-dehydrocholesterol contain a high proportion of a sterol which has a retention time, on gas-liquid chromatography, similar to that of the tetraethenoid sterol which is a major component of *Sacch. cerevisiae* NCYC 366 [15]. A particularly interesting finding is that, although half of the sterols in cells grown in cholesterol-supplemented medium are esterified, cells grown on ergosterol contain very little sterol ester. It would seem that cells grown anaerobically on ergosterol could be used to study the kinetics of sterol-ester formation when they are incubated under aerobic conditions, in a way similar to that in which anaerobically grown yeast cells are used to investigate mitochondrial biogenesis.

D. *Use of Lipid-requiring Mutants*

An attractive alternative to using anaerobically induced lipid requirements of *Sacch. cerevisiae* is to isolate mutant strains of the yeast that are auxotrophic for a membrane lipid component or which have an altered lipid composition. Suitable medium supplementation will then permit controlled manipulation of the lipid composition of the yeast membranes. A number of mutants of this type have recently been described in the literature, and already some of these have been used to probe basic problems in membrane function. So far, just two classes of these mutants have been examined.

Unsaturated Fatty Acid-requiring Mutants. The first report of mutants of *Sacch. cerevisiae* with mandatory requirement for an unsaturated fatty acid for growth was by Resnick and Mortimer [30], who used ultraviolet irradiation of a wild-type haploid culture. Certain of these mutants have since been subject to extensive biochemical and genetical examination [31, 32]. In particular, a detailed examination has been made of the specificity of the requirement of these mutants for unsaturated fatty acids. All of those examined in detail are desaturase mutants, and must therefore be supplied with an unsaturated acid. The specificity of the requirement depends on the strain, but in general it can be met by any one of a wide range of acids, with chain lengths ranging from 14-20 carbon atoms, and with one, two or three double bonds in a variety of locations along the chain. Moreover, the double bond need not be *cis* [33]. The requirement for an unsaturated fatty acid by these mutants is therefore broader than the requirement by anaerobically grown *Sacch. cerevisiae.* The extent of enrichment of the cell lipids by the fatty acid supplied varies between 50 and 75% of the total fatty acid [33].

A second series of unsaturated fatty acid-requiring mutants of *Sacch. cerevisiae* has been described, but as yet they have received only cursory examination. These are chain-elongation mutants [34, 35, 36] and have a nutritional requirement for a saturated fatty acid. They cannot grow when supplied with a highly purified unsaturated fatty acid.

So far, unsaturated fatty acid-requiring mutants of *Sacch. cerevisiae* have been employed mainly to study the dynamics of lipid alkyl chains in biological membranes using the technique of spin labelling [33, 37]. Reports have not yet appeared on formation of sphaeroplasts from these mutants, or on the effect, if any, which incorporating different unsaturated fatty acids into the cell lipids has on synthesis of the various classes of lipid. The use of these mutants in membrane studies has advantages, as well as disadvantages, compared with the use of anaerobically induced lipid requirements in *Sacch. cerevisiae.* Since they can be grown aerobically, they clearly are easier to use in the laboratory. And they can readily be subjected to genetical analysis. However, unsaturated fatty acid-requiring mutants cannot be used to study interactions in membranes between different sterols and unsaturated fatty acids. Nevertheless, there can be little doubt that the studies being made with these fatty acid-requiring mutants will help to elucidate interactions between individual components in biological membranes.

Sterol Mutants. Several groups of zymologists have attempted to isolate sterol-requiring mutants of *Sacch. cerevisiae,* but without success [38].

Recent research on polyene resistance in yeasts has revealed a way in which the sterol content and composition of *Sacch. cerevisiae* can be varied, and these findings could be useful to workers interested in composition-function relationships in yeast membranes. Polyene antibiotics complex with membrane-bound sterols, an interaction which severely disrupts the permeability of the plasma membrane [39]. Mutants of *Sacch. cerevisiae* that are resistant to one or more polyene antibiotics (etruscomycin, filipin, nystatin, pimaricin, rimocidin) have been isolated and, as expected, have an altered sterol composition compared with the parent strain [40, 41]. The possibility of exploiting these sterol mutants, particularly when coupled with a fatty-acid requirement, is attractive and could prove a valuable probe in membrane studies.

IV. Distribution of Lipids in Subcellular Structures of *Saccharomyces cerevisiae*

Establishing the nature and extent of environmentally induced changes in the lipid composition of *Sacch. cerevisiae* is only the first of a series of investigations, the aim of which is to explain relationships between composition and function in cellular membranes. If *Sacch. cerevisiae* is to be used as a model organism in membrane studies, it is essential first of all to establish the distribution of lipids in the various types of membrane (nuclear, mitochondrial, vacuolar, plasma), since it is a eukaryotic organism. When specific alterations in the lipid composition of *Sacch. cerevisiae* have been effected, it is then necessary to discover whether the lipid composition of all of the types of membrane in the cell are changed to the same extent, or whether the changes are confined to just one or two types of membrane.

Several preliminary studies have been reported on the separation of subcellular organelles in *Sacch. cerevisiae* [15, 42]. The most popular method is to prepare sphaeroplasts by digesting the cell wall, and to separate subcellular organelles from sphaeroplast lysates by differential, density-gradient or zonal centrifugation.

Having established methods for effecting specific alterations in the lipid composition of *Sacch. cerevisiae* NCYC 366, recent work in our Laboratory has been directed towards developing techniques for separating subcellular organelles from sphaeroplast lysates. As we are interested in both composition-function relationships in the plasma membrane, and the physiological roles of triacylglycerols and sterol esters, our work has concentrated on isolating plasma membranes and on establishing the subcellular location of triacylglycerols and sterol esters, using sucrose density gradients.

In order to establish rigorously the relative purity of subcellular fractions isolated by density-gradient centrifugation, it is essential to have available marker components or enzymes that are uniquely associated with a particular subcellular structure. These markers can then be used to monitor the distribution of individual subcellular structures on the gradient. Nuclear DNA and mitochondrial cytochromes are used to locate nuclei and mitochondria respectively. Marker compounds or enzymes for surface or plasma membranes are less well documented. In studies involving subcellular fractionation of animal cells, several enzyme activities have been used as markers for the surface membrane, including 5′-nucleotidase and [$Na^+ + K^+$]-stimulated ATPase [43, 44]. However, 5′-nucleotidase activity could not be detectable in subcellular fraction of *Sacch. cerevisiae* NCYC 366.

Because of the obvious need to isolate fractions rich in plasma membrane from *Sacch. cerevisiae* NCYC 366, one of us (G. E. Wheeler) investigated criteria for establishing the distribution of plasma membrane upon density-gradient centrifugation of lysed sphaeroplasts of *Sacch. cerevisiae* NCYC 366. Since there are no reports of enzyme activities specifically associated with the plasma membrane of yeast, initially an attempt was made to label chemically the plasma membrane of sphaeroplasts before lysis and fractionation. Marinetti and Gray [45] found that the fluorescent compound 4-acetamido-4′-isothiocyanate-stilbene-2,2′-disulphonic acid (SITS) can bind specifically to the surface membrane of liver cells if the label is introduced before the cells are homogenized. Intact sphaeroplasts of *Sacch. cerevisiae* NCYC 366 were treated with SITS, excess reagent removed, and the sphaeroplasts lysed. The lysate was then submitted to density-gradient centrifugation, and the material at each interface of the discontinuous gradient collected. A qualitative determination of the concentration of SITS extracted from each fraction with potassium hydroxide indicated the distribution of plasma membrane on the gradient.

In our Laboratory, R. J. Diamond (unpublished observations) had previously obtained evidence for the presence of a [Na^+ + K^+]-stimulated ATPase in crude plasma membranes from *Sacch. cerevisiae* NCYC 366; these membranes were obtained by differential centrifugation of lysed sphaeroplasts [17]. Subsequent experiments (G. E. Wheeler, unpublished observations) showed that while this activity can be found in crude plasma membrane fractions, and in fractions from density-gradient separations providing that the yeast cells have not been treated with a thiol reagent before or during sphaeroplast formation, nevertheless the activity could not consistently be demonstrated. Therefore, the activity is not a suitable marker enzyme for plasma membranes.

The enzyme phosphatidylinositol kinase has been shown to be located predominantly on the surface membranes of liver cells [46], and this activity was investigated as a possible plasma-membrane marker for *Sacch. cerevisiae* NCYC 366. The activity was detected in homogenates prepared either by lysing sphaeroplasts or by mechanical disruption of intact cells by shaking with glass beads. Phosphatidylinositol kinase activity was largely concentrated in material that collected at the interface between densities 1.13 and 1.16 on sucrose gradients [47]. When [Na^+ + K^+]-stimulated ATPase activity could be detected, it too was largely concentrated in material with the same density as that containing phosphatidylinositol kinase with the highest specific activity. As surface membranes from mammalian cells have a density in this range [43], material derived from *Sacch. cerevisiae* NCYC 366 with densities in the range 1.13-1.16 was considered to be mainly plasma membrane.

A preliminary analysis of a fraction rich in plasma membrane from density gradients of lysed sphaeroplasts of *Sacch. cerevisiae* NCYC 366 (Table IV) shows that the lipid extracts contain predominantly phospholipid and sterol. Compared with whole-cell lipids, it is relatively deficient in sterol ester. These data indicate that the plasma membrane of *Sacch. cerevisiae* may have a lipid composition similar to that of animal cell-surface membranes [48].

TABLE IV
Lipid composition of fractions from a sucrose density-gradient of a sphaeroplast lysate of *Saccharomyces cerevisiae* NCYC 366

Density	Phospholipids	Sterols	Sterol esters
<1.08	27	29	52
1.08-1.13	11	11	11
1.13-1.16	12	14	1
1.16-1.18	20	18	13
>1.18	30	27	22

Values indicate percentage of total

Analyses of material from other density bands on sucrose gradients show that sterol esters are concentrated in organelles with a density below that of the plasma membrane-rich fraction (K. Hunter: Ph.D. Thesis; University of Bath). Analyses have been made of the low-density organelles (Table V) and these reveal that their composition resembles that of low-density lipid-containing vesicles from mammalian cells and plasma [49], that is they contain a large amount of lipid which is largely sterol ester and triacylglycerol. Low-density vesicles from *Sacch. cerevisiae* NCYC 366, like those from animal sources, are also precipitated by heparin in the presence of divalent cations.

Further work is required to establish the nature and physiological function of low-density lipid-containing vesicles from *Sacch. cerevisiae*. They could conceivably be associated with subcellular

TABLE V

Properties of low-density vesicles from *Saccharomyces cerevisiae* NCYC 366 compared with vesicles from mammalian systems

	High density lipoprotein	Low density lipoprotein	Very low density lipoprotein	Chylomicrons	Yeast vesicles
Lipid composition (%)					
Phospholipids	53	28	20	9	32
Triacylglycerols	14	14	42	89	26
Free sterols	33	10	8	0.8	3
Sterol esters		47	30	1.7	38
Density range	1.06-1.20	1.00-1.06	0.9-1.00	0.94	$<$1.08

Data on mammalian vesicles are from Margolis [49].

structures that have been identified as vacuoles [50]. Alternatively, they may be similar to the vesicles which may be involved with release of material from the plasma membrane during wall growth [51] or secretion of internally synthesized enzymes such as invertase [52]. It is attractive to assign such a role to these vesicles, at least provisionally, since an enzyme that catalyses hydrolysis of sterol esters has been located in the periplasmic space of *Sacch. cerevisiae* NCYC 366 [53], an activity which may be concerned with dissociation of enzymes and wall material from the vesicles.

Acknowledgements

Work carried out in this Laboratory was supported by a grant from the Science Research Council (U.K.), No. B/RG/182, and by the competent technical assistance of Mrs. Diane Arkell and Mrs. Zena Newman. We also wish to acknowledge the patient secretarial help of Mrs. Hazel Ellis.

References

1. Rouser, G., Nelson, G. J., Fleischer, S. and Simon, G. (1968). *In:* Biological Membranes, p. 5 (Chapman, D. ed.). Academic Press, London and New York.
2. Silbert, D. F. and Vagelos, P. R. (1967). *Proc. natn. Acad. Sci. U.S.A.*, **58**, 1579.
3. Razin, S. (1973). *Adv. microbial Physiol.* **10**, 1.
4. Hunter, K. and Rose, A. H. (1971). *In:* The Yeasts, Vol. 2, p. 211 (Rose, A. H. and Harrison, J. S., eds). Academic Press, London and New York.
5. Goldfine, H. (1972). *Adv. microbial Physiol.*, **8**, 1.
6. Shaw, N. (1970). *Bact. Rev.*, **34**, 365.
7. Wilkinson, S. G. (1969). *Biochim. biophys, Acta*, **187**, 492.
8. Baraud, J., Maurice, A. and Napias, C. (1970). *Bull. Soc. chim. Biol.*, **52**, 421.
9. Brennan, P. J., Flynn, M. P. and Griffin, P. F. S. (1970). *FEBS Letters*, **8**, 322.
10. Suomalainen, H. and Keränen, A. J. A. (1968). *Chem. Phys. Lipids*, **2**, 296.
11. Steiner, S. and Lester, R. L. (1972). *J. Bact.*, **109**, 81.
12. Lester, R. L. and Steiner, M. R. (1968). *J. biol. Chem.*, **243**, 4889.
13. Steiner, S., Smith, S., Waechter, C. J. and Lester, R. L. (1969). *Proc. natn. Acad. Sci. U.S.A.*, **64**, 1042.
14. Farrell, J. and Rose, A. H. (1967). *A. Rev. Microbiol.*, **21**, 101.
15. Hunter, K. and Rose, A. H. (1972). *Biochim. biophys. Acta*, **260**, 639.
16. Kates, M. and Baxter, R. M. (1962). *Can. J. Biochem. Physiol.*, **40**, 1213.

17. Diamond, R. J. and Rose, A. H. (1970). *J. Bact.*, **102**, 311.
18. Letters, R. (1966). *Biochim. biophys. Acta*, **116**, 489.
19. Kennedy, E. P. and Weiss, S. B. (1956). *J. biol. Chem.*, **228**, 195.
20. Waechter, C. J., Steiner, M. R. and Lester, R. L. (1969). *J. biol. Chem.*, **244**, 3419.
21. Waechter, C. J. and Lester, R. L. (1971). *J. Bact.*, **105**, 837.
22. Ratcliffe, S. J., Hossack, J. A., Wheeler, G. E. and Rose, A. H. *J. gen. Microbiol.*, **76**, 445.
23. Andreasen, A. A. and Stier, T. J. B. (1954). *J. cell. comp. Physiol.*, **43**, 271.
24. Light, R. J., Lennarz, W. J. and Bloch, K. (1962). *J. biol. Chem.*, **237**, 1793.
25. Alterthum, F. and Rose, A. H. *J. gen. Microbiol.*, **77**, 371.
26. Bauer, H., Bush, D. A. and Horisberger, M. (1972). *Experientia*, **28**, 11.
27. Mager, J. (1959). *Biochim. biophys. Acta*, **36**, 529.
28. Demel, R. A., Geuerts Van Kessel, W. S. M. and Van Deenen, L. L. M. (1972). *Biochim. biophys. Acta*, **266**, 26.
29. Proudlock, J. W., Wheeldon, L. W., Jollow, D. J. and Linnane, A. W. (1968). *Biochim. biophys Acta*, **152**, 434.
30. Resnick, M. A. and Mortimer, R. K. (1966). *J. Bact.*, **92**, 597.
31. Keith, A. D., Resnick, M. R. and Hayley, A. B. (1969). *J. Bact.*, **98**, 415.
32. Wisnieski, B. J., Keith, A. D. and Resnick, M. R. (1970). *J. Bact.*, **101**, 160.
33. Wisnieski, B. J. and Kiyomoto, R. Y. (1972). *J. Bact.*, **109**, 186.
34. Schweizer, E. and Bolling, H. (1970). *Proc. natn. Acad. Sci. U.S.A.*, **67**, 660.
35. Schweizer, E., Kühn, L. and Castorph, H. (1971). *Z. Physiol. Chem.*, **352**, 384.
36. Henry, S. A. and Keith, A. D. (1971). *Chem. Phys. Lipids*, **7**, 284.
37. Eletr, S. and Keith, A. D. (1972). *Proc. natn. Acad. Sci. U.S.A.*, **69**, 1353.
38. Adelberg, E. A., Roslansky, P. F., Myers, J. W. and Coughlin, C. A. (1955). *J. Bact.*, **69**, 733.
39. Lampen, J. O. (1966). *Symp. Soc. gen. Microbiol.*, **16**, 111.
40. Woods, R. A. (1971). *J. Bact.*, **108**, 69.
41. Molzahn, S. W. and Woods, R. A. (1972). *J. gen. Microbiol.*, **72**, 339.
42. Matile, Ph. (1970). *FEBS Symposium*, **20**, 39.
43. Wheeler, G. E., Coleman, R. and Finean, J. B. (1972). *Biochim. biophys. Acta*, **255**, 917.
44. Emmelot, P. and Bos, C. J. (1962). *Biochim. biophys. Acta*, **58**, 374.
45. Marinetti, G. V. and Gray, G. N. (1967). *Biochim. biophys. Acta*, **135**, 580.
46. Michell, R. H., Harwood, J. L., Coleman, R. and Hawthorne, J. N. (1967). *Biochim. biophys. Acta*, **144**, 649.
47. Wheeler, G. E. and Rose, A. H. (1972). *Biochem J.*, **127**, 64P.
48. Skipski, V., Barclay, M., Archibald, F. M., Terebuo-Kekish, O., Reichman, E. S. and Good, J. J. (1965). *Life Sci.*, **4**, 1673.
49. Margolis, S. (1969). *In:* Structural and Functional Aspects of Lipoproteins In Living Systems, p. 369 (Tria, E. and Scanu, A. M., eds). Academic Press, London and New York.
50. Matile, Ph., and Wiemken, A. (1967). *Arch. Mikrobiol.*, **56**, 148.
51. Sentandreu, R. and Northcote, D. H. (1969). *J. gen. Microbiol.*, **55**, 393.
52. Beteta, P. and Gascon, S. (1971). *FEBS Letters*, **13**, 297.
53. Wheeler, G. E. and Rose, A. H. (1973). *J. gen. Microbiol.*, **74**, 189.

Enzyme Synthesis, RNA Synthesis and the Effects of Actinomycin D in Protoplasts of the Fission Yeast *Schizosaccharomyces pombe*

J. M. MITCHISON, J. CREANOR and MARIA LUISA SARTIRANA

Department of Zoology, University of Edinburgh, Scotland

In earlier papers, we have shown that the basal rate of synthesis of three enzymes (sucrase, acid and alkaline phosphatase) doubles sharply at a point in the cell cycle of the fission yeast *Schizosaccharomyces pombe* which we have called the "critical point" [1]. Measurements were made on synchronous cultures produced by selection synchronization and it was assumed that rises in enzyme activity could be equated with enzyme synthesis. We have also measured the rate of synthesis of sucrase and maltase in samples taken at intervals from synchronous cultures and derepressed for these enzymes by lowering the glucose concentration in the medium [1, 2]. This rate of enzyme synthesis on derepression, also called enzyme "potential", stays constant through the cycle until the critical point is reached when it increases sharply.

Our interpretation of these results has been that the doubling in the rates of basal synthesis and of potential are gene-dosage effects and that the critical point is the time in the cycle when the number of functional genes doubles. Surprisingly, it is appreciably later in the cycle than the period of DNA synthesis and we have suggested that it may be related to changes in chromosome structure. This interpretation assumes that regulatory control is being exerted at the level of RNA transcription. Van Wijk *et al.* [3], however, have suggested that derepression by lowering the glucose concentration is an effect controlled at the translational level in the case of α-glucosidase in protoplasts of the budding yeast *Saccharomyces carlsbergensis.* In a key experiment, they showed that actinomycin D at 80 μg/ml did not inhibit the derepression of this enzyme over a period of 200 min. In order to test whether or not this effect occurs in *S. pombe,* we have prepared protoplasts and measured the synthesis of three enzymes (sucrase, maltase and acid phosphatase) and of RNA, with and without the addition of actinomycin D.

I. Materials and Methods

A. *Preparation of Protoplasts*

S. pombe strain NCYC 132 (new strain [2]) was maintained and grown at 32°C in EMM (2), a minimal medium with glucose as the carbon source [4]. Cells in the exponential phase of growth were converted into protoplasts by a modification of the technique of Duell *et al.* [5]. In a typical experiment, the cells from 500 ml culture at a concentration of $2 \cdot 10^6$ cells/ml were collected by filtration, resuspended in 20 ml of pretreatment medium, and incubated at 32°C in a shaking water bath for 30 min. This medium contained 0.14 M 2-mercaptoethanol, 40 mM EDTA, and 20 mM citric acid/Na_2HPO_4 buffer at pH 5.6. After incubation, the cells were centrifuged, washed twice with the buffer above, resuspended in 5 ml of protoplasting medium and incubated as before for 30 min. The protoplasting medium was made up by mixing 4 ml of EMM (2) containing 1.5 M $MgSO_4$ and 30 mM 2-mercaptoethanol with the contents of a 1 ml ampoule of snail enzyme ("suc d'Helix pomatia" from Industrie Biologique Francaise, 92 Gennevilliers, France). High magnesium proved much more effective than sorbitol or mannitol in making protoplasts of this yeast. On the other hand, it inhibited the synthesis of the enzymes which were being investigated. The magnesium concentration was therefore reduced slowly (to prevent osmotic bursting) from 1.2 M to 0.3 M. After the first 30 min, 2 ml of EMM (2) were added every 10 min until the final volume of the protoplasting suspension was 20 ml. The protoplasts were then centrifuged, washed twice with EMM (2) containing 0.5 M mannitol, and resuspended in this growth medium.

The final suspension contained protoplasts which were easily distinguishable from intact cells because the former were spherical and the latter cylindrical. Many of the protoplasts were in clumps which made it difficult to get accurate counts. In one experiment, measurement of the DNA content of a suspension of protoplasts showed that the yield of protoplasts was 71% of the initial cells. There was, however, considerable variation in the apparent yield from one experiment to the next, and this figure should not be relied on. As well as the protoplasts, there were a small number of cells with intact walls but with an abnormal appearance. Tests showed that these cells did not produce any significant quantity of the enzymes after the protoplasts had been burst by omitting mannitol from the growth medium.

B. *Preparation of Samples*

All three of the enzymes studied are external enzymes [6] in the sense that they are situated outside the cell membrane and are bound to or contained by the cell wall in intact cells. In protoplasts, however, some but not all of each of these enzymes are released into the medium. It was necessary, therefore, to assay both protoplasts and medium in order to follow the patterns of enzyme synthesis. Samples, usually 0.7 to 1.0 ml, were taken, frozen and thawed, dialysed overnight against water at 4° C and then split into smaller aliquots for enzyme assay. The dialysis removed glucose and phosphate from the medium and also actinomycin, all of which would have interfered with the assays.

Samples for radioactivity measurements were collected on glass fibre filters (Whatman GF/A), washed with 0.5 M mannitol, dried and then counted in a liquid scintillation counter. Samples for uridine and adenosine incorporation were washed twice with 5% trichloracetic acid and then with water before drying.

C. *Enzyme Assays and Activity*

Enzymes were assayed by techniques described earlier [1, 2]. In summary, sucrase (β-D-fructofuranoside fructohydrolase, EC 3.2.1.26) and maltase (α-D-glucoside glucohydrolase, EC 3.2.1.20) were assayed by measuring, with Glucostat Special reagent (Worthington), the amount of glucose released from sucrose and maltose as substrates. Acid phosphatase (orthophosphoric monoester phosphohydrolase, EC 3.1.3.2) was assayed with p-nitrophenyl phosphate as the substrate.

The enzyme activity values for protoplasts have not been given in the text figures because an accurate value for the protoplast yield in each experiment was not known. An estimate of the activity, however, can be given from the following figures. If U is the activity in μmole substrate/min, then the value of the last point on the first control curve for sucrase in Fig. 2 is 1.67 U/10^{10} original intact cells. The same point in Fig. 3 for maltase is 1.19 U/10^{10} original cells, and for acid phosphatase in Fig. 4 is 3.65 U/10^8 original cells. If the protoplast yield was 70%, the values for protoplasts would be 1.4 times higher.

D. *Preparation and Electrophoresis of RNA*

Protoplasts were lysed at 0°C in a mixture of 2% tri-isopropyl naphthalenesulphonic acid, sodium salt (Eastman Kodak), 0.1 M NaCl and 0.5 M Tris-HCl buffer at pH 7.5. The lysate was shaken with an equal volume of phenol mixture [7], the phenol phase removed and the aqueous phase, made 0.4 M with respect to NaCl, shaken again with one volume of fresh phenol mixture. The nucleic acids were precipitated from the aqueous phase by adding two volumes of 96% ethanol.

Nucleic acids were separated by electrophoresis on 2.6% polyacrylamide gels according to the technique of Loening [8]. The gels were scanned for absorption at 265 nm and fractionated into slices for radioactivity measurements. Tritium-labelled RNA in the slices was hydrolysed and counted by scintillation [8]. RNA labelled with ^{32}P was counted under a thin end window Geiger tube [9].

II Results

A. *Properties of Maltase*

The properties of sucrase and acid phosphatase in *S. pombe* have been described in detail in an earlier paper [1]. Maltase, however, has only been described briefly [2], so we give a fuller account here.

Maltase is an external enzyme since the activity of living cells was more than 90% of that of freeze-dried cells. Michaelis-Menten plots were linear and gave a K_m of 4.9 mM with maltose as substrate. This K_m did not change when the enzyme was derepressed. An activity/pH curve with a maltose substrate was shallow and symmetrical with a maximum activity at pH 4.5. The activity with *p*-nitrophenyl α-glucoside was low—less than 5% of the activity with maltose.

Maltase could be derepressed by lowering the glucose concentration in the medium (Fig. 1). A similar result has been found with sucrase in *S. pombe* [1] and with α-glucosidase in budding yeast [3]. The addition of maltose at concentrations up to 2% had no significant effect on the course of derepression (Fig. 1). This is in sharp contrast to the situation with α-glucosidase in budding yeast where the presence of maltose caused a considerable increase in the rate of derepression [3]. There are two other clear differences between these two enzymes. Maltase showed little activity against nitrophenol glucoside; and it was external whereas the α-glucosidase is internal [6]. It might be argued that sucrase and maltase are the same enzyme in *S. pombe*. Evidence against this is that these enzymes had

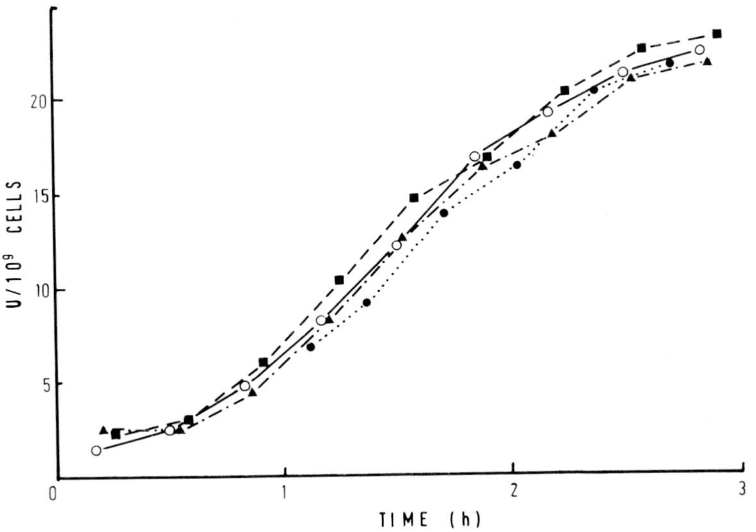

Fig. 1. Derepression of maltase in intact cells with and without maltose. At 0 time, samples were taken from an exponential phase culture in EMM (2) with $2.6 \cdot 10^6$ cells/ml and diluted with 19 times their volume of glucose-free EMM (2). This lowered the glucose concentration from 1% (w/v) to 0.05%. The cultures were incubated at 32°C and aliquots taken for maltase assays. U = μmole substrate/min. ○——○ control without maltose. ▲–•–▲ 0.1% maltose added at 0 h. ■— —■ 0.5% maltose added at 0 h. ● . . . . . ● 2.0% maltose added at 1 h.

different synthetic patterns in synchronous cultures [10], that it took longer for the derepression of maltase to become linear (40 min for maltase, 20 min for sucrase), and that the K_m of maltase remained unchanged after derepression whereas that of sucrase fell by a factor of four [1].

B. *Enzyme Synthesis and Actinomycin*

The effect of actinomycin on the derepression of sucrase and maltase was tested in the following way. Protoplasts were kept in growth medium with 80 μg/ml actinomycin D (Sigma). After periods of 0, 2 and 3.5 h, samples were removed and the protoplasts added to medium similar to that above except that the glucose concentration in the EMM (2) was 0.05% instead of the normal 1.0%. The low glucose concentration derepressed both enzymes, and the progress of this derepression was followed for 2 h. Control cultures contained no actinomycin.

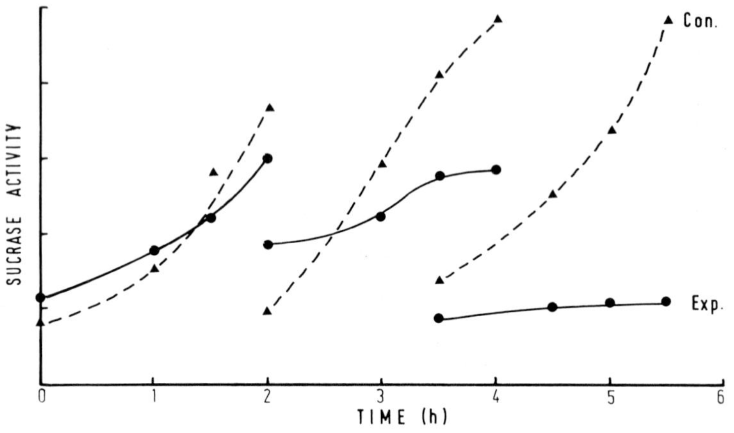

Fig. 2. Derepression of sucrase in protoplasts with and without actinomycin. At 0 time, a freshly-prepared suspension of protoplasts at 32°C was split into two subcultures, the control being in EMM (2) + 0.5 M mannitol and the experimental being in the same medium + 80 μg/ml actinomycin D. At 0, 2 and 3.5 h, samples were taken from each subculture and derepressed by transfer to low glucose (0.05%) EMM (2) + 0.5 M mannitol with and without 80 μg/ml actinomycin D. The samples were incubated at 32°C and aliquots taken for sucrase assay. ▲— —▲ control; ●——● experimental with actinomycin. The curves for the 3.5 h samples are from a different experiment to those for 0 and 2 h.

Results for sucrase are shown in Fig. 2. In the 0 h samples, the actinomycin-treated culture showed slightly less derepression than the control. This effect was more marked in the 2 h samples, and in the 3.5 h samples there was practically no apparent enzyme synthesis. Very similar results for maltase are shown in Fig. 3.

Experiments with acid phosphatase were similar except that normal EMM (2) was used for the samples rather than a low glucose medium. What was being tested, therefore, was the normal rate of synthesis of the enzyme rather than the rate of derepression. Results are presented in Fig. 4 and are similar to those with sucrase and maltase. Note that the basal rate of synthesis in the controls falls off as the culture ages.

In a second set of experiments, the procedures were the same except that the cells were left in growth medium for 2 h before actinomycin was added and the initial samples removed. The results for maltase (Fig. 5) showed that the inhibitory effects of the actinomycin appeared sooner than in the first set of experiments (cf. Fig. 3). There is practically no derepression after 2 h treatment with

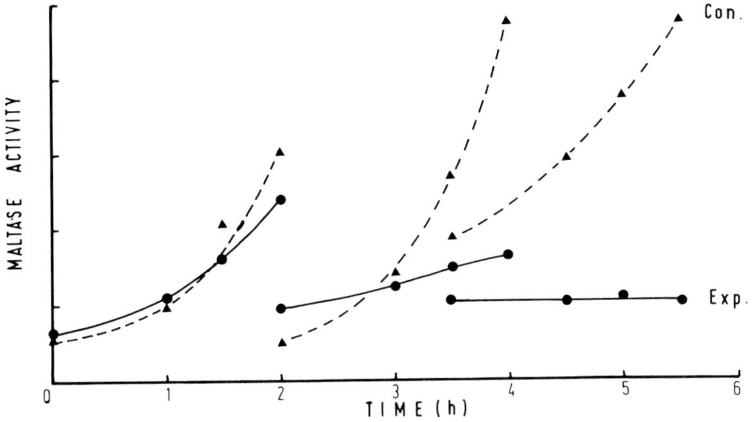

Fig. 3. Derepression of maltase in protoplasts with and without actinomycin. The procedure was the same as for Fig. 2, except that the assay was for maltase.

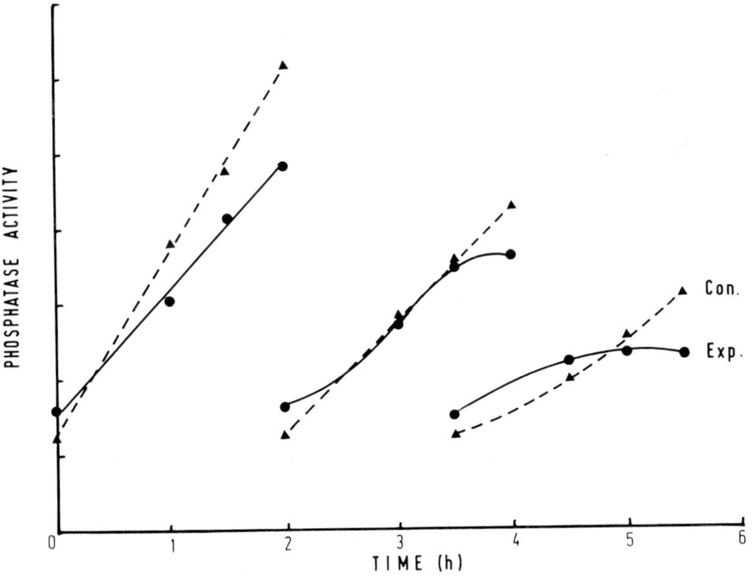

Fig. 4. Synthesis of acid phosphatase in protoplasts with and without actinomycin. The procedure was the same as for Fig. 2, except that the samples were transferred to media with ordinary EMM (2) instead of low-glucose EMM (2), and the assay was for acid phosphatase.

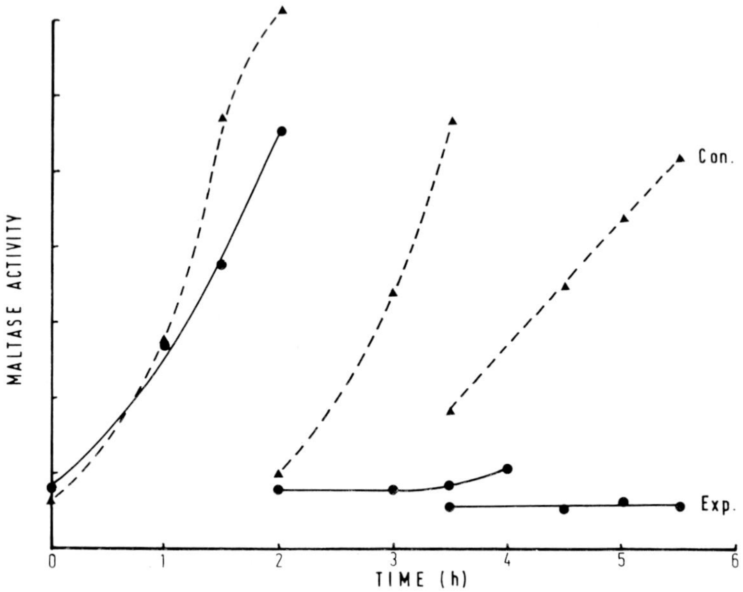

Fig. 5. Derepression of maltase in protoplasts with and without actinomycin. The procedure was the same as for Fig. 3, except that the protoplasts had been in EMM (2) + 0.5 M mannitol for 2 h after preparation.

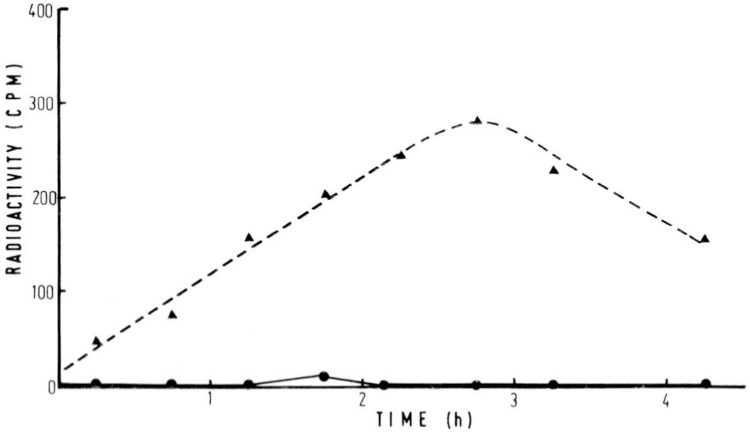

Fig. 6. Incorporation of uridine pulses by protoplasts with and without actinomycin. Freshly-prepared protoplasts from $6 \cdot 10^8$ cells at 32°C were split into two 6 ml subcultures at 0 time. Control (▲— —▲) was in EMM (2) + 0.5 M mannitol, and experimental (●——●) was in the same medium + 80 μg/ml actinomycin D. At 0 time and every 30 min thereafter until 4 h, a sample of 0.5 ml was removed and incubated for 15 min with 0.1 ml of $[^3H_{5,6}]$ uridine (Amersham, 24 Ci/mmole, 10 μCi/ml). Each point is the radioactivity with the time as the end of the pulse. The background count was 146 cpm.

the inhibitor. Very similar results were obtained for sucrase and acid phosphatase.

C. *RNA Synthesis*

Although we have not been able to measure net increase of RNA in protoplasts by direct measurements, synthesis does appear to go on as judged by the incorporation of labelled precursors. Figure 6 shows the rate of uridine incorporation in 15 min pulses for the first 4 h after suspending protoplasts in growth medium. The control showed a rising rate for the first 2.5 h, after which it declined. In contrast, treatment with actinomycin prevented any significant incorporation. This might suggest that actinomycin produced an immediate and drastic inhibition of RNA synthesis. It seems, however, that the inhibition is one which applies to uridine uptake or incorporation rather than to RNA synthesis since the results with adenosine were different. Adenosine is a legitimate label for RNA in this yeast since the DNA/RNA ratio is unusually low (1/100) and incorporated adenine is 97% digestible with ribonuclease [11]. Figure 7 shows the results of labelling with adenosine at 0, 2 and 3.5 h after transfer of protoplasts to growth medium with and without actinomycin, a timing which is identical with the enzyme assays in Figs 2-4. The controls showed a recovery of RNA synthesis after an initial low value when the protoplasts were transferred from protoplasting medium to growth medium (similar to the pattern for uridine incorporation in Fig. 6). With actinomycin, the initial rate of incorporation was similar to that in the controls, but there was no recovery, rather a decline until the rate after 3.5 h was less than half that at the start of the experiment. In another set of experiments (Fig. 8) the procedure was the same except that the protoplasts were suspended in growth medium for 2 h before the addition of actinomycin and the start of the labelling (a timing identical with that of the enzyme assays in Fig. 5). This allowed the controls to recover fully and they showed, if anything, a slight fall-off in rate as the culture aged. As with the enzymes, the inhibitory effect of actinomycin was faster and more complete. After 2 h, there was little or no incorporation.

Acrylamide gel electrophoresis was used to investigate the types of RNA that were being synthesized in these experiments. RNA was labelled with tritiated adenosine for 10 min after the protoplasts had been in growth medium for 2 h both with and without actinomycin. The radioactivity profiles after electrophoresis are shown in Fig. 9.

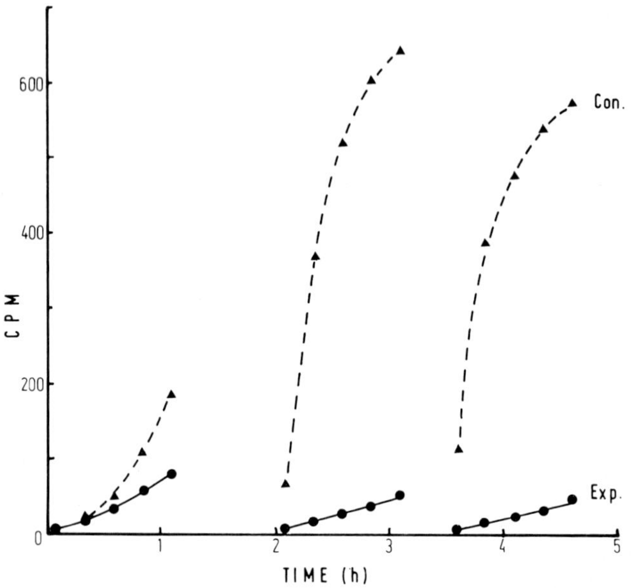

Fig. 7. Incorporation of adenosine by protoplasts with and without actinomycin. Freshly-prepared protoplasts from $1.3 \cdot 10^9$ cells at 32°C were split into two 10 ml subcultures at 0 time. Control (▲— —▲) was in EMM (2) + 0.5 M mannitol, and experimental (●——●) was in the same medium + 80 μg/ml of actinomycin D. At 0 time, 3 ml was removed from each and incubated with 0.1 ml of [$^3H_{2,8}$] adenosine (Amersham, 15.5 Ci/mmole, 10 μCi/ml). 0.5 ml samples of this incubation mixture were removed at the times indicated up to 65 min and prepared for counting. The same procedure was repeated at 2 and 3.5 h. The background count was 250 cpm.

The control shows two major peaks of ribosomal RNA (26S and 18S; 1.3 m and 0.7 m M_r) coincident with peaks on the absorbance profile. There are also peaks representing ribosomal precursors which will be described in a forthcoming paper. With actinomycin, there is a broad spread of labelled material but no clear-cut ribosomal peaks. We conclude therefore that the majority of the RNA that is synthesized after actinomycin treatment is non-ribosomal.

Gels were also run with RNA labelled with [$^{32}PO_4^{3-}$]. These were not satisfactory for protoplasts treated for some time with actinomycin because of a long "tail" of low molecular weight compounds in the radioactivity profile which were not digestible by ribonuclease (possibly polyphosphates). They showed, however, that control cultures labelled for 10 min soon after transfer to growth medium (e.g. the first curve in Fig. 7) had ribosomal precursor peaks but no

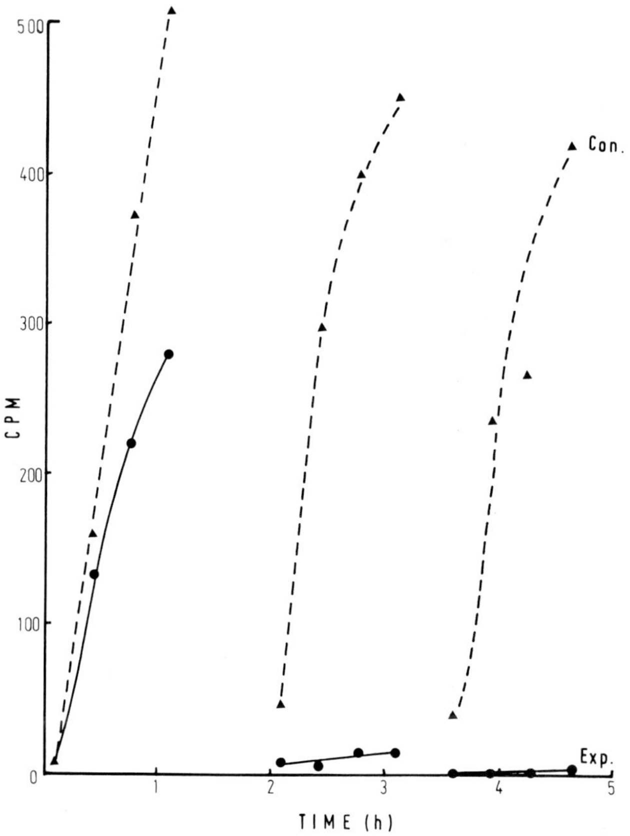

Fig. 8. Incorporation of adenosine by protoplasts with and without actinomycin. The procedure was identical with that for Fig. 7, except that the protoplasts had been for 2 h after preparation in EMM (2) + 0.5 M mannitol, and had been prepared from $6 \cdot 10^8$ cells.

peaks of fully formed ribosomal RNA. The recovery of RNA synthesis in the control cultures is presumably due to recovery of the capacity to synthesize ribosomal RNA at the normal rate.

D. *Uptake of Actinomycin*

The fact that the effect of actinomycin on enzyme synthesis increases as the period of pretreatment lengthens could be caused by slow penetration of the inhibitor into the protoplasts. We therefore investigated the uptake of labelled actinomycin D into protoplasts

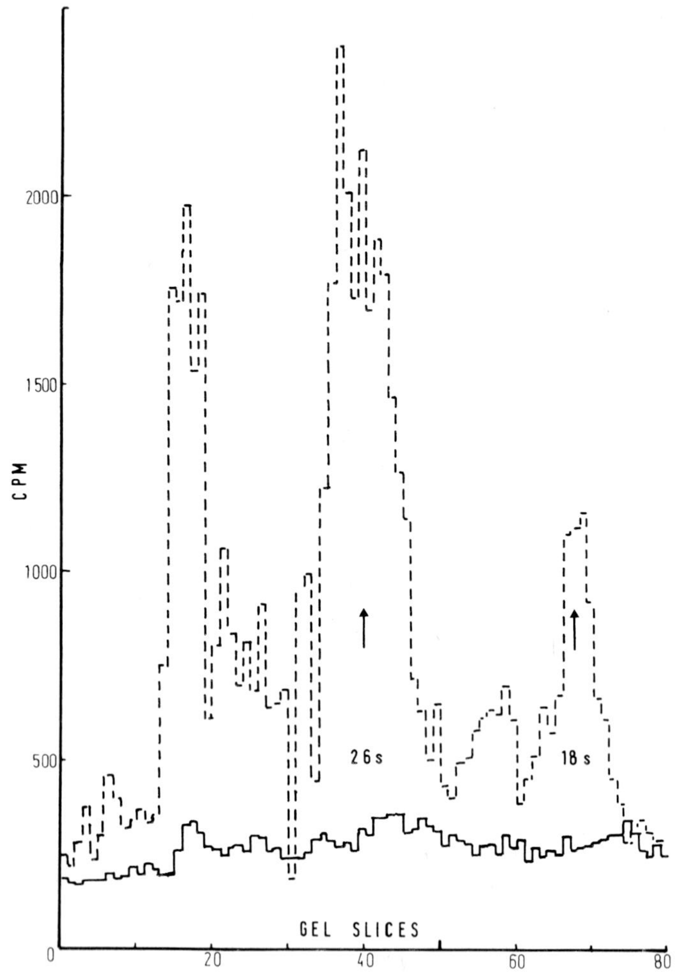

Fig. 9. Radioactivity profiles of acrylamide gels of RNA from protoplasts with and without actinomycin. Freshly-prepared protoplasts from $1.1 \cdot 10^9$ cells were split into two subcultures of 4 ml. Control (– – –) was in EMM (2) + 0.5 M mannitol, and experimental (——) was in the same medium + 80 μg/ml actinomycin D. After 2 h, each culture was labelled for 10 min with 100 μCi of $[^3H_{2,8}]$ adenosine (Amersham, 15.5 Ci/mmole). The protoplasts were then centrifuged down and treated with 5 ml of the lytic mixture described in the Methods section above. The arrows under two of the control peaks indicate the peaks of the ribosomal components (26S and 18S) from the absorbance scan which is not shown otherwise. The background count was 50 cpm.

and into intact cells. The patterns of uptake are shown in Fig. 10. Although the rate of uptake into protoplasts diminished a little after the first 30 min, it did not stop but continued at a constant rate for 1.5 h. Other experiments showed that the rate remained constant for a further 2 h, and that the same uptake curve was obtained if the protoplasts had been suspended for 2 h in growth medium before the addition of the actinomycin.

The uptake into intact cells was much less than that into protoplasts, and presumably accounts for the relative insensitivity of normal yeast cells to actinomycin.

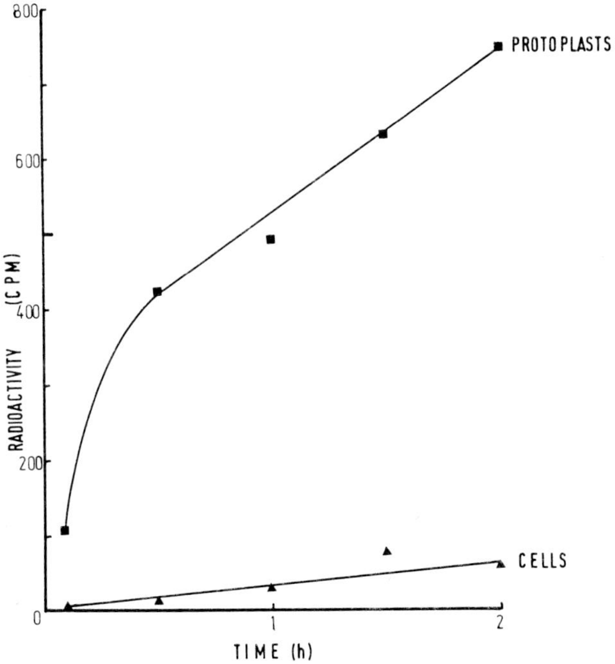

Fig. 10. Uptake of actinomycin by protoplasts and intact cells. At 0 time, freshly-prepared protoplasts from $8.6 \cdot 10^8$ cells were suspended in 2 ml of EMM (2) + 0.5 M mannitol + [G – ^{3}H] actinomycin D (Amersham, to give 80 μg/ml at 390 μCi/mmole). 0.2 ml samples taken for counting at times indicated. The same procedure was followed for $2.6 \cdot 10^8$ intact cells. Background count was 100 cpm. The two curves are comparable if the yield of protoplasts is assumed to be 30%. If the yield was greater, the difference between the curves would be decreased.

E. *Cycloheximide and Enzyme Synthesis*

In earlier work on intact yeast cells, we added cycloheximide under conditions of normal growth and during derepression [1, 2]. Cycloheximide is an effective inhibitor of protein synthesis at the ribosome level and, at 100 μg/ml, it reduced leucine incorporation after 10 min to 8% of its normal value. With the four enzymes tested (sucrase, maltase, acid and alkaline phosphatase), the enzyme activity stopped rising and remained constant for at least 2 h. This suggests that the enzymes are relatively stable and that the activity rises are due to enzyme synthesis rather than activation. In addition, however, there was an appreciable time before the activity of some of the enzymes stopped rising after the addition of the inhibitor (30 min for alkaline phosphatase and 15 min for sucrase and maltase). We have called this "precursor delay" and have suggested that it might be due to delayed assembly of subunits or conformational change, or perhaps to a delay in the positioning of the enzyme in the cell. Since the situation in protoplasts might be different from that in intact cells, we repeated these cycloheximide experiments with protoplasts. The results were identical. Figure 11 shows a 15 min precursor delay for maltase, and no delay for acid phosphatase.

III. Discussion

Actinomycin D has been widely used in the last decade to distinguish between transcriptional and translational control of protein synthesis [12, 13]. If the assumptions are made that actinomycin stops RNA synthesis and has no other effect, and that the messenger RNA is unstable, then transcriptional control is indicated if the effect of actinomycin is to cause a rapid decrease of protein synthesis. If, on the other hand, protein synthesis continues or increases, this suggests a stable message and translational control. There can of course be added complexities such as specific regulatory proteins [14] or stabilization of the message by actinomycin [15].

On a simple model, our results are consistent with transcriptional control by an unstable message. Once the protoplasts have recovered by a 2 h period in growth medium, the fall-off in enzyme synthesis matches the fall-off in RNA synthesis (cf. Figs 5 and 8). The match is not quite so good if actinomycin is added as soon as the protoplasts have been removed from the protoplasting medium and before they have recovered the capacity to synthesize ribosomal RNA. The synthesis of enzymes on derepression has ceased after 3.5 h, but

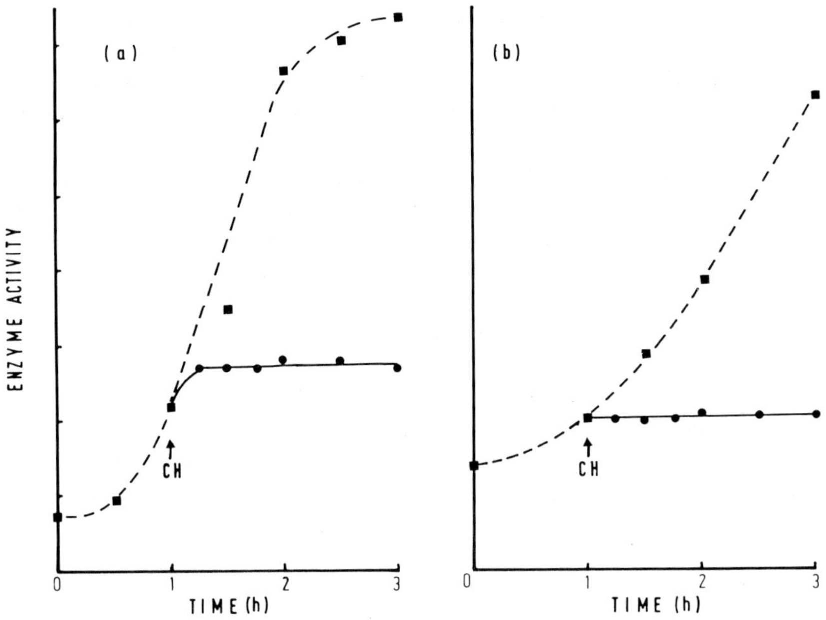

Fig. 11. Effect of cycloheximide on enzyme activity. (a) At 0 time, freshly-prepared protoplasts were suspended in low glucose (0.05%) EMM (2) + 0.5 M mannitol and samples taken thereafter for maltase assay (■— —■). A portion of the culture was removed at 1 h, treated with cycloheximide at 100 μg/ml, and samples taken for maltase assay (●—●). (b) The same procedure as (a) except that the assays were for acid phosphatase and the medium was EMM (2) + 0.5 M mannitol.

there is still some RNA synthesis though at a reduced level (compare Figs 2-4 and 7). We do not know, however, how much of the non-ribosomal RNA which is synthesized in the presence of actinomycin is true message and how much is nuclear RNA which never leaves the nucleus.

Whatever the exact relations between the syntheses of enzymes and of messenger RNA, we do not find a situation in which enzyme synthesis continues in the long-term presence of actinomycin. In this, our results differ from those of Van Wijk *et al.* [3] who found that the same concentration of actinomycin did not inhibit the derepression of α-glucosidase in budding yeast over a period of 200 min. It would be interesting to know the effect of pretreatment of their protoplasts with actinomycin since a longer exposure to actinomycin might have shown an effect on enzyme synthesis. It must be remembered, however, that there are substantial differences, which have been described earlier, between their α-glucosidase and our maltase.

Any detailed analysis of the control mechanisms is made difficult by the relatively slow effect of the actinomycin on RNA and enzyme synthesis. This may be due to slow penetration into the protoplasts, an effect which is known for other systems, e.g. sea urchin eggs [16]. The curves for the uptake of labelled actinomycin are consistent with slow penetration since they do not reach a stable equilibrium value within 3.5 h. A very rough calculation shows that the concentration of actinomycin within the protoplasts is about 1.5 μg/ml at 1.5 h.

The results of gel electrophoresis show that non-ribosomal RNA continues for a time to be synthesized in the presence of actinomycin, while ribosomal RNA synthesis is reduced or absent. This differential action of actinomycin on ribosomal RNA synthesis confirms in a lower eukaryote what has been known for some time in mammalian cells [17, 18]. It is also interesting that freshly-prepared protoplasts behaved like those treated with actinomycin in synthesizing what was mainly non-ribosomal RNA, and only recovered their capacity to synthesize ribosomal RNA after a period in growth medium. This is similar to the behaviour of intact cells of this yeast after a "step-down" in medium [19].

The fact that uridine incorporation was almost completely suppressed by actinomycin whereas adenosine incorporation continued gives a warning that the absence of uridine incorporation in other cellular systems treated with actinomycin does not necessarily imply that RNA synthesis has stopped. There are reports that uracil uptake is reduced by actinomycin in tobacco leaves [20], and that uridine uptake is reduced in rat cells by another inhibitor, phenyl ethanol [21].

We can conclude that our results are consistent with transcriptional control and a short-lived message for the three enzymes investigated. There is also one bit of independent evidence that acid phosphatase is controlled by a short-lived message (May, quoted in Mitchison and Creanor [1]). Because of the slow action of actinomycin, however, we cannot rule out the possibility of messages of moderate stability (e.g. half-life of 30 min) or of some degree of translational control.

Acknowledgements

We gratefully acknowledge the technical assistance of Mr. R. J. Pitcairn and Miss Yvonne Pryde. This work was supported by a research grant from the Science Research Council.

References

1. Mitchison, J. M. and Creanor, J. (1969). *J. Cell Sci.*, **5**, 373.
2. Mitchison, J. M. and Creanor, J. (1971). *Expl Cell Res.*, **69**, 244.
3. Van Wijk, R., Ouwehand, J., Van Den Bos, T. and Koningsberger, V. V. (1969). *Biochim. biophys. Acta*, **186**, 178.
4. Mitchison, J. M. (1970). *In:* Methods in Cell Physiology, Vol. 4, p. 131 (Prescott, D. M., ed.). Academic Press, New York and London.
5. Duell, E. A., Inoue, S. and Utter, M. F. (1964). *J. Bact.*, **88**, 1762.
6. Lampen, J. O. (1968). *Antonie van Leeuwenhoek*, **34**, 1.
7. Kirby, K. S. (1965). *Biochem. J.*, **96**, 266.
8. Loening, U. E. (1967). *Biochem. J.*, **102**, 251.
9. Loening, U. E., Jones, K. and Birnstiel, M. L. (1969). *J. Molec. Biol.*, **45**, 353.
10. Bostock, C. J., Donachie, W. D., Masters, M. and Mitchison, J. M. (1966). *Nature, Lond.*, **210**, 808.
11. Mitchison, J. M. and Lark, K. G. (1962). *Expl Cell Res.*, **28**, 452.
12. Gross, P. R. (1968). *Ann. Rev. Biochem.*, **37**, 631.
13. Martin, D. W. and Tomkins, G. M. (1970). *Proc. natn. Acad. Sci. U.S.A.*, **65**, 1064.
14. Bechet, J. and Wiame, J. M. (1965). *Biochem. biophys. Res. Commun.*, **21**, 226.
15. Endo, Y., Tominaga, H. and Natori, Y. (1971). *Biochim. biophys. Acta*, **240**, 215.
16. Gross, P. R. and Cousineau, G. H. (1964). *Expl Cell Res.*, **33**, 368.
17. Perry, R. P. (1967). *In:* Progress in Nucleic Acid Research and Molecular Biology, Vol. 6, p. 220 (Davidson, J. N. and Cohen, W. E., eds). Academic Press, New York and London.
18. Penman, S., Fan, H., Perlman, S., Rosbash, M., Weinberg, R. and Zylber, E. (1970). *Cold Spring Harb. Symp. Quant. Biol.*, **35**, 561.
19. Mitchison, J. M. and Gross, P. R. (1965). *Expl Cell Res.*, **37**, 259.
20. Babos, P. and Shearer, G. B. (1969). *Virology*, **39**, 286.
21. Plagemann, P. G. W. (1970). *J. Cell Physiol.*, **75**, 315.

Use of a Spheroplast-like *Neurospora crassa* Slime Mutant for Investigations of Subcellular Structures

ROLAND R. THEIMER

Botanisches Institut der Universität, München, Germany

For many well known reasons fungi serve as good subjects for regulatory studies in eucaryotic cells. A major problem, however, results from the rigidity of the cell walls. For satisfactory breakage of cells rather powerful disruption techniques have to be applied which in turn destroy most of the subcellular structures. Especially fragile cell organelles such as microbodies (peroxisomes, glyoxysomes) are extremely susceptible to damage by rough extraction procedures. Treatment of mycelia with snail gut enzymes is widely employed in order to achieve a gentle digestion of cell walls [1] and to liberate easily disruptable spheroplasts. Yet long term incubation with wall degrading enzymes is likely to introduce artificial changes which, in regulation studies, may confuse the biochemical interpretation. Besides, such treatment may have severe effects on intracellular enzyme activities to an extent not yet understood. Since spheroplasts tend to regenerate cell walls they are not suitable for long term experiments.

To date subcellular fractionation of fungal cells has yielded contradictory reports about the existence of microbodies in fungi although the widespread occurrence of this type of organelle in higher plant cells [2] and certain animal cell types [3] is well documented. Some investigators failed to detect microbodies in yeast cell extracts [4, 5] while others [6, 7] succeeded in demonstrating particulate catalase activity. Kobr *et al.* [8], attempting to isolate microbodies (glyoxysomes) from *Neurospora crassa* wild type, met with difficulties in the separation from mitochondria. The presence in *N. crassa* of typical microbody enzymes has already been reported [9, 10, 11] yet little is known about the intracellular distribution. Enzymes common to almost all microbodies are catalase and uricase [2, 3, 12, 13]; in addition, in animal microbodies D-amino acid oxidase [3] and in plant microbodies isocitrate lyase are found [2]. In this paper the growth and subcellular fractionation of a *N. crassa* "slime" mutant, previously isolated and described by Emerson [14], is reported.

I. Material and Methods

Neurospora crassa (fz; sg; os-l; slime) was a generous gift from the Fungal Genetics Stock Center, Arcata, California/U.S.A. (FGSC no. 1118). The fungus was cultivated at 28° C in the dark. Stock cultures were grown on petri dishes containing Vogel's medium [15] supplemented with 0.75% peptone, 0.5% yeast extract and 2% sucrose, solidified with 2% agar. The mold was grown in bulk essentially as described previously [16] except that only reciprocal shakers were employed. In the liquid media the sucrose was replaced by 0.2 M ethanol and 0.233 M sorbitol. After 48 h the cells from approx. 450 ml growth medium were harvested by centrifugation at 250 x g for 15 min. The sediment was washed twice and finally suspended in grinding medium. This solution was the same as previously described [12] with the addition of 0.06% Ficoll. Extraction of the organelles was achieved by gentle grinding with mortar and pestle for 10 min. The resulting brei which still contained a portion of unbroken cells was subjected to differential centrifugation. After sedimentation of the debris and whole cells at 1,000 x g for 10 min the supernatant suspension was decanted and spun for 30 min at 27,000 x g to yield a supernatant solution S_{II} and a crude particulate fraction P_{II}. The latter was carefully resuspended in grinding medium in which the sucrose concentration had been increased to 0.8 M, and an aliquot (2-3 ml) was layered on top of a continuous sucrose density gradient (38 ml) ranging from 30% (w/w) to 57% sucrose in 0.001 M EDTA solution. After centrifugation for 5 h at 65,000 x $g_{av.}$ in a Spinco L2-50B ultracentrifuge the gradient was fractioned and each sample was examined for sucrose concentration, protein content and enzyme activities. Protein contents were determined by the procedure of Lowry *et al.* [17]. Enzyme activities were determined spectrophotometrically using a Gilford recording spectrophotometer Model 2400-S as described in the literature [12, 18, 19, 20]. Since the addition of 0.02% (v/v) Triton X-100 to the test mixtures increased the enzyme activities up to 20%, the detergent was included routinely. The assay mixture for D-amino acid oxidase contained in 1.0 ml : 0.09 M $Na_2P_2O_7$ buffer, p H 8.7, saturated with O_2, 30 mM D-alanine, pH 8.7, 10 mM phenylhydrazinium chloride, pH 7.0, and 20-80 μg protein. The increase in absorbance at 324 nm due to the formation of the pyruvatephenylhydrazone was recorded, the molar extinction coefficient was 1.7 cm^2/mol.

II. Results and Discussion

A. *Growth and Some Properties of* N. crassa *slime*

The slime mutant grows as unicellular spheres which lack any evident cell walls when cultured in a medium of suitable osmotic strength. Duplication seems to take place by a budding-like process (Fig. 1).

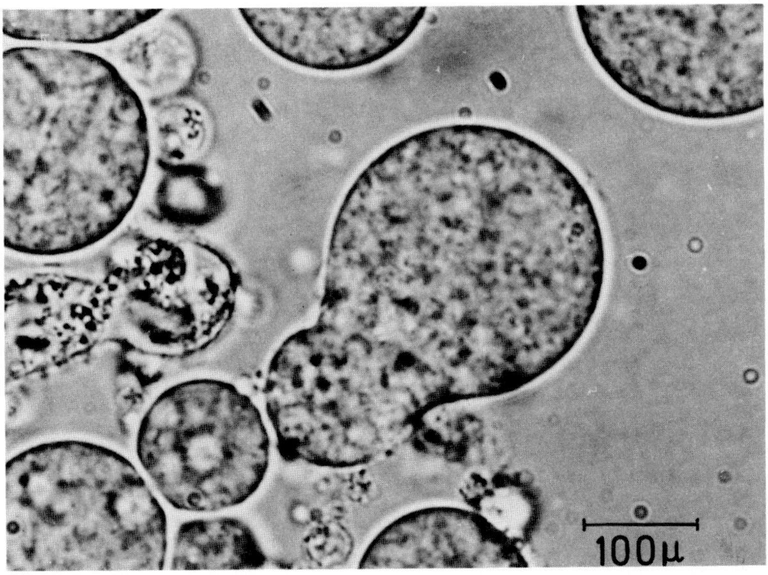

Fig. 1. A "budding" cell of the *N. crassa* "slime" mutant.

The cells are very sensitive to osmotic changes and readily burst upon dilution of the medium. This is illustrated in Fig. 2. The micrographs were taken at 30 sec. intervals after addition of buffer. In this respect the slime mutant cells closely resemble fungal spheroplasts. After 2 min practically all cells have burst leaving only granular debris of yet unknown composition. Under the conditions employed the mutant grows logarithmically from approximately 12 to 48 h after inoculation (Fig. 3), producing per ml growth medium about 3 mg of protein (precipitable with trichloracetic acid). The number of cells per ml increases by two orders of magnitude in 48 h to approx. 4×10^8 cell/ml growth medium, and then decreases due to the

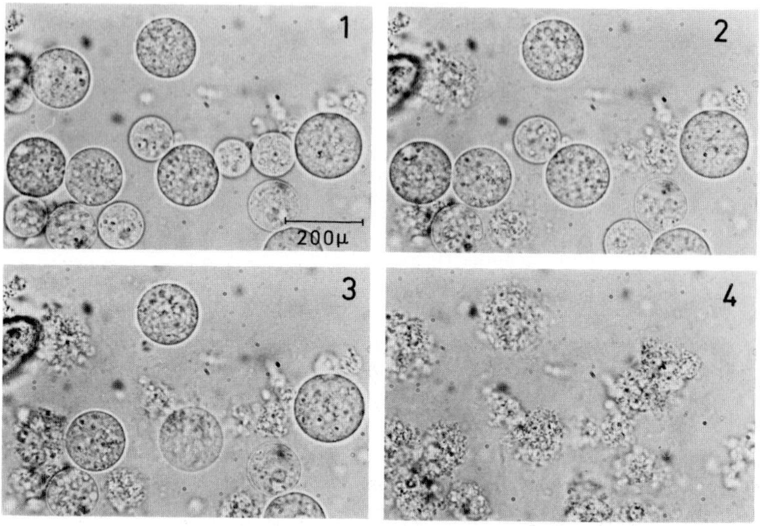

Fig. 2. A few drops of buffer were added to a drop of cell suspension to dilute the culture medium [1]. Subsequently, pictures were taken at 30 sec. intervals.

damage of the cells caused by agitation of the medium on reciprocal shakers (Fig. 4).

These results render the slime mutant a promising subject for studies on regulation of subcellular structures. It grows quickly and can easily be transferred from a culture medium into an induction medium.

The lack of rigid cell walls permits disruption of the cells by gentle methods and avoids prolonged incubation with snail gut enzymes. The slime phenotype of the mutant seems to be genetically stable; under our experimental conditions no wild type regenerates could be observed.

B. *Isolation of Subcellular Structures*

Table I lists a series of enzyme activities commonly used as markers for mitochondria or microbodies from plant and animal cells [2, 3, 12]. Fumarase, a mitochondrial enzyme, apparently tends to leak out of the particles since only 50% of its activity is recovered in the crude particulate fraction (P_{II}). For the first time the presence of remarkable amounts of activity of the enzymes catalase, uricase, D-amino acid oxidase and isocitrate lyase in particulate fractions of *N. crassa* can be reported. These enzymes are generally believed to be

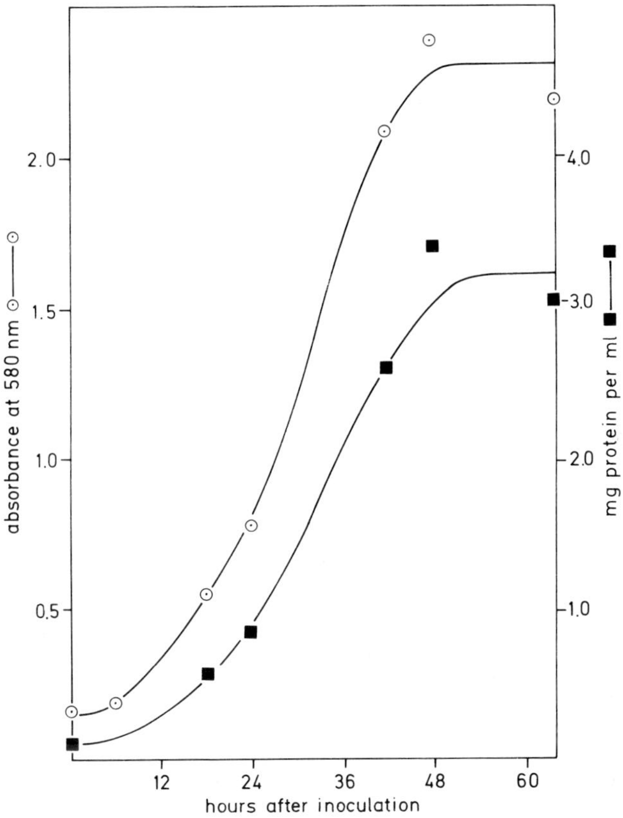

Fig. 3. Increase of absorbance and protein contents of a liquid culture of a *N. crassa* slime mutant.

located in microbodies in plant and animal cells [2, 3]. When the P_{II} -fraction is subjected to equilibrium density centrifugation two distinct and visible protein bands are obtained. A typical enzyme distribution pattern of such a sucrose density gradient is shown in Fig. 5.

The upper protein band sedimenting at a density of 1.16 ± 0.01 g/cm^3 coincides with the peak of fumarase activity indicating the presence of mitochondria in these fractions. No appreciable activity of any of the other enzymes sedimented in this region of the gradient. Catalase, D-amino acid oxidase, uricase and isocitrate lyase activities band together at a density of 1.20 ± 0.01 g/cm^3 corresponding to the second protein peak. Since

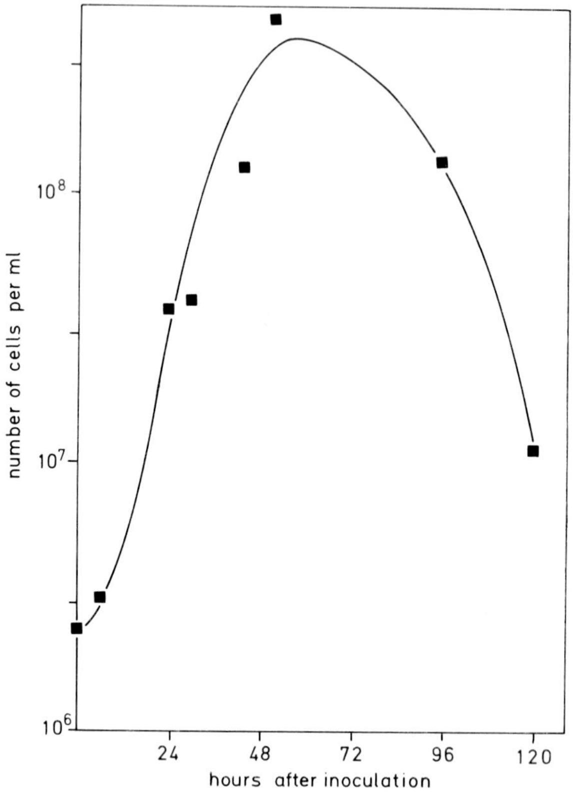

Fig. 4. Number of cells per ml growth medium in a liquid culture of a *N. crassa* slime mutant.

the enzymatic equipment as well as the density of these particles are in fair agreement with the properties of microbody fractions isolated from higher plants and animals [2, 3], this band is referred to as microbodies. The minor peak in this region of fumarase activity probably reflects a trapping of mitochondrial fragments by the microbodies or the adherence to the organelles of enzyme leaked out of the mitochondria.

In contrast to *N. crassa* wild type [8] the slime mutant clearly facilitates the ready separation of mitochondria from microbodies. Moreover, our results are interpreted to mean that there may exist a type of fungal microbody carrying peroxisomal enzymes such as catalase, uricase and D-amino acid oxidase as well as the glyoxysomal enzymes isocitrate lyase. Our findings also provide biochemical

TABLE I
Distribution of enzyme activities among subcellular fractions of a *N. crassa* slime mutant

	1,000 x g supernatant S_I	27,000 x g supernatant S_{II}*	27,000 x g pellet P_{II}*
Protein (mg)	311.6	156.6 (50.5)	152.9 (49.5)
Fumarase (μmoles/min)	5.74	2.9 (48.2)	3.11 (51.8)
Catalsae (μmoles/min)	4,025.0	623.0 (12.5)	4,347.0 (87.5)
Uricase (nmoles/min)	1,720.0	908.0 (60.2)	600.0 (39.8)
D-amino acid oxidase (μmoles/min)	72.2	34.1 (42.5	45.9 (57.5)
Isocitrate lyase (μmoles/min)	195.0	92.1 (55.0)	75.0 (45.0)

* Values in parentheses give the percentage of S_{II} + P_{II}.

confirmation of a preliminary observation by van Winkle [21] whose electron micrographs of the slime mutant show abundant numbers of microbody-like organelles. Ultimate proof of the identity of the organelles is currently being sought by electron micrography of the isolated particles.

In summary, our results demonstrate that *N. crassa* slime is a useful organism for the investigation of subcellular structures in eucaryotic cells. Employing a nutritionally inert osmoticum which allows the undisturbed growth of the mutant *N. crassa* "slime" should also be ideal for the study of the induction and regulation of cellular organelles. Since little is known about the membrane or possible cell wall relicts surrounding the mutant's cells a detailed analysis may shed some light on the mechanisms of fungal cell wall synthesis. Experiments along these lines are currently being conducted in this laboratory.

Acknowlegements

The skilled and dedicated technical assistance of Frl. Gisela Mundorf is appreciated. This work was supported by a grant from Deutsche Forschungsgemeinschaft.

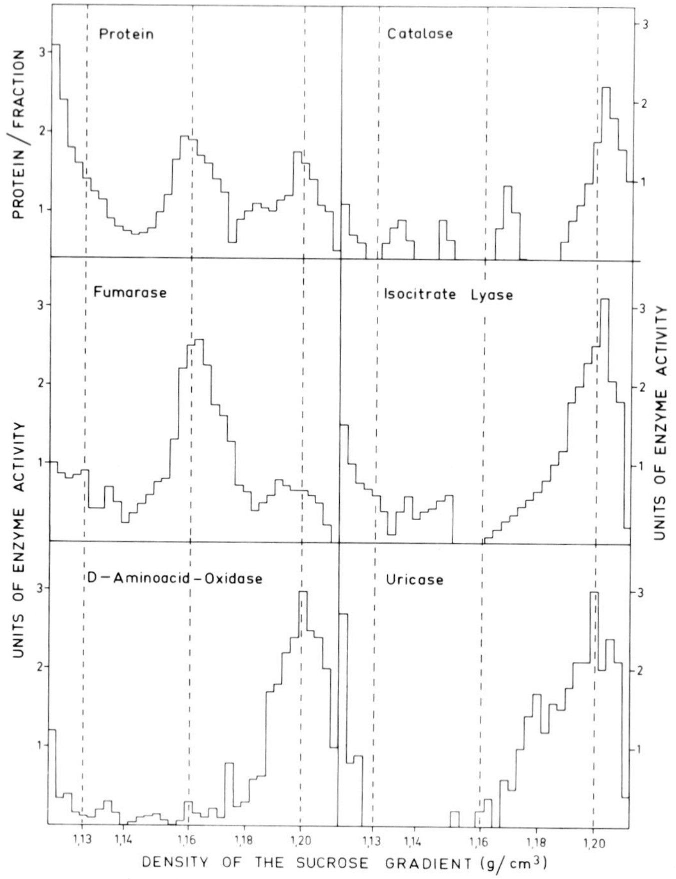

Fig. 5. Isopycnic density gradient centrifugation of a crude particulate fraction from *Neurospora crassa* slime mutant. Enzyme activity is given in the following units (per fraction): Fumarase, nmoles (x12) fumarate formed/min; D-amino acid oxidase, μmoles (x0.84) pyruvatephenylhydrazone formed/min; catalase, μmoles (x8.4) H_2O_2 destroyed/min; isocitrate lyase, μmoles (x0.28) glyoxylate-phenylhydrazone formed/min; uricase nmoles (x0.28) urate utilized/min. Protein is given in μg (x31)/fraction.

References

1. Duell, E. A., Inoue, S. and Utter, M. F. (1964). *J. Bact.*, **88**, 1762.
2. Tolbert, N. E. (1971). *A. Rev. Pl. Physiol.*, 22, 45.
3. De Duve, Ch. (1969). *Proc. Roy. Soc.*, **B173**, 71.
4. Duntze, W., Neumann, D., Gacedo, J. M., Atzpodien, W. and Holzer, H. (1969). *Eur. J. Biochem.*, **10**, 83.

5. Perlman, P. S. and Mahler, H. R. (1970). *Arch. Biochem. Biophys.*, **136**, 245.
6. Szabo, A. S. and Avers, Ch. J. (1969). *Ann. N.Y. Acad. Sci.*, **168**, 302.
7. Cartledge, T. G. and Lloyd, D. (1972). *Biochem. J.*, **126**, 126.
8. Kobr, M. J., Vanderhaeghe, F. and Combepine, G. (1969). *Biochem. biophys. Res. Commun.*, 37, 640.
9. Autrich, H. (1966). *Wiss. Z. Karl-Marx-Uni. Lpz.*, **15**, 359.
10. Greene, R. C. and Mitchell, H. K. (1957). *Arch. Biochem. Biophys.*, **70**, 603.
11. Sjogren, R. E. and Romano, A. H. (1967). *J. Bact.*, **93**, 1638.
12. Theimer, R. R. and Beevers, H. (1971). *Pl. Physiol.*, **47**, 246.
13. Huang, A. H. C. and Beevers, H. (1971). *Pl. Physiol.*, **48**, 637.
14. Emerson, S. (1963). *Genetica*, **34**, 162.
15. Vogel, J. H. (1956). *Microb. Genet. Bull.*, 13, 42.
16. Woodward, V. W. and Woodward, C. K. (1968). *Neurospora Newsl.*, 13, 18.
17. Lowry, O. H., Rosebrough, N. J., Farr, A. L. and Randall, R. J. (1951). *J. biol. Chem.*, **193**, 265.
18. Bergmeyer, H. U. (ed.) (1970). Methoden der enzymatischen Analyse I, p. 640. Verlag Chemie, Weinheim.
19. Dixon, G. H. and Kornberg, H. L. (1959). *Biochem. J.*, 72, 3.
20. Racker, E. (1950). *Biochem. biophys. Acta*, **4**, 211.
21. Van Winkle, B. E. (1969). *Neurospora Newsl.*, **14**, 5.

PART IV

Protoplasts of Moulds

The Enzymes of Polyphosphate Metabolism in Protoplasts and some Subcellular Structures of *Neurospora crassa*

I. S. KULAEV

Department of Plant Biochemistry, Moscow State University and Institute of Biochemistry and Physiology of Microrganisms of the USSR Academy of Sciences, Pustchino-on-the-Oka, Moscow Region, USSR

During recent years my laboratory has been engaged in studying the physiological role of high molecular inorganic polyphosphates in the metabolism of lower organisms, especially in fungi. According to what we know today [1-3], these compounds play an important role in conservation of phosphate and energy, and regulation of the level of intracellular orthophosphate (Fig. 1).

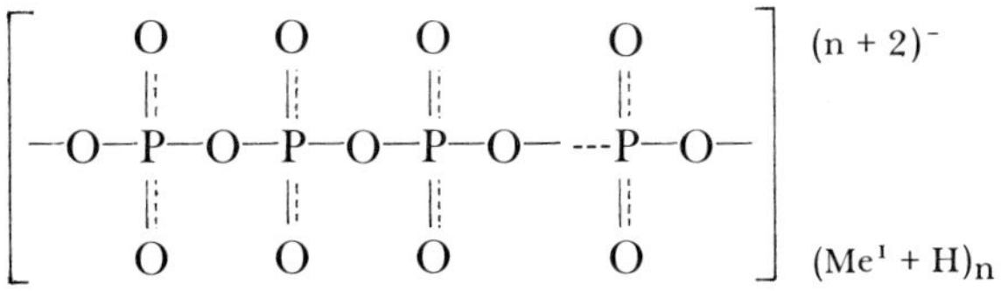

Fig. 1. The structure of inorganic polyphosphates. Me^1 = univalent cations, n = the number of condensed molecules of orthophosphate.

Also, they aparently participate, as a source of energy, in the processes of translocation of ions and substances across the plasma membrane of the cells of microorganisms.

To study in detail the physiological role of polyphosphates, we have investigated the intracellular localization of both the polyphosphates themselves and the enzymes participating in their metabolism in various fungi.

The most useful method of isolating intact cellular structures from fungal cells is the one involving various enzymatic preparations capable of lysing rigid cell walls of these organisms and giving protoplasts as an obligatory intermediate.

Using this method for studying the localization of different fractions of polyphosphates in the cells of *Neurospora crassa,* we demonstrated previously [4] the existence of three different pools of polyphosphates (Fig. 2).

One of them is an alkali-soluble high polymer fraction localized between the cell wall and the plasma membrane. The second includes less polymer salt-soluble polyphosphates and is to be found in the cell nucleus. Finally, the third pool consists of the least polymer acid-soluble polyphosphates localized exclusively in volutine-like granules.

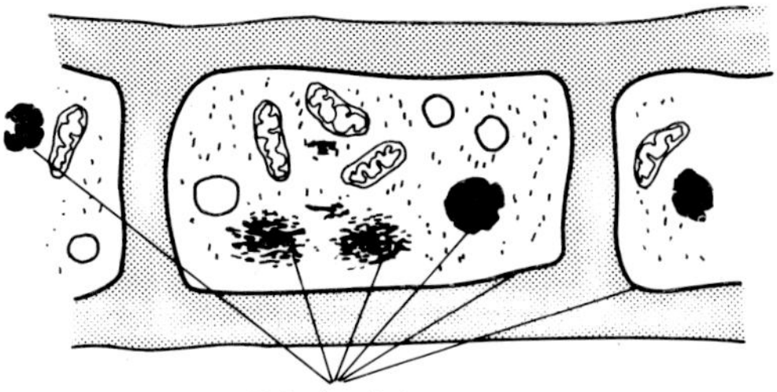

Fig. 2. Localization of inorganic polyphosphates in fungal cell.

Thus, these data have disproved the concept [2] that the high molecular polyphosphates in fungi and apparently in other microorganisms can be found only in the metachromatic volutine granules.

Our recent findings associated with the localization of the enzymes of polyphosphate metabolism in different cellular structures of *N. crassa* will be reported in this paper.

For this purpose, G. I. Konoshenko, A. M. Umnov, and myself [5], obtained protoplasts from 18-h mycelium of *N. crassa Abbot 4A* with the help of an enzymic preparation from *Helix pomatia* as described by Bachmann and Bonner [6].

Using differential centrifugation and ultracentrifugation in sucrose density gradient [7-11], as is shown in Fig. 3, we obtained highly purified preparations of nuclei, mitochondria, volutine-like granules, microsomes and hyaloplasma from the protoplasts.

The nuclei, mitochondria and microsomes were obtained from protoplasts in independent experiments, in which three different methods of fractionation and purification were used. The purity and intactness of the subcellular structures obtained were controlled by electron microscopy as well as by some biochemical tests. The data of an electron-microscope study of intact cells, of *N. crassa* protoplasts, nuclei and mitochondria are presented in Figs 4-7.

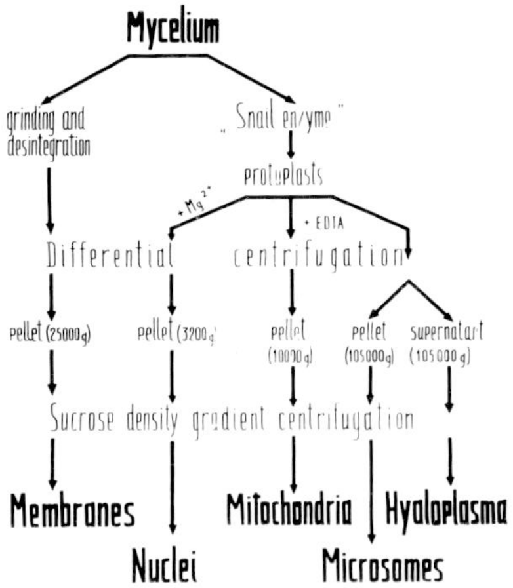

Fig. 3. Isolation of some subcellular structures of *N. crassa* (General scheme).

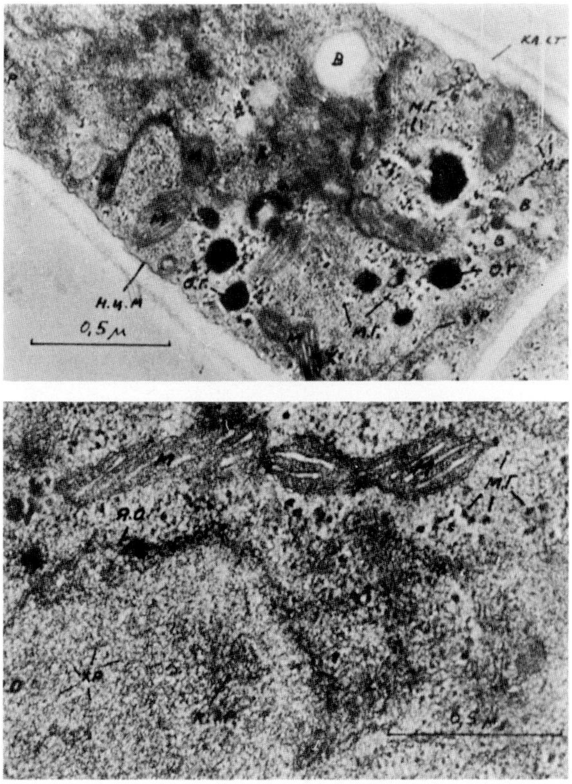

Fig. 4. Electron micrograph of *N. crassa* mycellium.

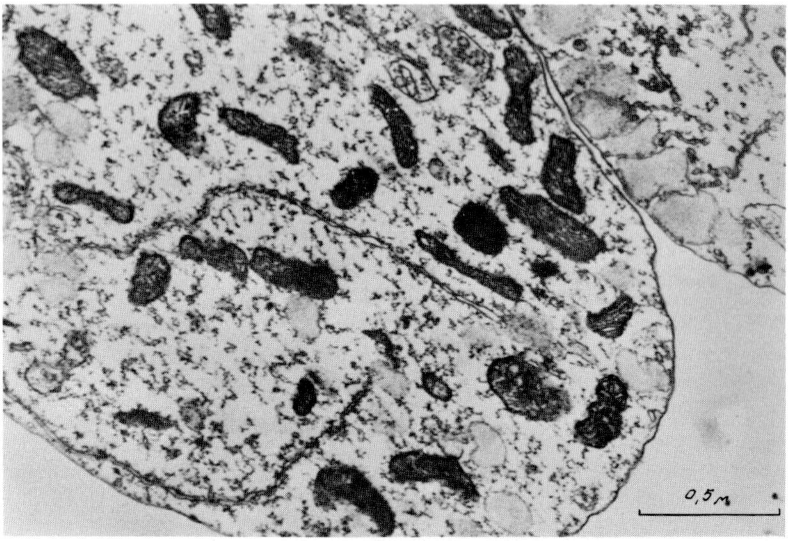

Fig. 5. Electron micrograph of *N. crassa* protoplasts.

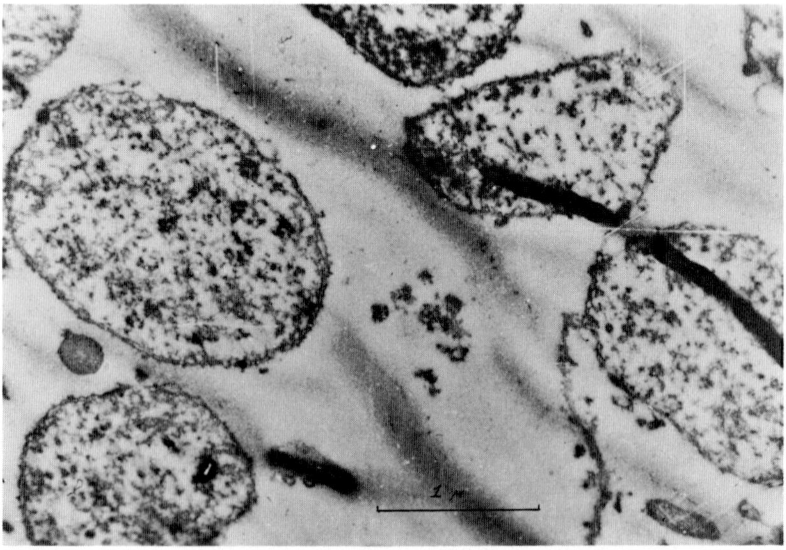

Fig. 6. Electron micrograph of *N. crassa* nuclei.

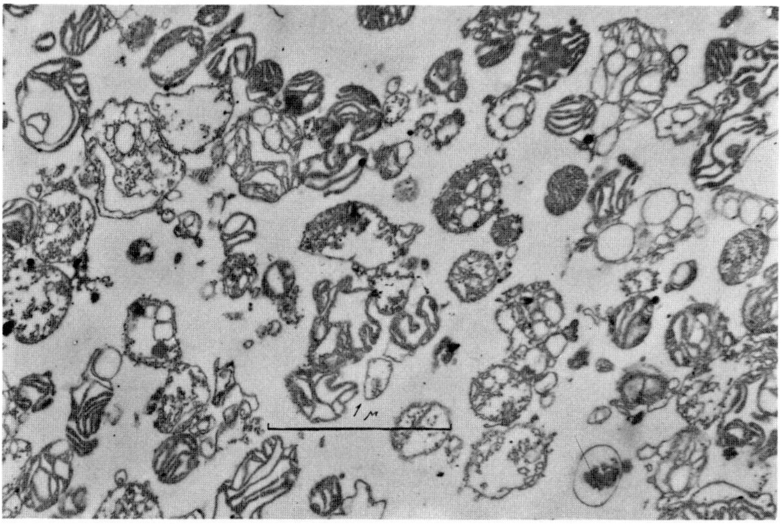

Fig. 7. Electron micrograph of *N. crassa* mitochondria.

We also made an attempt to obtain the preparations of *N. crassa* plasma membranes from protoplasts by using the methods of Boulton [12], Mendoza and Villanueva [13-14] and Longley *et al.* [15] for yeast cells. For this purpose the protoplasts were disrupted by osmotic shock and in a high-speed disintegrator [16]. However, we did not succeed in isolating plasma membranes from protoplasts of *N. crassa* because of the low yield of protoplasts from the mycelium. The second (and the main) reason is that isolation of plasma membranes from protoplasts of *N. crassa* resulted in formation of very long fragments, which always sedimented together with mitochondria and were hard to separate. So, we used the method proposed by Matile *et al.* [17] for obtaining sufficient quantities of highly purified fractions of plasma membranes of *N. crassa.* As can be seen in the fractionation scheme (Fig. 3), in the latter case we first crushed the *N. crassa* mycelium by grinding it in a mortar in the isolation medium together with glass beads No. 15 and then in a high-speed disintegrator [16]. After the differential centrifugation and purification in the sucrose density gradient we obtained a sufficiently pure membrane fraction, the electron micrograph of which is presented in Fig. 8.

The data on the character of the ATPase activity of these preparations, as shown in Table I, also testified to the fact that in

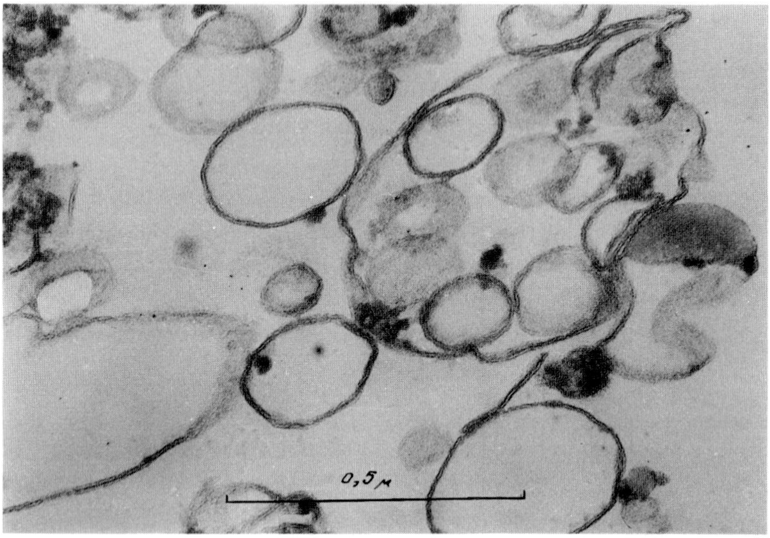

Fig. 8. Electron micrograph of purified fraction of *N. crassa* plasma membrane.

TABLE I
Properties of ATPase of *N. crassa* plasma membranes

Additions	Activity (mE)
Control	0
+3 µmoles $MgCl_2$	276
+6 µmoles $MgCl_2$	241
+10 µmoles KCl	39
+3 µmoles $MgCl_2$ + 10 µmoles KCl	426
+3 µmoles $MgCl_2$ + 10 µmoles KCl + 1 µg oligomycine	411
+3 µmoles $MgCl_2$ + 10 µmoles KCl + 10 µg oligomycine	399

Incubation medium: 2mM ATP, 0.025 mM Tris-HCl, pH 7.3. Total volume—1.0 ml.

this case we dealt with a fraction of plasma, and not of mitochondrial or nuclear membranes.

We could see that the ATPase of the membrane preparation obtained was K^+, Mg^{2+}-dependent and oligomycin inhibited it very little. This means that the contamination of the fraction with mitochondrial membranes was negligible.

In the plasma membranes thus obtained and also in intact cells, protoplasts, main subcellular structures and fractions, the activity of the following enzymes of polyphosphate metabolism were studied: polyphosphate phosphohydrolase, polyphosphate polyphosphohydrolase, tripolyphosphate phosphohydrolase and 1,3-diphosphoglycerate : polyphosphate phosphotransferase. According to our data, other enzymes known to take part in the polyphosphate metabolism—ATP: polyphosphate phosphotransferase and polyphosphate: glucose 6-phosphate phosphotransferase—are not present in *N. crassa*.

$$(\text{poly-P})_n + H_2O \rightarrow (\text{poly-P})_{n-1} + P_i$$

Fig. 9. The scheme of polyphosphate phosphohydrolase reaction.

First of all, we investigated the localization of polyphosphate phosphohydrolase (polyphosphatase) which splits off one phosphate group from the end of the chain as is shown in Fig. 9. As can be seen from the data obtained in several independent experiments and shown in Table II, the polyphosphatase activity of protoplasts is approximately twice as low as that of intact cells.

TABLE II
Specific activity of polyphosphatase in cells, protoplasts and some subcellular structures of *N. crassa*

Structures	Polyphosphatase (mE/mg of protein)
Cells	30-50
Protoplasts	15-30
Nuclei	0
Mitochondria	0
Plasma membrane	5
Microsomes	2
Hyaloplasma	40

Incubation medium: Tris-HCl, pH 7.2—50 μmoles; $MgCl_2$—0.5 μmole; KCl—100 μmoles; polyphosphate ($\bar{n}$ = 290)—0.1-0.15 mg; protein—100-200 mg. Total volume—0.5 ml. Incubation time at 37° C—20 min.

This activity is entirely absent in the nuclei and mitochondria. It can be observed in the membranes, but its intensity is relatively low. Nearly all the polyphosphatase activity remaining in protoplasts was found to be present in the fraction of hyaloplasma upon further fractionation. These data allowed us to conclude that at least a part of polyphosphatase in *N. crassa* is located between the cell wall and the plasma membrane and the other part is connected with the protoplast in a rather labile way.

The next series of experiments was aimed at elucidating the effect of short-time washing (the structures preserve their intactness) with the isolation medium containing a small quantity (0.02%) of detergents, on polyphosphatase activity of protoplasts. Table III shows that the washing of protoplasts with a deoxycholate solution produced no effect on their polyphosphatase activity, whereas a short-time washing with a solution of Triton X-100 taken in the same concentration led to a considerable decrease in the activity.

TABLE III
The effect of detergents on the polyphosphatase activity of protoplasts

	Enzyme activity	
	In mE per gram of intact cells	In per cent from enzyme activity in intact cells
Intact cells	1020	100
Protoplasts	700	70
Protoplasts + sodium deoxycholate	700	70
Protoplasts + Triton X-100	280	27
Protoplasts + sodium deoxycholate + Triton X-100	320	31

Incubation time at room temperature—15 min. Detergent concentration—0.02%.

To establish whether Triton X-100 inactivates polyphosphatase or washes the enzyme off the surface of the protoplast, S. E. Mansurova and G. I. Velichko carried out an experiment, the results of which are shown in Table IV. Protoplasts of *N. crassa* were incubated in the isolation medium for 15 min without any additions for 0.02% and 0.1% of Triton X-100 were added. Then the protoplasts were separated from the incubating medium by centrifugation for 10 min. After that the polyphosphatase activity was assayed in all the samples both in control and the incubation medium and in the protoplasts themselves The resulting data revealed that the enzyme was almost completely washed out of the protoplasts in the course of incubation with the 0.02% solution of Triton X-100.

The fact that, according to the results obtained, such incubation did not affect the intactness of protoplasts allows the conclusion to be made that polyphosphatase in the cells of *N. crassa* is localized entirely outside the plasma membrane and seems bound with it loosely.

Inactivation of this enzyme, as follows from the data of the same experiment (Table IV), does take place in the presence of Triton X-100. However, this detergent has a significant inactivating effect only in concentrations exceeding 0.1%.

TABLE IV
The effect of Triton X-100 on the release of polyphosphatase from *N. crassa* protoplasts

Additions	Total activity of enzyme (mE)			
	In protoplasts	In incubation medium	Sum	Loss of activity in per cent
Control (without detergent)	210	20	230	
+0.02% Triton X-100	10	160	170	26
+0.1% Triton X-100	0	40	40	83

Incubation time at room temperature—25 min.

It is important to mention that, according to our preliminary data, the removal of Triton X-100 from the solution, containing polyphosphatase inactivated by it, leads to reactivation of the enzyme.

Then we [5] studied the localization of another enzyme, tripolyphosphate phosphohydrolase (tripolyphosphatase), hydrolysing in a specific manner inorganic tripolyphosphate [18] via the reaction presented in Fig 10.

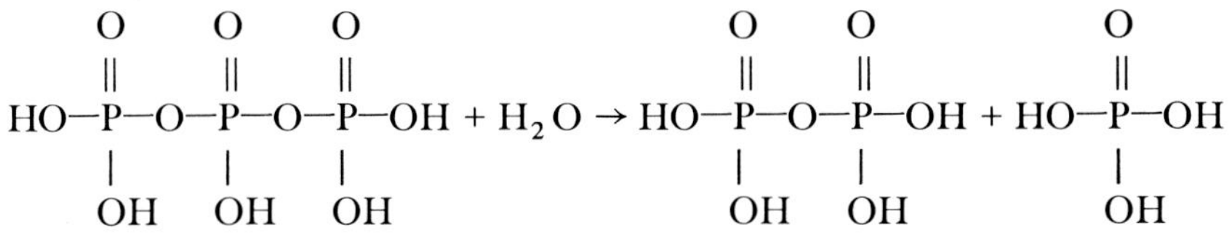

Fig. 10. The scheme of tripolyphosphate phosphohydrolase reaction.

It was shown (Table V) that lysis of the cell wall by "snail enzyme" preparation in the case of protoplasts does not lead to any loss of tripolyphosphatase activity of *N. crassa* cells.

It is interesting to point out that, apart from hyaloplasma, the

TABLE V

Specific activity of tripolyphosphatase in cells, protoplasts and some subcellular structures of *N. crassa*

Structures	Tripolyphosphatase (mE/mg of protein)
Cells	90-150
Protoplasts	100-160
Nuclei	0
Mitochondria	150-200
Plasma membrane	6
Microsomes	20
Hyaloplasma	160

Incubation medium: Tris-HCl, pH 7.2–50 μmoles; $MgCl_2$–0.5 μmole; tripolyphosphate–0.1-0.15 mg; protein–100-200 mg. Total volume–0.5 ml. Incubation time at 37°C–20 min.

activity of the enzyme was observed only in mitochondria and not in other subcellular structures obtained from protoplasts. The data in Table VI show that mitochondria possess 70% to 100% of tripolyphosphatase activity per gram of cells (calculated on the basis of cytochrome "a" content).

TABLE VI

Tripolyphosphatase activity of *N. crassa* mitochondria

Structures	Total activity of enzyme	
	In mE per gram of intact cells	In per cent from activity of intact cells
Cells	3400-3800	100
Protoplasts	3400-4400	100-117
Mitochondria	2300-3800	70-100

Even the most thorough purification of the preparation of mitochondria in the sucrose density gradient (Fig. 11) does not make it possible to separate the tripolyphosphatase activity from mitochondrial protein.

Thus the data obtained allow us to conclude that the specific enzyme which splits tripolyphosphate in *N. crassa* is mainly, if not entirely, localized in mitochondria and apparently plays some

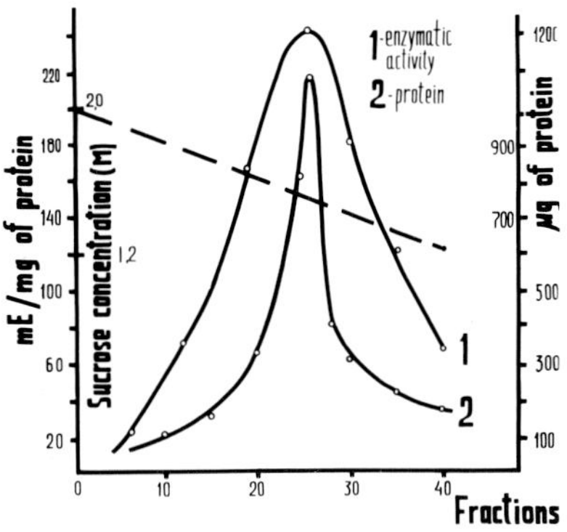

Fig. 11. Distribution of tripolyphosphatase in sucrose gradient of crude preparation of mitochondria.

specific role in their phosphorus and energy metabolism. It is not excluded, for instance, that tripolyphosphatase, like ATPase and pyrophosphatase [19], may play an important role in translocation of ions across mitochondrial membrane and in other energy-dependent processes.

The next stage of our work conducted together with G. I. Konoshenko, E. K. Chernyshova and M. S. Kritsky was to study the localization of polyphosphate polyphosphohydrolase (polyphosphate depolymerase) in *N. crassa* cells [20].

This enzymes which breaks phosphoanhydride bonds in the middle part of high polymer polyphosphates catalyses the reaction shown in Fig. 12. This work was carried out with two different strains of *N. crassa.* Since the data obtained in both cases were identical we shall confine ourselves only to those obtained with one strain.

$$(\text{poly-P})_n + H_2O \rightarrow (\text{poly-P})_{n\text{-}x} + (\text{poly-P})_x$$

Fig. 12. The scheme of polyphosphate polyphosphohydrolase reaction.

Table VII shows that the removal of the cell wall of *N. crassa* with the help of "snail enzyme" leads to a nearly complete disappearance of the enzymatic activity from protoplasts. The polyphosphate depolymerase activity was not detected in mitochondria. Trace amounts of it revealed in nuclei and microsomes are likely to be the result of contamination of the preparations of the cellular structures with fragments of plasma membrane.

TABLE VII

Specific activity of polyphosphate depolymerase in cells, protoplasts and some subcellular structures of *N. crassa*

Substrates	Polyphosphate depolymerase activity (in mE/mg of protein)					
	Intact cells	Proto-plasts	Nuclei	Mito-chondria	Micro-somes	Hyalo plasma
Polyphosphate (ñ = 290)	8,3	0,6	0,2	0,0	2,2	0,0
Polyphosphate (ñ = 180)	9,4	1,3	0,2	0,0	–	0,0

Incubation medium: sodium acetate, pH 3.2–3.4-0.3 μmole; polyphosphate–25 μmole of acid-liable phosphorus; protein–0.5-1.0 mg. Total volume–3.0 ml. Incubation time at 30° C–25-30 min.

Thus the findings obtained testify to the fact that polyphosphate depolymerase as well as polyphosphatase are localized in the periphery of the *N. crassa* cell between the cell wall and the plasma membrane, just where the most high polymer polyphosphates are known to be situated [4] and where their synthesis is most likely to take place. Since polyphosphate depolymerase takes part in the breakdown of high polymer polyphosphates into fragments of lower polymericity, which are known to be localized inside *N. crassa* cells (in the nuclei and in volutine-like granules), it is not excluded that these enzymes may also participate in translocation of smaller fragments of polyphosphates, obtained with their participation across the plasma membrane inside the cell.

We have studied the localization of another enzyme in *N. crassa* cells in collaboration with G. I. Konoshenko [21]. It is 1,3-diphosphoglycerate : polyphosphate phosphotransferase, a new enzyme that we have recently detected in *N. crassa* [22]. It catalyses the reaction shown in Fig. 13.

$$\begin{array}{lcl} \mathrm{COO{\sim}\textcircled{P}} & & \mathrm{COOH} \\ \vert & & \vert \\ \mathrm{CHOH} + (\mathrm{poly\text{-}P})_n & \rightarrow & \mathrm{CHOH} + (\mathrm{poly\text{-}P})_{n+1} \\ \vert & & \vert \\ \mathrm{CH_2O\text{-}\textcircled{P}} & & \mathrm{CH_2O\text{-}\textcircled{P}} \end{array}$$

Fig. 13. The scheme of 1,3-diphosphoglycerate : polyphosphate phosphotransferase reaction.

TABLE VIII

Specific activity of 1,3-dip hosphoglycerate polyphosphate-phosphotransferase in cells, protoplasts and some subcellular structures of *N. crassa*

Structures	Enzymatic activity (In mE/mg of protein)
Cells	0,170
Protoplasts	0,180
Nuclei	0
Mitochondria	0,050
Microsomes	0,037
Hyaloplasma	0,028

Incubation medium: glycil-glycine buffer, pH 7.0—15 μmoles; $MgCl_2$-6 μmoles; polyphosphate $\bar{n}$ = 75)—0.015 μmole; glyceraldehyde phosphate dehydrogenase—14 mg of protein; aldolase—12 mg of protein; fructose-1,6-diphosphate—5,2 μmoles; $K_2H^{32}PO_4$—50 mC; KH_2PO_4—8.0 μmoles; protein—0.2-1 mg. Total volume—0.6 ml; incubation time at 37°C—30 min.

Table VIII shows that the specific activities of the enzymes in intact cells and in protoplasts are identical. This allows us to assume that their localization is intracellular.

Further fractionation of protoplasts showed that the enzyme is absent in the cell nuclei, whereas it was always present in the other subcellular structures studied. In all the experiments the largest quantity of the enzyme was detected in crude fractions of mitochondria. It became clear that about 30% to 40% of the activity observed in the extracts of intact cells was contained in those preparations. However, the evidence obtained after the purification of the preparation of mitochondria from *N. crassa* in the sucrose density gradient (Fig. 14) clearly indicates that the 1,3-diphosphoglycerate : polyphosphate phosphotransferase activity is localized not in the mitochondria themselves but in the particles which sediment together with them. It should be noted that polyphosphates were also found in the same fractions. According to their behaviour during fractionation and their characteristics, these particles were identical to volutine-like granules that we had earlier found in the crude fraction of *N. crassa* mitochondria [4].

The occurrence of 1,3-diphosphoglycerate : polyphosphate phosphotransferase in polyphosphate-containing volutine-like granules may mean that they play a role as regulators of the intracellular orthophosphate level in the cells and control the rate of glycolysis.

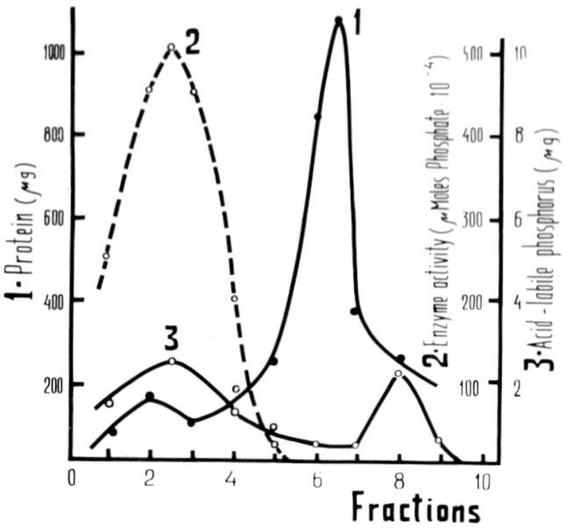

Fig. 14. Distribution of 1,3-diphosphoglycerate : polyphosphate phosphotransferase in sucrose gradient of crude preparation of mitochondria.

Thus all mentioned above permit us to conclude that different fractions of high polymer polyphosphates are metabolized in *N. crassa* cells with the help of various enzymatic systems, and they carry out specific functions in every case.

Polyphosphatase and polyphosphate depolymerase play the most important role in the metabolism of the most polymer polyphosphates localized between the cell wall and the plasma membrane. We have definite indications that this fraction of polyphosphates in fungi is closely connected with the sugar uptake and biosynthesis of cell wall polysaccharides [23].

Apparently the metabolism of polyphosphates localized in volutine-like granules is primarily connected with glycolysis.

The mechanisms of biosynthesis and transformations of high-molecular polyphosphates localized in the nuclei is still an enigma. In the case of *N. crassa,* none of the known enzymes of polyphosphate metabolism have been found in those cellular structures However, this fraction of polyphosphate in *N. crassa* is very closely associated with biosynthesis of nucleic acids, according to the data recently obtained in our laboratory [24].

As to the mitochondria of *N. crassa*, it seems to us that those cellular structures have nothing to do with the metabolism of high molecular polyphosphates. However, our evidence permits us to assume that inorganic tripolyphosphate may play an important role in the metabolic processes taking place in mitochondria.

References

1. Kulaev, I. S. and Belozersky, A. N. (1962). *Isv. Acad. Nauk. SSSR*, **3**, 354; **4**, 502.
2. Harold, F. M. (1966). *Bact. Rev.*, **30**, 772.
3. Kulaev, I. S. and Afanasieva, T. P. (1969). *Antonie van Leewenhoek*, **35**, suppl. B29.
4. Kulaev, I. S., Afanasieva, T. P., Krasheninnikov, I. A. and Mansurova, S. E. (1970). *Acta Facult. Med. Univer. Brunensis*, 37, 81.
5. Kulaev, I. S., Konoshenko, G. I. and Umnov, A. M. (1972). *Biokhimiya*, 37, 227.
6. Bachmann, B. J. and Bonner, D. M. (1959). *J. Bact.*, **78**, 550.
7. Kulaev, I. S., Krasheninnikov, I. A. and Kokurina, N. A. (1966). *Biokhimiya*, **31**, 850.
8. Kulaev, I. S., Krasheninnikov, I. A. and Polyakov, V. Yu. (1970). *In:* Cell Nucleus and Its Ultrastuctures, p. 310 (Zbarsky, I. B., ed.) "Nauka", Moscow.
9. Rozijn, T. H. and Tonino, G. I. M. (1964). *Biochim. biophys. Acta*, **91**, 105.
10. Luck, D. J. L. (1963). *J. Cell Biol.*, **16**, 483.
11. Krasheninnikov, I. A., Kulaev, I. S., Konoshenko, G. I. and Biryusova, V. I. (1968). *Biokhimiya*, **33**, 800.
12. Boulton, A. A. (1965). *Expl Cell Res.*, **37**, 343.
13. Mendoza, C. G. and Villanueva, J. R. (1962). *Microbiol. Espan.*, **15**, 139.
14. Mendoza, C. G. and Villanueva, J. R. (1967). *Biochim. biophys. Acta*, **135**, 189.
15. Longley, R. P., Rose, A. H. and Knights, B. A. (1968). *Biochem. J.*, **108**, 401.
16. Kulaev, I. S., Umnov, A. M. and Konoshenko, G. I. (1972). *Cytologiya*, **14**, 1038.
17. Matile, Ph., Moor, H. and Muhlethaler, R. (1967). *Arch. Microb.*, **58**, 201.
18. Kulaev, I. S. and Konoshenko, G. I. (1971). *Biohimiya*, **36**, 1175.
19. Baltscheffsky, H., Baltscheffsky, M. and Stedingk, L.-V. von (1969). *In:* Progress in Photosynthesis Research, Vol. 3, p. 1313 (Metzner, H., ed.).
20. Kulaev, I. S., Konoshenko, G. I., Chernyshova, E. K. and Kritsky, M. S. (1972). *Dokl. Akad. Nauk SSSR*, **206**, 1.
21. Kulaev, I. S. and Konoshenko, G. I. (1971). *Dokl. Acad. Nauk SSSR*, **200**, 477.
22. Kulaev, I. S. and Bobyk, M. A. (1971). *Biokhimiya*, **36**, 426.
23. Kulaev, I. S., Vagabov, V. M. and Tziomenko, (1972). *Dokl. Akad. Nauk SSSR*, **204**, 734.
24. Kritsky, M. S., Chernyshova, E. K. and Kulaev, I. S. (1970). *Dokl. Akad. Nauk SSSR*, **192**, 1166.

Formation and Regeneration of Protoplasts of *Pythium* PRL 2142

J. H. SIETSMA* and W. R. DE BOER

Laboratory of Microbiology, University of Amsterdam
Amsterdam, The Netherlands

The regeneration of fungal protoplasts is of current interest because cell wall formation is probably an important factor in morphogenesis. To study this process it is necessary to have a system for making protoplasts in large quantities.

It was known from previous studies [6] that the main components of the cell wall of *Pythium* were cellulose, a β-(1-3) glucan and the almost ubiquitous present branched β-(1-3) glucan with β-(1-6) linkages at the branching points. As minor components lipoid material was detected for almost 10% and some protein. Protoplasts could quite easily be obtained by incubating the mycelium with a combination of two enzymes, cellulase and exo-laminaranase. Not essential for the protoplast forming were lipase and endo-laminaranase, but the latter two enzymes stimulated this process progressively. With the same enzyme complex protoplasts could be obtained from several other Oomycete species. The cell wall composition of these species seems to be quite similar.

The enzyme complex for preparing protoplasts was usually obtained by growing *Streptomyces satsumaensis* with dead *Pythium* mycelium as carbon source. The disadvantages of this method are that it takes a long incubation time (almost 24 h) to get a sizeable number of protoplasts, and the preparation and purification of the enzyme complex is very laborious. A better method was found in using commercially available helicase in combination with *Aspergillus niger* cellulase in order to make large quantities of protoplasts of *Pythium.* Although helicase itself contained a considerable amount of cellulase activity, addition of this particular cellulase was essential for the protoplast formation. Presumably this could be due to a difference in specificity of both cellulases. Chromatographic analysis of the products obtained by hydrolysing carboxy methyl cellulose with helicase or with *A. niger* cellulase supports this conclusion. The

*Present address: Botanical Laboratory, Rijksuniversiteit Groningen, Haren, The Netherlands.

former enzyme complex seemingly contains a cellulase with a more exo character and the latter one with a more endo type activity. Although lipase activity was not detected in the *A. niger* cellulase preparation, it is still possible that the latter enzyme preparation contains an unknown substance, which promotes protoplast formation. Further it is quite likely that cellulose in the cell wall of *Pythium* has a crystalline nature and very specified enzymes are needed to attack the crystalline structure of a substance [9]. For effective protoplast formation not only an adequate enzyme system able to hydrolyse the cell wall is needed, but also a good osmotic stabilized medium, to prevent the protoplasts from bursting. Several factors affecting protoplast formation are seen in Table I. Inorganic salts, like $MgSO_4$, NaCl, KCl, NH_4Cl were very effective osmotic stabilizers when added in the suitable concentration. Organic stabilizers strongly inhibited the protoplast formation. The osmolarity of all the solutions was kept equal.

Several workers have shown that the sulfhydryl compounds like 2-mercaptoethanol, cysteamine, dithiotreitol, stimulate the degradation of yeast and fungal walls [1, 3, 7, 8]. *Pythium* cell wall was not sensitive for this treatment. Neither chelating agents, like EDTA or citrate, nor variation of the pH or temperature influenced protoplast formation. On the other hand, short incubation (30 min) of the mycelium with a detergent strongly influenced the protoplast formation. Very young mycelium (not older than 24 h) was not affected by this treatment, but older mycelium could not be transformed into protoplasts without pretreatment with a detergent. The best protoplast preparation was obtained by pretreating the mycelium with Triton X 100 at a concentration of 0.05% in 0.4 M NaCl for 30 min, then collecting the mycelium by filtration; it was thoroughly washed and suspended in 0.4 M NaCl and finally helicase (1.2 mg/ml) and *A. niger* cellulase (2 mg/ml) were added. After 4 h of incubation, numerous protoplasts were obtained, which could be separated from mycelium debris by filtration through glass wool, precipitated by gentle centrifugation and washed. The fact that Triton X 100 strongly stimulated protoplast forming, especially in older mycelium, and also that lipase enhanced this process, indicate that the cell wall could be covered with a protective layer of lipoid material. The sequence of events happening during protoplast formation is shown in Fig. 1.

For some other fungi and yeasts, like *Geotrichum candidum*, *Saccharomyces cerevisiae* and others, protoplast formation is stimulated by sulfhydryl compounds and in some cases also by

TABLE I

Factors effecting protoplast formation

Osmotic stabilizer in the incubation medium	No pretreatments No additions	Pretreatments of the mycelium		Additions to the incubation medium	
		SH-compounds[a]	detergents[b]	SH-compounds	detergents
KCl (0.4 M)	+++[c]	+	++++	+	—[d]
NaCl (0.4 M)	+++	+	++++	+	—
NH_4Cl (0.4 M)	++	—	++++	+	—
$MgSO_4$ (0.65 M)	++	+	++++	+	—
Sorbitol (0.65 M)	+	—	+	—	—
Mannitol (0.65 M)	—	—	+	—	—
Sucrose (0.65 M)	—	—	—	—	—

[a] SH-compounds used were 2-mercaptoethanol, cysteamine or dithiothreitol in a concentration of 250 mM during pretreatment, or of 10 mM when added to the incubation medium. [b] Detergents used were desoxycholate, laurylsulfate or Triton X 100 in the same concentrations as the SH-compounds. [c] —: no protoplasts after 24 h; +: few protoplasts after 8 h; ++: numerous protoplasts after 6 h; +++: numerous protoplasts after 4 h; ++++: numerous protoplasts after 4 h also with older mycelium. [d] Addition of detergents to the incubation medium caused lysis of the protoplasts.

pronase. This may suggest that an outer proteinous layer is present. These protective layers of proteinous or lipoid nature apparently change as the organism grows older, since the ease with which protoplasts can be formed decreases progressively with mycelial age.

I. Regeneration of Protoplasts

Protoplasts suspended in a medium which provided suitable osmotic protection plus adequate nutrients were able to regenerate into new mycelium. After 3 h of incubation some protoplasts started to form one or more buds, which grew out to normal hyphae (Fig. 2a, b c).

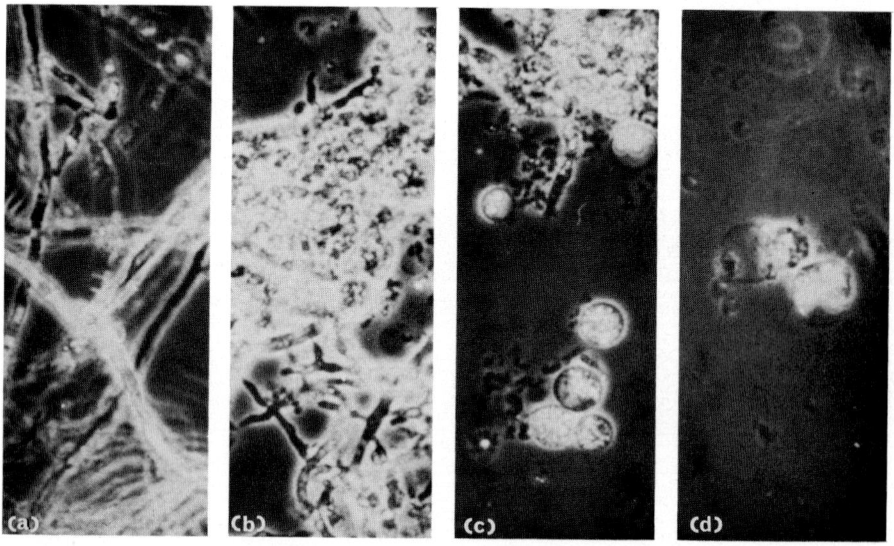

Fig. 1. Protop last formation of *Pythium*. Magnification 800 ×; Phase contrast. (a) Unchanged mycelium; (b) Several swellings are seen on the hyphae after 2 h of incubation with the enzyme complex; (c) Protoplasts and mycelial debris after 4 h of incubation; (d) Protoplasts which burst when the medium was diluted with water.

Some protoplasts extended to a structure with several swellings before a normal hypha arose (Fig. 2d). However a great number of protoplasts were still unchanged after 18 h of incubation.

The regeneration process was also followed under the fluorescence microscope, after incubating samples with Calcofluor, an optical brightener [5]. This substance adheres specifically to glucans and makes them fluoresce strongly when illuminated with near U.V.

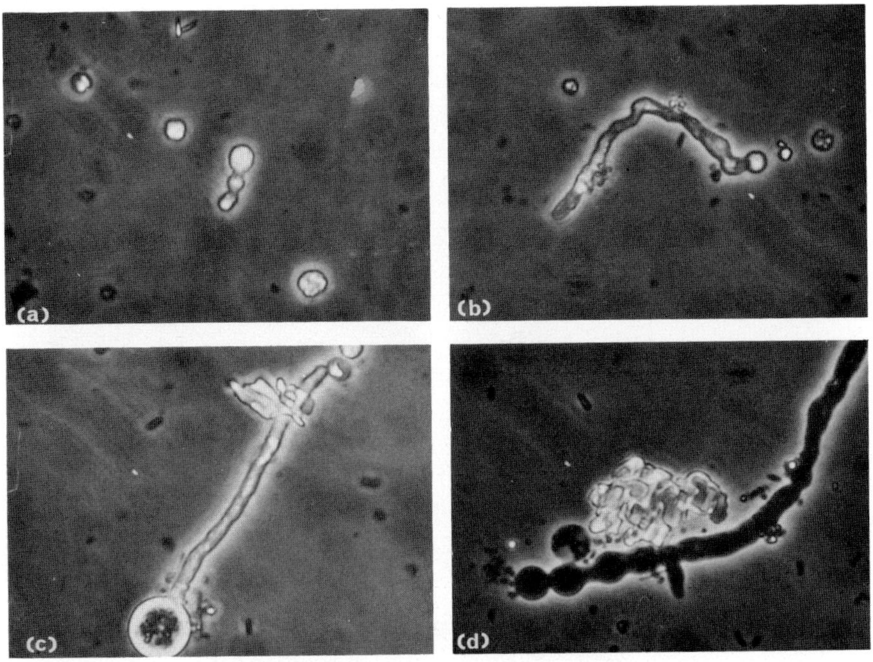

Fig. 2. Regeneration of protoplasts of *Pythium*. Magnification 800 ×; Phase contrast. (a) Buds are formed after 2 h of regeneration; (b) Protoplasts grown into a hypha after 8 h of regeneration; (c) As (b) but after 10 h of regeneration; (d) Hypha formed from protoplast after several swellings.

light. Freshly prepared protoplasts did not fluoresce at all, but after 1 h of incubation in the regeneration medium most of them showed a weak fluorescence, that rapidly became brighter as the regeneration proceeded. Newly formed buds and hyphae fluoresced very strongly. The tips of the hyphae generally fluoresced more strongly than the rest of the mycelium (Fig. 3a, b, c, d). The absence of fluorescence after incubating protoplasts with Calcofluor is a small indication that *Pythium* protoplasts lack most of the wall components. After 1 h of incubation all protoplasts did fluoresce.

In order to quantify the results a protoplast was considered to be able to regenerate when it could grow out into a visible colony on an agar plate provided with a suitable osmotic stabilizer and some nutrients. For determining the percentage of regeneration it did not make any difference whether the protoplasts were suspended in, or equally spread over the surface of, the agar. Gelatine could not be used as a solidifier since it was readily hydrolysed by *Pythium*.

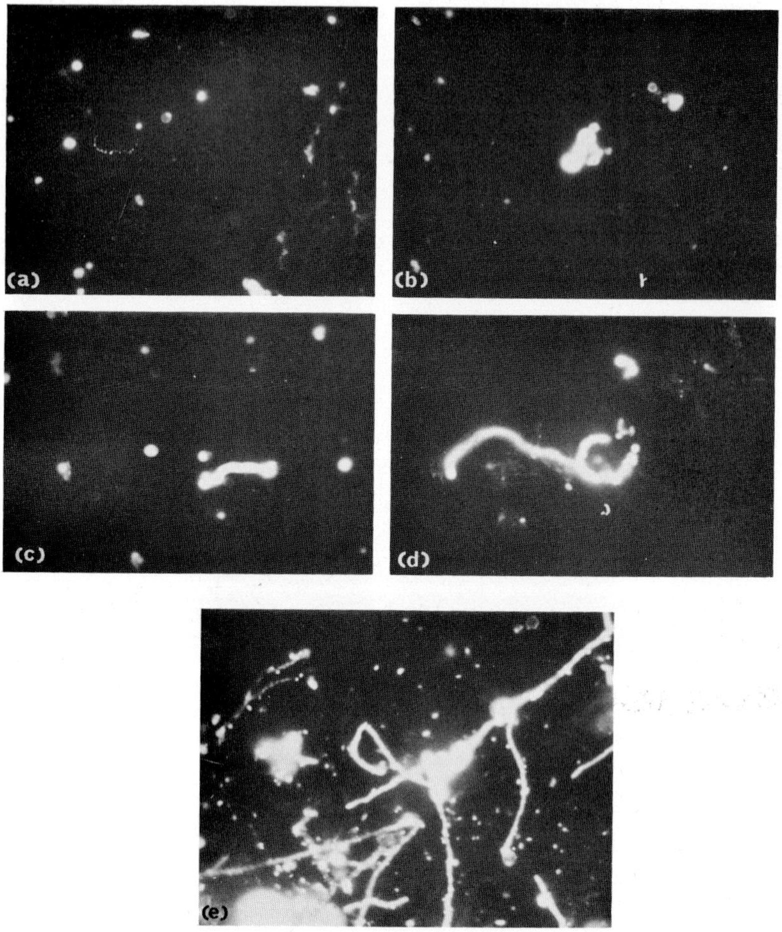

Fig. 3. Regeneration of protoplasts followed by fluorescence microscopy. Magnification 500 x. (a) Weak fluorescent protoplast after 1 h of regeneration; (b) Several strong fluorescent buds after 3 h of regeneration; (c) Same as (b) after 5 h of regeneration; (d) Newly formed hypha after 10 h of regeneration; (e) Newly formed mycelium after 16 h of regeneration.

It is shown in Table II that successful regeneration requires the use of a suitable osmotic stabilizer. Inorganic salts like NaCl, KCl, NH_4Cl completely inhibited the regeneration, whereas organic compounds like sucrose, mannitol, sorbose, did not inhibit this process. The percentage of protoplasts able to regenerate in a new mycelium appears to depend on the mycelial age. Protoplasts from 18 h cultures showed 18% regeneration and protoplasts from 72 h old

TABLE II

Factors effecting protoplast regeneration into a new mycelium

Age of the mycelium from which protoplasts are formed	Percentage of protoplasts[a] capable of regeneration	Osmotic stabilizer in regeneration medium	% Regeneration
18 h	18%	KCl (0.4 M)	0%
24 h	20%	NaCl (0.4 M)	0%
42 h	27%	NH_4Cl (0.4 M)	0%
48 h	25%	$MgSO_4$ (0.65 M)	0%
66 h	30%	Mannitol (0.65 M)	15%
72 h	30%	Sorbitol (0.65 M)	12%
		Sucrose (0.65 M)	11%

[a] Protoplasts were formed from mycelium pretreated with Triton X 100.

TABLE III
Regeneration in liquid medium

Additions to the regeneration medium	Dry weight (mg/ml) after 24 h of regeneration of 10^6 protoplasts/ml[a]	
	Osmotic stabilizer NaCl (0.4 M)	Osmotic stabilizer Mannitol (0.65 M)
No additions	0.7	0.4
1% glucose (gl)	0.4	1.5
0.1% asparagine (asp)	0.5	0.8
0.1% yeast extract (ye)	0.6	4.2
gl + asp	0.5	2.7
gl + ye	0.3	3.5
gl + asp + ye	0.7	9.0

[a] Protoplasts (10^6/ml) were suspended in regeneration medium with the additions mentioned and incubated at 30° C in a shaking water bath. After 24 h the suspension was centrifuged (3,000 x g; 10 min) and the dry weight of the pellet was estimated.

mycelium showed 30% regeneration. In studies with liquid medium glucose was essential for good regeneration. Nitrogen sources like asparagine, yeast extract or peptone were not necessary but stimulated the process of regeneration considerably (Table III).

With the fluorescent staining technique two types of regeneration could be observed. In the first type protoplasts formed one or more strongly fluorescent buds, that grew out into normal hyphae. These protoplasts probably contained one or more nuclei including all the information for the development of a new mycelium. In the second type protoplasts started to fluoresce after 1 or 2 h, but then the process stopped and no buds were formed. These protoplasts probably still contained a cell wall synthesizing system but lacked a nucleus. The later protoplasts might be formed out of the hyphal tips. Electron-microscopic techniques of Bracker and co-workers [2] have shown that no nuclei were present within 20 μm from the apex, whereas an active cell wall synthesizing system was still detectable. Given now the facts that the average diameter of a protoplast is 5 μm and of a hypha is 3.5 μm then one protoplast contains protoplasmic material from 7 μm of the hypha. When protoplasts are obtained from mycelium not treated with a detergent, one can presume that the majority of the protoplasts are formed from the apical region, because it is most vulnerable to attack by the glucanases. Then it is to be expected that not very many protoplasts contain a nucleus. This is especially true for younger mycelium. Protoplasts derived from mycelium that is pretreated with a detergent, are also formed from intercalary locations where the nuclei are more evenly distributed over the hyphal region. This might be an explanation for the higher percentage of regeneration of protoplasts derived from older mycelium than from younger cultures.

In conclusion, protoplasts could be obtained from *Pythium* by incubating the mycelium with a combination of helicase and cellulase, in the presence of an inorganic stabilizer. Pretreatment of the mycelium with a weak detergent made mycelium older than 24 h vulnerable to the enzyme complex.

Protoplasts suspended in a medium which provided suitable osmotic protection plus adequate nutrients were able to regenerate into new mycelium. The percentage of protoplasts able to regenerate varied from 15-30%.

Regeneration was carried out both in liquid and solid medium.

References

1. Duell, E. A., Inoue, S. and Utter, M. F. (1964). *J. Bact.*, **88**, 1762-1773.
2. Grove, S. N., Bracker, C. E. and Morre, D. J. (1970). *Am. J. Bot.*, **57**, 245-266.
3. Kovac, L., Bednarova, H. and Greksak, M. (1968). *Biochim. biophys. Acta*, **153**, 32-42.
4. Necas, O. (1971). *Bact. Rev.*, **35**, 149-170.
5. Preece, T. F. (1971). *In:* Methods in Microbiology, Vol. 4, pp. 509-516 (Booth, C., ed.). Academic Press, London and New York.
6. Sietsma, J. H., Eveleigh, D. E. and Haskins, R. H. (1969). *Biochim. biophys. Acta*, **184**, 306-317.
7. Sommer, A. and Lewis, M. J. (1971). *J. Gen. Microbiol.*, **68**, 327-335.
8. Villanueva, J. R. and Garcia Acha, I. (1971). *In:* Methods in Microbiology, Vol. 4, p p. 665-718 (Booth, C., ed.). Academic Press, London and New York.
9. Reese, E. T. and Mandels, M. (1963). *In:* Advances in Enzymatic Hydrolysis of Cellulose and Related Materials, pp. 197-234 (Reese, E. T., ed.). Pergamon Press, New York.

Studies on the Regeneration of *Aspergillus nidulans* Protoplasts

J. F. PEBERDY, C. E. BUCKLEY and R. K. GIBSON

Department of Botany, University of Nottingham
Nottingham, U. K.

Regeneration of protoplasts has been observed in several filamentous fungi [1, 2] and a variety of morphological forms of the process have been described. Mycelial protoplasts of *Aspergillus nidulans* will readily regenerate in certain liquid media. Initially the protoplast produces an outgrowth which has the appearance of a chain of cells separated by septa, or of an irregular hypha with randomly positioned septa. From this primary structure one or more apparently normal hyphae are produced. The regenerating protoplast of this fungus is of interest as a system for the study of wall synthesis as well as the cause of the morphogenetic changes that occur.

Recently we have described a method in which fluorescent brightener dyes, which adsorb specifically to β-linked polysaccharides, were used in an attempt to quantify regeneration [9]. In this paper we discuss further experiments based on this technique.

I. Materials and Methods

Preparation and Regeneration of Protoplasts. Protoplasts were obtained from 36 h mycelium of *A. nidulans* grown in shake cultures using Streptomyces lytic enzyme [3, 4]. For regeneration experiments the protoplasts (10^6/ml) were suspended in basal salts medium [3] supplemented with various carbon sources.

Adsorption of Fluorescent Brightener. Protoplasts were examined by UV microscopy using the methods described previously [9]. For estimations of the amount of brightener adsorbed, the protoplasts were sampled from the regenerating cultures at various time intervals. After washing and homogenization, the debris was resuspended in 5 ml aqueous Tinopal BOPT (500 mg/ml; Ciba-Geigy (UK) Ltd., Manchester, U.K.) for 5 min. The amount adsorbed by the disrupted protoplasts was estimated by measurement of UV absorbance of the solution at 350 nm [9].

Gas-Liquid Chromatography of Sugars. The residual carbohydrate content of the samples removed from the regeneration cultures was estimated by the method of Clamp *et al.* [5]. Trimethylsilyl ethers of the sugars were analysed using a Perkin-Elmer F11 Gas Chromatograph with Silicone SE30/HMDS Chromosorb (5 : 95, w/w) columns. The injection block temperature was 240° C and the chromatograph was temperature programmed from 130° to 200°C at 1°/min. Hydrogen and air pressures were 10^5 Nm^{-2} and 2×10^5 Nm^{-2} respectively and the nitrogen carrier gas flow was 60 ml/min. Sugar concentrations, based on peak areas, were calculated according to the procedure of Clamp *et al.* [5].

Electron Microscopy. The preparation and examination of thin sections of protoplasts has been described previously [4]. Freeze-etched preparations were obtained by fixing protoplasts in 2% glutaraldehyde in 0.2 M phosphate, pH 7, containing 0.4 M mannitol. After washing, six times in fresh glass distilled water, droplets were frozen on to gold stubs and fractured at −150° C in a Balzer Freeze Etcher. The specimens were etched at −100° C, shadowed at 45° with platinum carbon, and finally carbon replicas were made. These were floated on hypochlorite overnight and then for 4 h on 70% H_2SO_4. After washing and transference to Formvar coated grids they were viewed in an AEI EM6B electron microscope.

II. Results

The Nature of the Protoplast Surface. Freshly isolated protoplasts of *A. nidulans* stained with Tinopal BOPT showed no fluorescence when viewed by UV microscopy suggesting the absence of wall polysaccharides. This observation was confirmed by the examination of thin sections of protoplasts in which the plasmalemma was found to be completely free of cell wall material (Fig. 1).

Observations on freeze-etched protoplasts, however, have revealed that the plasmalemma is not completely smooth but has irregular masses of material distributed over the surface (Fig. 2).

Fig. 2. Freeze-etched protoplast of *A. nidulans.* The plasmalemma has been fractured in the central region where the outer layer of the membrane has been removed revealing characteristic granules adhering to the surface of the inner layer. The outer surface of the membrane has irregularly distributed masses of material.

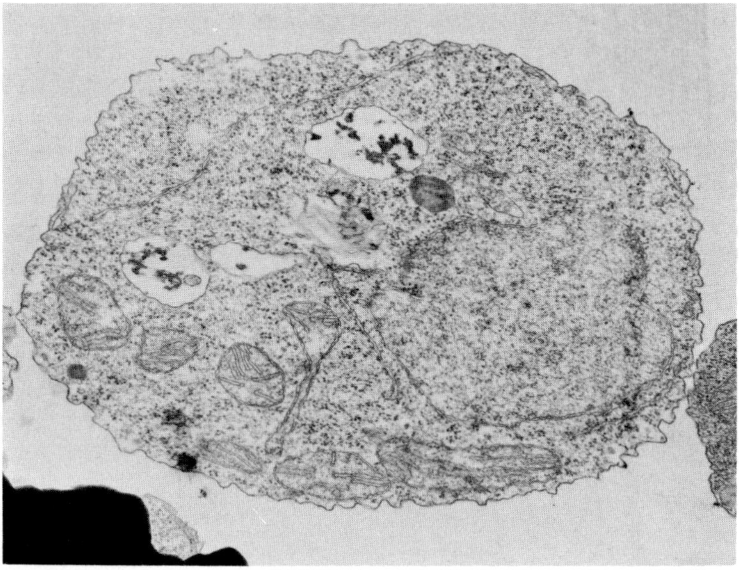

Fig. 1. Thin section of protoplast from *A. nidulans.* The plasmalemma is highly corrugated but lacks any residual wall material. Most organelles normally seen in hyphae also present.

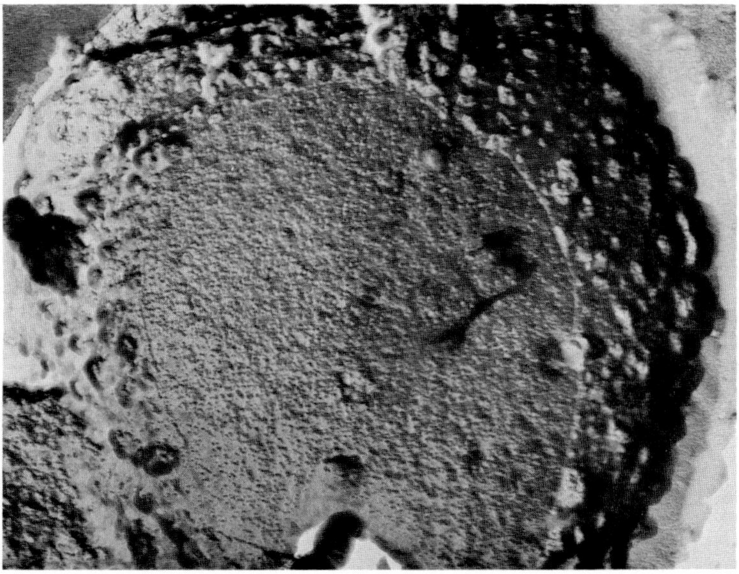

Regeneration of Protoplasts and Brightener adsorption. Protoplasts were cultured in media containing sugars, either singly or as mixtures, found in the hyphal walls of *A. nidulans,* i.e. N-acetylglucosamine (NAG), glucose, galactose and mannose [6]. The formation of new wall during regeneration could be observed by staining with fluorescent dye. Typical forms of regeneration are shown in Figs 3 and 4. The morphology and size of the primary

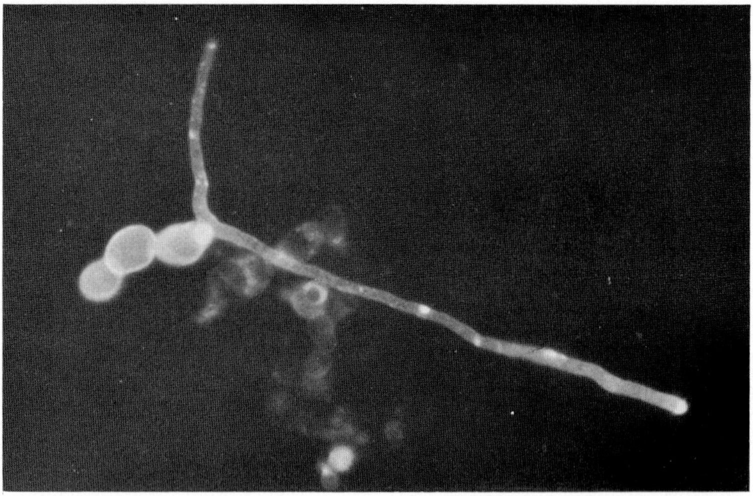

Fig. 3. Regenerating protoplast on NAG medium after 12 h, stained with Tinopal BOPT.

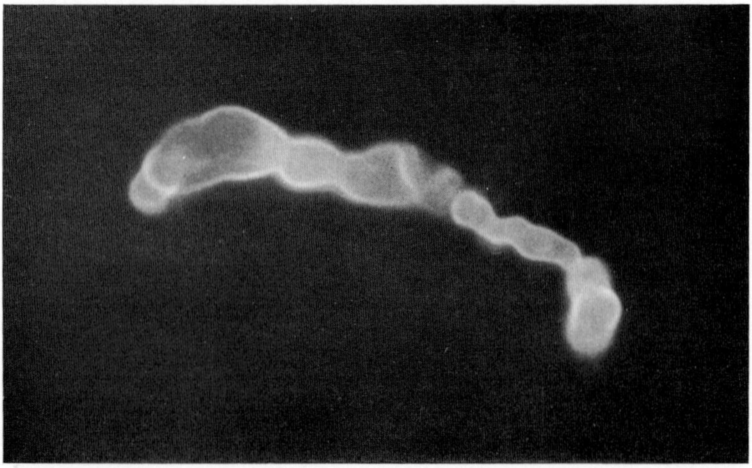

Fig. 4. Regenerating protoplast on glucose medium after 40 h, stained with Tinopal BOPT.

outgrowth was influenced by the nature of the carbon source in the regeneration medium. With NAG alone short cell chains or irregular shaped hyphae were formed. Normal hyphae developed after 12-16 h, much sooner than on the other media used.

On glucose medium the primary stage was a long chain of about 20 cells and hyphal development was delayed until 40-48 h. On the other media hyphal development occurred at 20-24 h on glucose : NAG and at 26-30 h on the mixed sugars.

The adsorption of optical brightener by protoplasts at various times during regeneration on different media is shown in Figs 5-8.

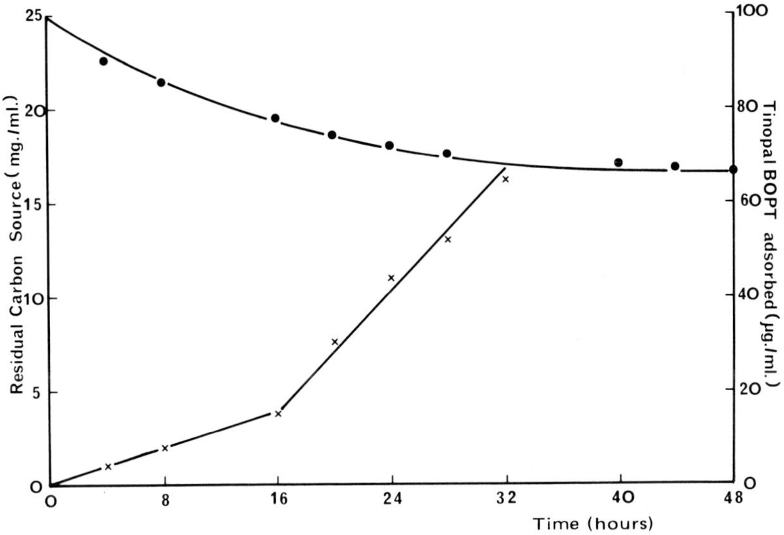

Fig. 5. Adsorption of Tinopal BOPT by protoplasts cultured in 2.5% NAG : salts medium.

In every case two distinct phases were observed, an initial one in which the rate of adsorption was low followed by a phase of rapid adsorption. Significant differences were observed in the time of transition between the two phases and in the rates of adsorption during both phases on the different media (Table I). The initial phase was shortest with NAG and longest with glucose. On mixed sugar media the transition point between the phases again occurred at intermediate times. Tinopal adsorption was most rapid with protoplasts in NAG medium during the initial phase and with protoplasts in glucose medium during the second phase.

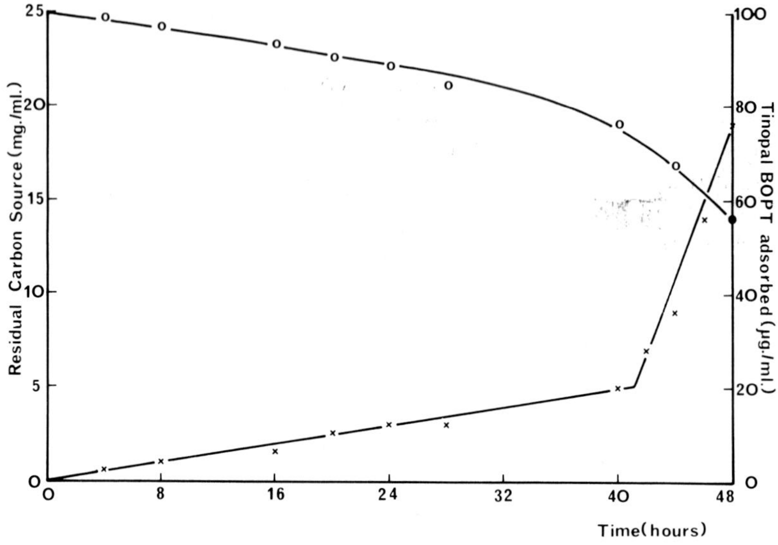

Fig. 6. Adsorption of Tinopal BOPT by protoplasts cultured on 2.5% glucose : salts medium.

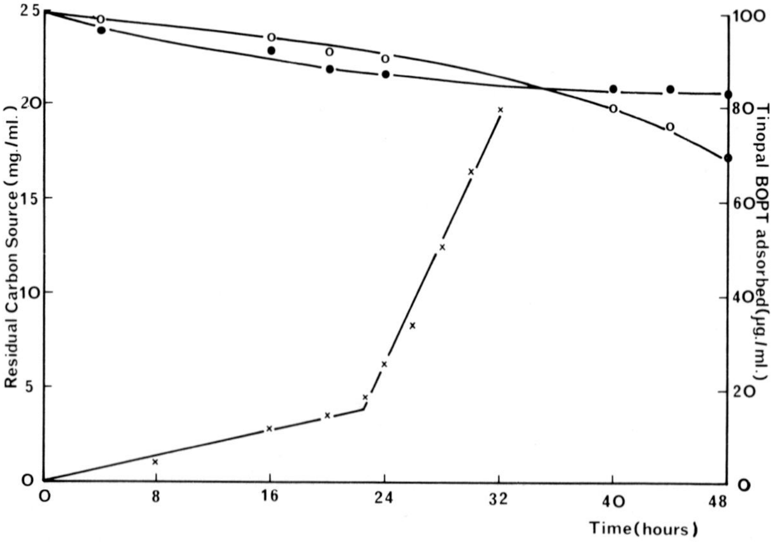

Fig. 7. Adsorption of Tinopal BOPT by protoplasts cultured on 2.5% NAG : 2.5% glucose : salts medium.

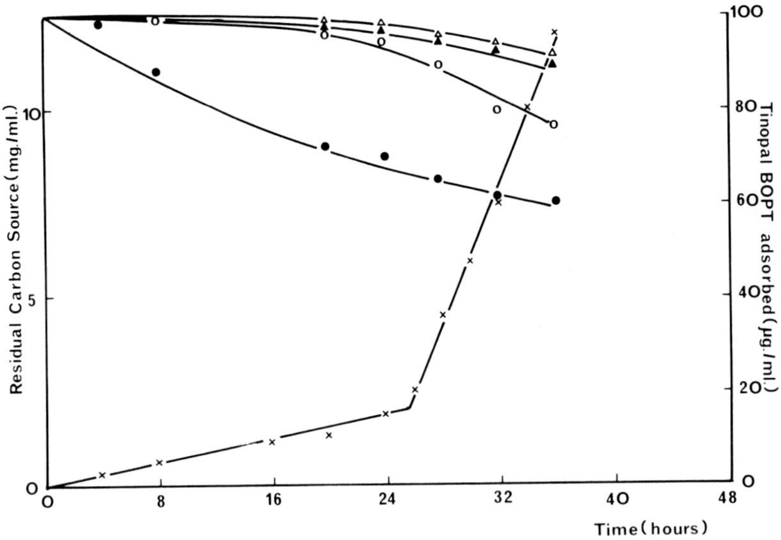

Fig. 8. Adsorption of Tinopal BOPT by protoplasts cultured on 1% NAG : 1% glucose : 1% galactose : 1% mannose : salts medium.

TABLE I
Tinopal BOPT removed by regenerating protoplasts

		Regeneration medium			
		2.5% NAG	2.5% Glucose	2.5% NAG 2.5% Glucose	1% NAG 1% Glucose 1% Galactose 1% Mannose
Brightener adsorbed μg/ml/h	Phase I	0.9	0.5	0.7	0.60
	Phase II	3.7	9.0	5.4	7.4

A difference in the pattern of uptake of NAG and glucose from regeneration media was also found. NAG uptake was most rapid during the initial phase slowing down in the second phase. The reverse situation was found for glucose.

III. Discussion

Our observations on freeze-etched protoplasts of *A. nidulans* have led us to query the designation of these structures. The nature of the material on the membrane surface is not yet known, although the lack of a fibrillar organization suggests that it is not polysaccharide. The lack of fluoresence of freshly produced protoplasts stained with optical brightener is also indicative of the absence of polysaccharide, but it is possible that the amount of material present, if it were β-linked polysaccharide, would not fluoresce. Investigations to determine the composition of this material are currently in progress.

The morphological stages of regeneration observed in *A. nidulans* are similar to those described in *Fusarium culmorum* by Garcia Acha *et al.* [7] although in our studies we have found that there are differences in the form of the primary stage which are associated with the nature of the carbon source used for regeneration. Although the regeneration process is not synchronous in all protoplasts the form of the brightener adsorption curves correlates with the morphological development. In each medium we found that the majority of regenerating protoplasts were producing hyphal initials at the time of transition in the adsorption rates.

The development of the primary stage from the protoplast and the ultimate formation of hyphae is a significant morphological change and it is of great interest to know the cause. The rapidity of regeneration on NAG alone as compared to glucose alone suggests that the composition of the newly formed wall may be of importance. The differences in the morphology of the primary stage on these two media may also reflect differences in wall composition. It could be assumed that on NAG medium the protoplasts might give rise to a primary structure which has a higher chitin content than the protoplasts on glucose medium and normal hyphal walls also. Gabriel [8] showed that *Rhizopus nigricans* protoplasts produced abnormal hyphae during regeneration in the presence of snail enzyme and concluded that unbalanced wall synthesis occurred. Our observations with washed protoplasts suggest that the primary stage of regeneration in *Aspergillus nidulans* may also be characterized by modified wall synthesis.

References

1. Villanueva, J. R. and Garcia Acha, I. (1971). *In:* Methods in Microbiology, Vol. 4, p. 665 (Booth, C., ed.). Academic Press, London and New York.
2. Peberdy, J. F. (1972). *Sci. Prog., Oxf.*, **60**, 73.
3. Peberdy, J. F. and Gibson, R. K. (1971). *J. gen. Microbiol.*, **69**, 325.
4. Gibson, R. K. and Peberdy, J. F. (1972). *J. gen. Microbiol.*, 72, 529.
5. Clamp, J. R., Bhatti, T. and Chambers, R. E. (1971). *In:* Methods of Biochemical Analysis, Vol. 19, p. 229 (Gluck, D., ed.). Interscience, New York.

6. Bull, A. T. (1970). *J. gen. Microbiol.*, **63**, 75.
7. Garcia Acha, I., Lopez-Belmonte, F. and Villanueva, J. R. (1966). *J. gen. Microbiol.*, **45**, 515.
8. Gabriel, M. (1968). *Fol. Microbiol.*, **13**, 231.
9. Peberdy, J. F. and Buckley, C. E. (1973). *J. gen. Microbiol.* **74**, 281.

Wall Structure, Wall Degradation, Protoplast Liberation and Wall Regeneration in *Schizophyllum commune*

J. G. H. WESSELS and O. M. H. DE VRIES

Department of Botany, University of Groningen
Haren, The Netherlands

Although many filamentous fungi have now been shown to release protoplasts after incubation with wall-degrading enzymes in a suitable osmotic stabilizer (cf. refs 1, 2), *Polystictus versicolor* [3] appears to be the only basidiomycetous species. Recently, the release of protoplasts from another basidiomycete, *Schizophyllum commune,* has also been reported [4]. The lytic enzyme system used was derived from the culture fluid of *Trichoderma viride* grown with *S. commune* walls as the carbon source. We have found that the same lytic enzyme system was also very successful for the production of protoplasts from a variety of other basidiomycetes [22] including *Collybia velutipes, Coprinus lagopus, Lentinus triginus, Pleurotus corticatus, Pholiota aurivella, Hericium coralloides, Polyporus biennis,* and *Polystictus versicolor.* The Trichoderma lytic enzyme system also released protoplasts from ascomycetes such as *Neurospora crassa,* various *Penicillium* species, and *Aspergillus nidulans.*

In some of the above-mentioned basidiomycetes viz. *Polyporus biennis, Hericium coralloides* and *Schizophyllum commune,* snail-gut enzymes were completely ineffective in releasing protoplasts. Therefore, an attempt was made to identify the essential component(s) in the *Trichoderma* enzyme system. The present report summarizes some data pertaining to the composition and structure of the hyphal wall of *S. commune* and shows the degradation of various wall components during protoplast liberation. The results indicate that it is the presence of an α-1,3-glucanase (S-glucanase) in the *Trichoderma* lytic system that accounts for its effectiveness in releasing protoplasts from *S. commune.* In fact, α-1,3-glucanase and chitinase appear to be the only enzymes required for this process [23].

I. Wall Composition and Structure

Most strains of *Schizophyllum commune* manufacture a mucilaginous polysaccharide which is present around the hyphae in high concentration but which can also be recovered from the culture medium. This mucilage was shown to consist of β-1,3-linked glucan chains with branches of single glucosyl units attached by β-1,6 linkage on every fourth unit, on average, along the chain [5]. When the hyphae are broken and the cytoplasm is extracted by extensive washing with buffers, the mucilage is also removed and what is left will be referred to as "complete walls".

Table I gives an analysis of the constituents of such a wall preparation. Quantitatively, the hexoses (mainly glucose) and hexosamine(s) are prominent but the presence of a relatively large amount of amino acids is also noteworthy.

TABLE I
Chemical composition of the hyphal wall of *Schizophyllum commune*[a]

Wall components	% of dry weight			% of complete wall	
	Complete walls	Fraction[b] A	Fraction[b] B	Fraction[b] A	Fraction[b] B
Glucose ($-H_2O$)	67.6	77.7	67.9	32.3	57.7
Mannose ($-H_2O$)	3.4[c]	12.1	—	—	—
Xylose ($-H_2O$)	0.2[c]	0.7	—	—	—
(ac) hexosamine ($-H_2O$)	12.5	0.2	22.2	0.1	102.1
Amino acids	6.4	2.2	9.2	3.0	83.3
Lipids (free)	1.0	—	—	—	—
Lipids (bound)	2.0	—	—	—	—
Sum	93.1	92.9	99.3		

[a] Strain 699 was grown in stationary culture at 25°C for five days. [b] The complete walls were extracted with 1N KOH at 60°C for 20 min, the extract was neutralized with acetic acid and the precipitate collected (fraction A). The residue left after alkaline extraction constitutes fraction B. [c] These figures were computed assuming that only fraction A contains these sugars.

An insight into the nature of the polysaccharides which make up the wall can be gained by fractionating the complete wall into an alkali-soluble and an alkali-insoluble portion. In effect, this procedure removes an electron-translucent outer wall layer that accounts for about 45% of the thickness of the wall [5]. About 66% of the material extracted by alkali can be precipitated by neutralizing the extract. Subsequent addition of ethanol only precipitates another 4% of the solubilized material, the rest (about 30%) remains soluble and probably arose from splitting of alkali-sensitive linkages.

Table I shows that this alkali-soluble but water-insoluble portion of the wall contains mainly glucose and small amounts of mannose and xylose. The glucose is derived from a glucan that contains a preponderance of α-1,3 linkages (S-glucan or pseudonigeran) and at least part of this glucan is present in the wall in a microcrystalline state displaying a typical rodlet structure when viewed in freeze-etched preparations [5]. The manner in which the mannose and xylose are present in the wall is unknown; either they are derived from a mannan and xylan, respectively, or these sugars are part of a heteropolymer possibly also containing glucose.

The alkali-insoluble portion of the wall (fraction B in Table I) contains a large amount of glucose, virtually all of the hexosamine and approximately 80% of the amino acids of the wall. At least part of the glucan in this fraction (R-glucan) can be characterized as bearing a similarity to yeast glucan, i.e. a glucan with both β1,3 and β1,6 linkages [6]. In contrast to yeast glucan, however, R-glucan is quite resistant to laminarinases and it probably contains chains with a structure similar to that indicated for the mucilage in which single glucosyl units are attached to the β-1,3-linked chain by β-1,6 linkage [5, 7]. We have found no indications for the presence of β-1,4 linkages as reported by other workers [8]. In fact, recent results from a methylation analysis suggests that in R-glucan almost exclusively 1, 3 and 1, 6 linkages occur (Sibers, Sietsma and Wessels, unpublished).

Undoubtedly, part of the hexosamine in fraction B is derived from chitin. Not only are chitin microfibrils visible in treated walls [9] but a typical X-ray diffraction pattern of chitin has also been obtained [6]. It appears, however, that not all of the hexosamine in this fraction originates from chitin because about 32% of the hexosamine can be solubilized by a treatment with 0.5 N HCl at 100°C for 1 h. Table II shows that enzymatic degradation of polymers containing this acid-soluble hexosamine is more rapidly effected by a glucanase than by a chitinase preparation, whereas the reverse is true for the polymer (chitin) containing the acid-insoluble hexosamine.

The presence of a glucan with various glycosidic linkages together with non-chitinous hexosamine and amino acids may indicate that the alkali-insoluble portion of the wall of *S. commune* contains one highly complex polymer in which hexosamine links polysaccharides and peptides. A similar picture is emerging from studies with other

TABLE II
Degradation of alkali-insoluble wall polymers in complete and KOH-extracted walls of *Schizophyllum commune* by glucanase and chitinase

Enzymes[a]	Complete walls			KOH-extracted walls		
	R-glucan	Hexosamine[b] insol.	Hexosamine[b] sol.	R-glucan	Hexosamine[b] insol.	Hexosamine[b] sol.
1. R-glucanase	22.6[c]	2.2	11.3	70.8	20.9	56.3
2. Chitinase	0.0	13.1	0.0	0.0	25.5	32.7
1 and 2	24.1	21.7	15.4	85.0	81.0	80.0

[a] The crude R-glucanase preparation was obtained from the culture fluid of *S. commune* strain K8 [7], the chitinase was a commercial preparation (NBCo). Degradation of pure crustacean chitin was six times faster with the chitinase than with the glucanase preparation. Walls were incubated with enzymes (1 mg/ml) in 0.05 N McIlvain buffer at 30°C for 4 h. [b] Hexosamine containing polymers were divided into two fractions: one solubilized by treatment with 0.5 N HCl at 100°C for 1 h and another remaining insoluble after this treatment (chitin). [c] Percentage of polymer degraded.

fungi (cf. 10, 11), suggesting that the presence of such complex peptidoglycans is a general feature of fungal walls.

The data of Table II also show that treatment of the walls with alkali greatly increases their susceptibility to glucanase and chitinase. Because the major component removed by this treatment is S-glucan, it is possible that this glucan shields the R-glucan-chitin complex against enzymatic degradation. Because hyphal walls were used in this experiment, the assumption has then to be made that the S-glucan is not only present as a relatively pure layer at the outer surface but also extends into the underlying R-glucan-chitin complex.

The data on the enzymatic breakdown of components of the alkali-insoluble portion of the wall (Table II) also indicate a synergistic effect of the combined action of chitinase and glucanase in the degradation of chitin. Such an effect is not evident for the breakdown of the glucan and of the acid-soluble portion containing hexosamine. It could be shown that a preliminary treatment with glucanase increases the susceptibility of chitin to chitinase. However, a preliminary treatment with chitinase does not increase the susceptibility to glucanase. Therefore, the R-glucan complex seems to protect the chitin against degradation, i.e. the chitin microfibrils may be embedded in a matrix substance containing glucan, hexosamine, and amino acids. Such an interpretation is substantiated by electron microscopic observations (Marchant and Wessels, unpublished) which reveal chitin microfibrils at both the inner and outer surface of alkali-extracted walls. This interpretation is at variance with suggestions put forward by Hunsley and Burnett [9] who envisage the wall of *S. commune* as a quadruple-layered structure with layers, from the inside to the outside, of chitin, protein, R-glucan, and S-glucan.

With respect to the surface localization of S-glucan [5, 9] it is noteworthy that this substance is apparently absent from the septa since these structures can be dissolved by treating isolated walls with enzyme preparations containing β-glucanase and chitinase but no S-glucanase [12].

II. Release of Protoplasts

Various enzyme preparations commonly employed for the liberation of fungal protoplasts were not successful in the case of *S. commune.* Among these were snail-gut enzymes (Helicase, Industrie Biologique Francaise, and Glusulase, Endo Labs Inc.) and extracellular enzymes from *Streptomyces* species [13] grown on *S. commune* hyphal walls. Abundant protoplast liberation was observed, however, by using an ammonium sulphate precipitate from the culture fluid of *Trichoderma viride* grown with isolated hyphal walls of *S. commune* as a carbon source [4].

For optimal release of protoplasts, 1-3-day-old mycelium of *S. commune* (3-7 mg dry wt/ml) is incubated with *Trichoderma* enzymes (1-2.5 mg protein/ml) in a solution containing 0.5 M $MgSO_4$ and 0.05 M maleic acid-NaOH buffer (osmotic potential −15.8 atm, pH 5.8). Other osmotic stabilizers such as KCl, NaCl, sucrose, mannitol, and sorbitol are also effective but only in KCl is the number of protoplasts as large as in $MgSO_4$. Initially, protoplasts are only released through ruptures near the hyphal apices (Fig. 1) but later also from older parts of the hyphae. If $MgSO_4$ is used as an osmotic stabilizer, the mycelium completely fragments and protoplasts are then also released from the open ends of hyphal compartments. Another phenomenon specifically exhibited when $MgSO_4$ is used is an enormous swelling of vacuoles in a number of protoplasts. These protoplasts float in the incubation mixture in contrast to another class of protoplasts, without these vacuoles, which sediment.

The released protoplasts of *S. commune* are osmotically sensitive. This, together with their release through ruptures in the wall, indicates that they lack a cell wall and thus can be considered as true protoplasts (cf. ref. 2). Indeed, electron micrographs (Fig. 2) do not

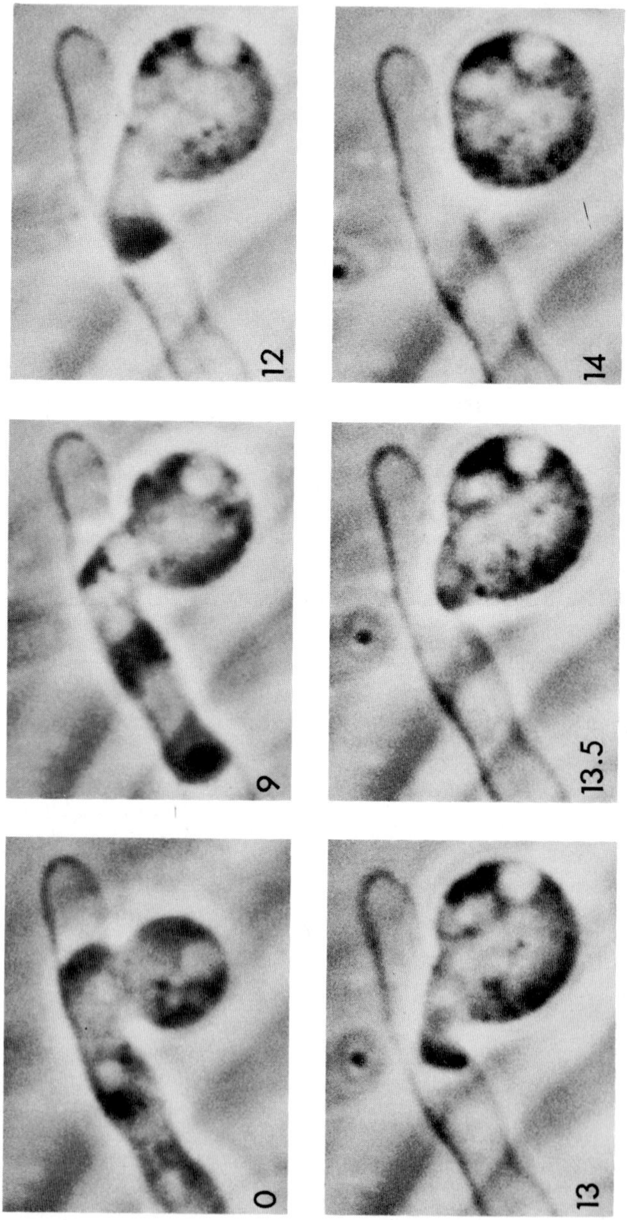

Fig. 1. Time-lapse photograph of the release of a protoplast from a hyphal tip of *Schizophyllum commune* strain 699 by *Trichoderma* enzymes in 0.5 M $MgSO_4$. The indicated numerals refer to minutes.

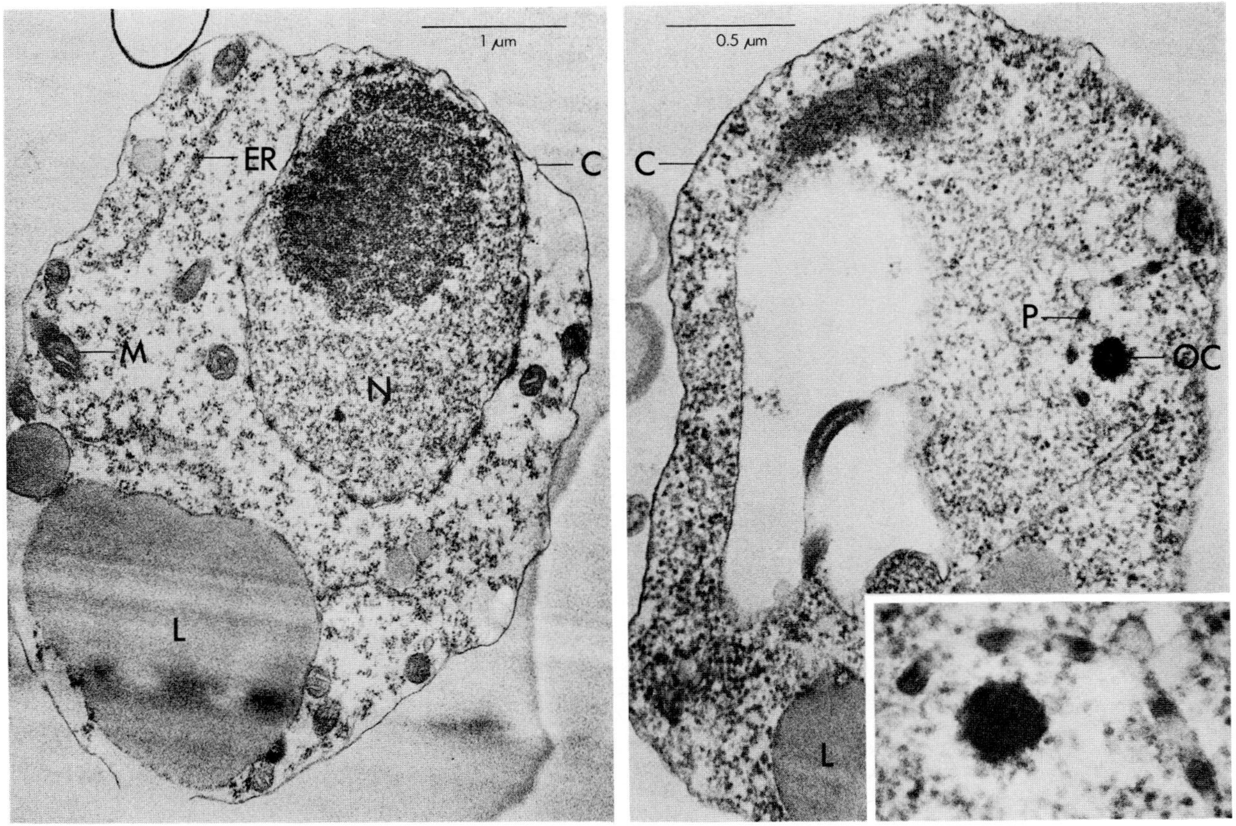

Fig. 2. Electron micrographs of protoplasts from *Schizophyllum commune* strain 699 released in 0.5 M $MgSO_4$. Fixation: glutaraldehyde-acrolein/OsO_4. Shown are a nucleus with nucleolus (N), lipid granula (L), mitochondria (M), endoplasmic reticulum with attached ribosomes (ER), cytoplasmic membrane (C), and a parenthesome (P) with pore occlusion (OC) (see inset at higher magnification). (Courtesy of P. van der Valk.)

show the presence of wall remnants. Figure 2 also shows the presence, in some of the protoplasts, of a typical basidiomycetous structure, a parenthesome. Apparently, this structure, together with the dense occlusion which is normally present in the septal pore became detached from the septum during protoplast liberation.

III. Wall Degradation During Protoplast Release and Identification of Essential Enzymes [23]

Trichoderma viride manufactures an α-1,3-glucanase when grown with an α-1,3-linked glucan as a substrate [14]. We found that a pure α-1,3-glucan is not necessary and that α-1,3-glucanase (S-glucanase) is also produced when hyphal walls of *S. commune* are used as a carbon source. Because S-glucanase is absent from the snail-gut enzyme preparations [4], which fail to induce protoplasts, the presence of this enzyme may be crucial for degrading the wall to such an extent that ruptures can occur.

Table III shows that during the liberation of protoplasts, degradation of all the three major wall polymers occurs. By the time that the maximum number of protoplasts is released, nearly all of the R-glucan and chitin has disappeared from the wall but about 50% of the S-glucan is still present and this amount only decreases very slowly during prolonged incubation.

If isolated S-glucan, precipitated by neutralizing alkaline solutions, is used as a substrate for the enzyme, only 40% can be broken down. The remaining 60% is resistant. Since it has been demonstrated that

TABLE III
Comparison of degradation of wall polymers in living mycelium and isolated hyphal walls of *Schizophyllum commune* by an extracellular enzyme preparation from *Trichoderma viride*

Wall polymer	Per cent degradation[a]					
	Living mycelium			Isolated walls		
	1 h[b]	4 h[b]	9 h	1 h	4 h	9 h
S-glucan	32	51	63	18	55	60
R-glucan	57	91	99	58	92	99
Chitin	23	84	97	30	83	98

[a] Living mycelium and isolated walls, standardized on the same amount of S-glucan (1 mg/ml), were suspended in 0.5 M $MgSO_4$ in 0.05 M maleic acid-NaOH buffer (pH 5.8) and incubated with *Trichoderma* enzymes at 25°C. [b] At 1 h the protoplasts were rapidly emerging and at 4 h the maximum number of protoplasts was attained.

both the precipitated S-glucan and the S-glucan in the native wall display a characteristic and sharp X-ray diffraction pattern [5] and since over 70% of the S-glucan is rapidly degraded after precipitation from alkali with ethanol, it is probable that the resistant S-glucan fraction represents that part which is present in crystalline condition.

Table III also shows that only minor differences exist between the rates at which the wall polymers are broken down in living mycelium and in isolated hyphal walls. These results do not provide any evidence that the S-glucan layer at the outside imposes a barrier to enzymes degrading the underlying R-glucan-chitin complex. However, by using snail enzymes on living mycelium and isolated walls it was found that such a barrier does exist.

The enzymes in the crude *Trichoderma* enzyme preparation were fractionated according to their activities towards laminarin, R-glucan, S-glucan, and chitin and the various fractions, alone or in combinations, were then tested for their effectiveness in releasing protoplasts from *S. commune* mycelium.

Table IV shows that, apparently, only S-glucanase and chitinase are needed to achieve sufficient wall degradation for protoplasts to

TABLE IV
Effectiveness of various enzyme preparations in releasing protoplasts from *Schizophyllum commune* mycelium

Enzyme preparation[a]	Enzyme activity[b] on S-glucan	R-glucan	Chitin	Number of protoplasts
I	+	—	—	0
II	—	+	—	0
III	—	—	+	0
I + II	+	+	—	0
I + III	+	—	+	92
II + III	—	+	+	0
I + II + III	+	+	+	101
Snail enzymes	—	+	+	0
Snail enzymes + I	+	+	+	200

[a] Fractions I-III refer to enzyme preparations obtained by purification of the crude *Trichoderma viride* enzyme mixture. The enzymes were adsorbed on CM cellulose at pH 3.8. Enzymes were eluted at the same pH with a linear gradient of NaCl (0-0.15 M). Laminarinase was eluted first. Fractions eluted around 0.08 M NaCl contained S-glucanase, R-glucanase, and chitinase but little laminarinase. Chitinase was then removed by adsorption on colloidal chitin and the S-glucanase and R-glucanase were completely separated by filtration through Sephadex G-100 (preparations I and II). After elution of the CM cellulose column with salt, a pure chitinase was eluted by raising the pH to 5.5 in the presence of 0.1 M NaCl (preparation III). [b] The minus sign refers to an activity of less than 2% of the activity present in the crude enzyme mixture.

emerge. However, when the mycelium was incubated in buffered 0.5 M MgSO without enzymes added, it was found that R-glucan can be degraded extensively by endogenous enzymes. Therefore, from the fact that no exogenous R-glucanase is necessary it cannot be concluded that R-glucan degradation is not a prerequisite for protoplast liberation.

Table IV also shows that snail-gut enzymes which themselves are ineffective in inducing protoplast release in *S. commune* are very effective in doing so when supplemented with the fraction containing S-glucanase. Because these snail enzyme preparations already contain a large variety of hydrolytic enzymes, except for S-glucanase, this adds to the evidence that it is actually the S-glucanase in enzyme preparation I which is responsible for the effect. On the other hand these experiments cannot completely exclude the possibility that the purified S-glucanase preparation contains an as yet unknown active factor.

At least one of the other basidiomycetes tested, viz. *Polyporus*, is known to contain S-glucan in its wall [15]. Like *S. commune, P. biennis* did not produce protoplasts with snail-gut enzymes but the *Trichoderma* system was effective. This suggests that in this case, too, the presence of S-glucanase was needed. On the other hand, we found that *Aspergillus nidulans,* which contains S-glucan in its wall [16], could produce protoplasts with snail-gut enzymes. Similar observations have been reported for *Aspergillus niger* [17] and *Schizosaccharomyces pombe* [18] both of which contain S-glucan [15, 19, 20]. This indicates that the mere presence of S-glucan in the wall not always necessitates the addition of an S-glucanase in order to release protoplasts. It might be that in these cases S-glucan is less important as a structural element in the wall. Also, some of these organisms might be able to manufacture an S-glucanase themselves as has been shown to be the case in *A. nidulans* [21].

IV. Wall Regeneration

After removal of the lytic enzyme system and re-incubation of the protoplasts of *S. commune* in osmotically stabilized growth medium, many protoplasts regenerated a new wall. Regeneration was observed in liquid medium as well as in medium solidified with gelatin but in the latter case the incidence of the formation of normally shaped hyphae was much higher. Up to 40% of the protoplasts could develop into normal mycelium (Fig. 3). This coincides well with

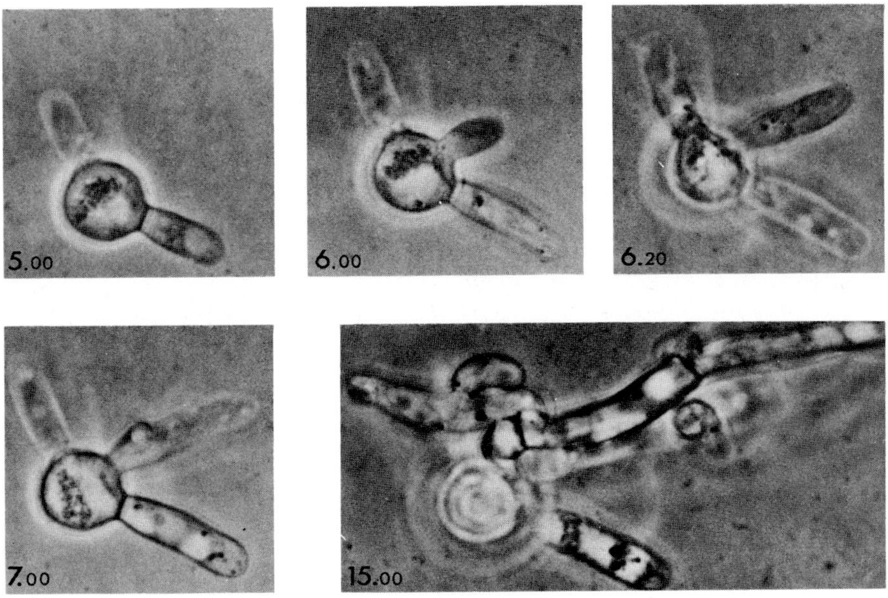

Fig. 3. Formation of normal hyphae from a protoplast of *Schizophyllum commune* dikaryon K8. Regeneration was done in minimal medium supplemented with 0.3% yeast extract, 0.05 M sorbitol, and 18% gelatin. The numerals refer to hours after inoculation of the protoplasts in growth medium.

estimates of the proportion of protoplasts that contain one or more nuclei [24].

When protoplasts were obtained from a dikaryon, most of the regenerated hyphae were also dikaryotic and showed clamp connections (Fig. 3). A smaller number contained only one of the nuclear types and thus developed into monokaryons. This provides a convenient method for recovering the two nuclear types from a dikaryon.

Further studies on the regeneration of protoplasts from *S. commune* may shed some light on the problem of wall synthesis. Experiments along these lines are in progress.

References

1. Strunck, C. (1970). *In:* Proc. Int. Symp. Yeast Protoplasts, p. 107 (Necas, O. and Svoboda, A., eds). Brno.
2. Villanueva, J. R. and Garcia Acha, I. (1971). *In:* Methods in Microbiology, Vol. 4, p. 665 (Booth, C., ed.). Academic Press, New York and London.

3. Strunk, C. (1965). *Biol. Resch,* 3, 242.
4. Vries. O. M. H. de and Wessels, J. G. H. (1972). *J. Gen. Microbiol.,* **73**, 13.
5. Wessels, J. G. H., Kreger, D. R., Marchant, R., Regensburg, B. A. and de Vries, O. M. H. (1972). *Biochim. biophys. Acta,* **273**, 346.
6. Wessels, J. G. H. (1965). *Wentia,* **13**, 1.
7. Wessels, J. G. H. (1969). *Biochim. biophys. Acta,* **178**, 191.
8. Wang, C. and Miles, P. G. (1966). *Am. J. Bot.,* **53**, 792.
9. Hunsley, D. and Burnett, J. H. (1970). *J. Gen. Microbiol.,* **62**, 203.
10. Ballesta, J. P. G. and Villanueva, J. R. (1971). *Trans. Brit. Mycol. Soc.,* **56**, 403.
11. Maret, R. (1972). *Arch. Mikrobiol.,* **81**, 68.
12. Janszen, F. H. A. and Wessels, J. G. H. (1970). *Antonie van Leewenhoek,* **36**, 255.
13. Sietsma, J. H., Eveleigh, D. E., Haskins, R. H. and Spencer, J. F. T. (1967). *Can. J. Microbiol.,* **13**, 1701.
14. Hasegawa, S. and Nordin, J. H. (1969). *J. biol. Chem.,* **244**, 5460.
15. Bacon, J. S. D., Jones, D., Farmer, V. C. and Webley, D. M. (1968). *Biochim. biophys. Acta,* **158**, 313.
16. Bull, A. T. (1970). *J. Gen. Microbiol.,* **63**, 75.
17. Musilková, M. and Fencl, Z. (1968). *Fol. Microbiol.,* **13**, 235.
18. Rost, K. (1969). *Z. Allgem. Mikrobiol.,* **9**, 289.
19. Johnston, I. R. (1965). *Biochem. J.,* **96**, 651.
20. Kreger, D. R. (1954). *Biochim. biophys. Acta,* **13**, 1.
21. Zonneveld, B. J. M. (1972). *Biochim. biophys. Acta,* **258**, 541.
22. Vries, O. M. H. de and Wessels, J. G. H. (1973). *Antonie van Leeuwenhoek,* **39** (In press).
23. Vries, O. M. H. de and Wessels, J. G. H. (1973). *J. Gen. Microbiol.,* **76**, 319.
24. Valk, P. van der and Wessels, J. G. H. (1973). *Protoplasma,* (In press).

PART V

Protoplasts of Plants

Isolation, Fusion and Development of Protoplasts of Higher Plants

E. C. COCKING

Department of Botany, The University of Nottingham
Nottingham Great Britain

Since the previous meeting on yeast protoplasts in Brno in 1968 work on higher plant protoplasts has steadily progressed, so much so that in many ways the spectrum of biological activities displayed by these higher plant protoplasts is, at the moment, greater than those in yeast protoplasts. Workers on yeast protoplasts are naturally interested in the detailed work on higher plant protoplasts and the reasons for the major advances in this field, and in this review I will try to highlight these advances and draw comparisons with work on yeasts and other micro-organisms where appropriate. Recently, several key reviews have been prepared. Cocking [1] has very comprehensively discussed the isolation and development of higher plant protoplasts and also their fine structure and cell wall regeneration [2]. The actual procedures to be adopted for the successful isolation of protoplasts for a wide range of physiological, virological and cultural studies have also been fully described by Evans and Cocking [3]; and very recently there has been the First International Symposium on Higher Plant Protoplasts in Versailles [4].

I. Isolation

The general enzymatic procedure for higher plant protoplast isolation using enzymes rich in β-1-4 glucanase is well established. Work has been greatly facilitated by the commercial availability of such cellulases as *Trichoderma viride* cellulase of various degrees of activity and as a result their use has superseded the use of *Myrothecium verrucaria* cellulase which is not available commercially and which was used in the first enzymatic isolations of higher plant protoplasts [5]. The fact that higher plant cells exist together in organized tissues complicates the isolation of protoplasts and for this reason certain workers, in particular the Japanese group [6], have advocated the initial separation of cells using polygalacturonase enzymes and then the digestion of the walls of these isolated cells with the cellulase enzyme complex. This procedure works well for tobacco leaves but is more time consuming and laborious than the mixed enzyme procedure developed by Power and Cocking [7] which results in the simultaneous separation of the cells and their wall degradation. There is considerable variation from one plant to another in the readiness with which the cells are separated and their walls degraded. In particular, variation in the amount and nature of the hemi-cellulose content of the cell walls affects the accessibility of β-1-4-glucanase to the cellulose fibrils of the wall and other factors may also be involved—for instance, the presence of protein enzyme inhibitors. Any lignification also prevents cellulase degradation of walls. As a result there is a considerable variation from leaf to leaf of different species in the release of protoplasts by these enzymatic methods. Evans, Keates and Cocking [8] also eliminated the necessity of removing the lower epidermis when working with cereal leaves—a procedure introduced by Takebe *et al.* [6] when working with tobacco leaves and since found unnecessary by Jensen, Francki and Zaitlin [9]. It may also be possible to apply a sequential treatment of enzymes without actually isolating the cells [10]; enzymatic degradation of the cuticle advocated by Schilde-Rentschler [11] is not usually necessary.

A particular advantage of enzymatic methods for the isolation of protoplasts (for a detailed discussion see Cocking [1]) over mechanical ones is that protoplasts can thereby be isolated from meristematic tissues. Higher plants, however, possess cells in which there are cytoplasmic connections between contiguous cells and during the actual wall degradation by the enzymes these cytoplasmic channels between cells can expand drawing the protoplasts together and causing them to fuse spontaneously [12]. As a result, multinucleate spontaneously fused protoplasts can be readily isolated from higher plant tissues, particularly meristematic tissues, when using the mixed enzyme procedure. Multinucleate spontaneously fused protoplasts have been cultured in suitable media and like uninucleate protoplasts they can in certain instances regenerate a new cell wall. During mitosis the nuclei aggregate together and good evidence has been obtained that the nuclei can fuse together at this stage [12, 13, 14]. These syncitia prior to division are perhaps near to the motile coencytic plasmodia of the Myxomycetes in their nuclear relationships.

At Nottingham we have given particular attention to the question of whether or not any fibrils of cellulose are present on the surface of isolated protoplasts. This is not an easy problem to resolve since the wall formed *de novo* around the isolated protoplasts also contains cellulose fibrils which are indistinguishable from cellulose fibrils of the original wall. Using the technique of deep etching in conjunction with freeze-etch technique it has, however, been possible to show that little, if any, cellulose is present (Fig. 1). The odd fibril of cellulose in no way affects the fusion and other characteristic properties of these isolated protoplasts [15].

Normally the cell walls of higher plant cells contain cellulose (β-1-4-glucan), hemi-celluloses and pectin as well as small amounts of special proteins. Pollen tetrads, however, have a callose wall within which the young uninucleate haploid microspores are enclosed. This callose wall is an unbranched β-1-3-glucan. Recently Bhojwani and Cocking [16] showed that the same enzyme which degrades the

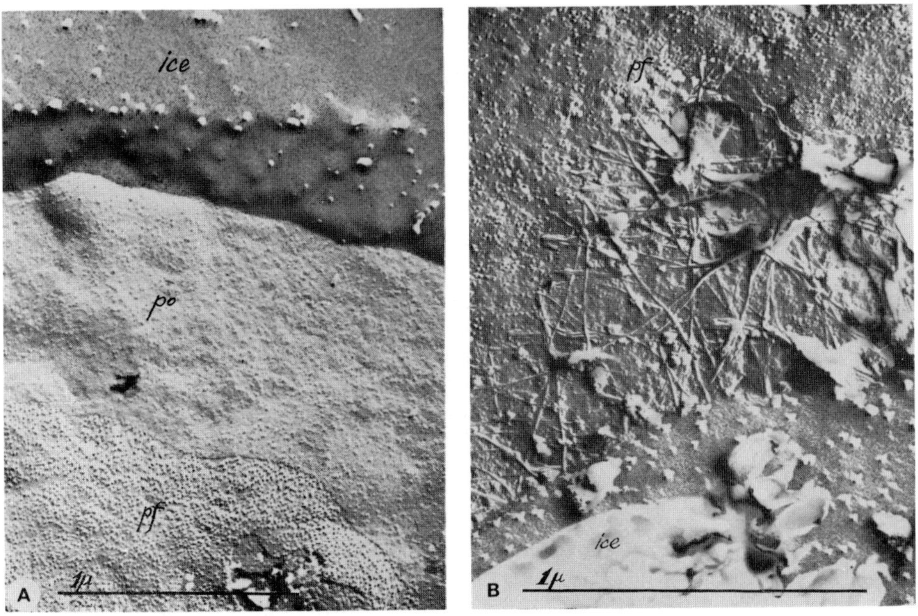

Fig. 1. (a) The outer face of the plasmalemma (po) of a freshly isolated tomato fruit protoplast is typically devoid of residual cell wall. The fracture face of the membrane (pf) is typically granular and serves as a reference layer for observations of wall formation. The etched surface of the enclosing ice (ice) is also visible. x60,000. (b) After a day or two cellulose microfibrils of the regenerated wall may be seen lying on and arising from the previously bare plasmalemma surface. x85,000.

yeast cell wall, liberating yeast protoplasts, will very readily degrade this callose wall enabling naked microspores to be isolated from pollen tetrads which have recently completed meiosis. Since the usual cellulases are inactive it would seem very likely that the ability of the snail enzyme complex to degrade the callose wall is due to the high level of β-1-3-glucanase present. Other enzymes capable of degrading the cell walls of fungi releasing protoplasts (see Villanueva [17]) may therefore also be suitable for pollen protoplast isolation. Using this new procedure it is possible to isolate naked, uninucleate haploid microspores from a wide range of species of flowering plants and, as will be discussed by Bajaj [18] at this Symposium, studies on the culture of these microspores may enable haploid plants to be produced from a wider range of species than is presently possible. The size of these microspore protoplasts is about the same as that of many fungal protoplasts. As will be described later, these naked spores can be readily fused together. Mature pollen grains have an extine which is impregnated with sporopollenin, a very chemically and enzymatically resistant material which incidently is present in the walls of certain algae such as Chlorella, making enzymatic isolation of protoplasts from Chlorella impossible [19]. Only after germination is the cellulose of the pollen grain susceptible to degradation by cellulase [20].

II. Fusion

As we have already seen, protoplasts of higher plants, which are present in a tissue while the walls of the tissues are being enzymatically degraded, can spontaneously fuse together as a result of the expansion of plasmodesmatal connections. This special type of intraspecies fusion is clearly different from the fusion of freely isolated protoplasts. The reproducible fusion of freely isolated protoplasts is required for interspecies fusion. Normally freely isolated protoplasts do not fuse together and it has become clear that one of the major difficulties is the fact that most isolated higher plant protoplasts are negatively charged. This prevents close aggregation—a required prelude to any membrane fusion and cytoplasmic mixing. The initial effect of a fusion inducing agent such as sodium nitrate [21] has been shown to largely eliminate this negative charge resulting in the clumping together of the isolated protoplasts [22, 23]. Membrane fusion is followed by the gradual mixing together of the cytoplasms, and the rapidity and extent to which this occurs is dependent on the extent of differentiation of the protoplasts including, for instance, the presence or absence of vacuoles and plastids and the development of the endoplasmic reticulum. With highly vacuolated isolated leaf protoplasts only a few of the fusing protoplasts may undergo extensive cytoplasmic mixing together; most may undergo limited cytoplasmic fusion to form a very closely aggregated fusing protoplast mass closely reminiscent of a chimaeral tissue. Moreover, any tendency to regenerate a new cell wall will tend to prevent continued cytoplasmic mixings. Consequently, nuclear fusion will therefore be greatly restricted. Under these circumstances further culture may yield very few nuclear hybrid cells. Less differentiated and less vacuolated more densely cytoplasmic protoplasts are more likely to yield on fusion heterokaryons made up of the cytoplasms and nuclei of the species being fused together with a close juxtaposition of nuclei. As far as can be judged from studies on homokaryons formed as a result of spontaneous fusion close juxtaposition of nuclei affords the opportunity for nuclear fusion as the fused protoplasts enter mitosis [2]. For a detailed discussion of nucleo-cytoplasmic interactions in the achievement of nuclear synchrony in DNA synthesis and mitosis in multinucleate cells see Johnson and Rao [24]. The ability to produce hybrid protoplasts and walled cells as a result of isolated protoplast fusion is a key step in the somatic hybridization of plants and it is clear that fusion needs to be induced by sodium salts such as sodium nitrate [21]. The key role that somatic hybridization by the fusion of these naked cells could play in improving the genetic variability of plants has been critically and comprehensively discussed by Power and Cocking [25]. The scientific and social implications could be profound. However, while the pattern of experimental approach is now well established (Fig. 2), it is clearly manifest that each interspecies fusion requires special handling in respect of the subsequent culture of the fusion products. It would seem likely that following fusion the fusion products regenerate a new cell wall. Cell wall regeneration involves pectin and cellulose synthesis, and fusion products capable of photosynthetic activity appear to resynthesize a wall more readily than non photosynthetic systems. The exploitation of the potentiality of somatic fusion necessitates the regeneration of the hybrid cells into whole plants which can then be subjected to an extensive breeding programme. Even for isolated higher plant cells which have not been subjected to enzyme treatment to remove cell walls, only a relatively few species of a particular cell type have been shown to undergo regeneration into whole plants. In particular, monocotyledons have proved to be particularly intransigent [26].

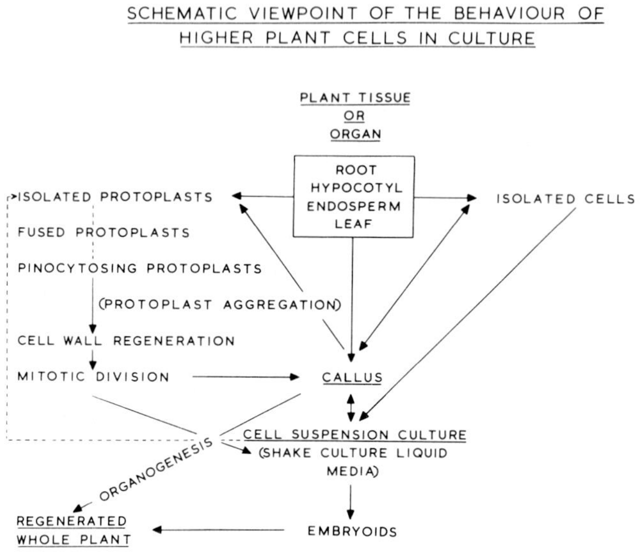

Fig. 2.

Slowly but steadily, however, the range of species from which it is possible to regenerate whole plants from isolated protoplasts is increasing (diploid tobacco [27, 28]; haploid tobacco [29]; carrot [30, 31]).

The difficulties in this respect should not, however, be underestimated. Even varieties of the same species differ greatly in the readiness with which regeneration takes place and the perfection of adequate reproducibilty of regeneration and a high plating effiency may take much time and energy. Each somatic interspecies cross requires a special separate investigation. The recent work with different but sexually compatible species of tobacco [32] was facilitated by knowledge of conditions required for culture and fusion, and regeneration. This is a further step towards establishing that somatic fusion, coupled with plant regeneration, will yield interspecies hybrids not capable of being produced by sexual means due to the already well recognized incompatibility barriers associated with pollen tube growth [33] and other factors. One of the major difficulties in this type of work is the adequate and unambiguous characterization of hybrid calluses and any hybrid plants which will always be expected to be produced at a very low frequency—often at the level of a few in a million. Since it is evident that novel cytoplasmic hybrids could be formed in a large proportion of the total hybrid-like product from any somatic fusion of protoplasts of different species isolated from highly vacuolated leaf and callus, claims for interspecies hybridization at the nuclear level need to be based on the karyotype of the resultant hybrid. The fusion of pollen tetrad microspore protoplasts, which fuse readily and extensively with rapid intermingling of the cytoplasms may help in increasing the frequency with which somatic fusions resulting in heterokaryons giving rise to nuclear hybrids take place. These haploid protoplasts may also facilitate the development of media for the selective growth of nutritionally dependent hybrid cell lines [4] ; it is clear that the development of general methods for the selection of somatic hybrid cells is of paramount importance for the further development of any parasexual cycle in plants.

III. Development

From what we have already seen it is evident that the further development of protoplasts is largely associated with their regeneration into cells as a result of wall regeneration. Wall regeneration by isolated higher plant protoplasts has been comprehensively described by Cocking [34] . In general, it would seem that while protoplasts in suitable media can undergo mitosis, cytokinesis is associated with the beginnings of wall regeneration. Of course, not all protoplasts which undergo wall regeneration immediately undergo cytokinesis, but cell wall regeneration and cytokinesis are frequently closely associated with each other (for a detailed analysis of the possible correlation see Cocking [2]).

The development of isolated protoplasts can also be influenced by the uptake of various macromolecules, particles such as virus particles and even whole micro-organisms. Isolated protoplasts undergo, to varying degrees, endocytosis [22, 34] and the ability to infect isolated higher plant protoplasts by viruses is probably related to this ability [35, 36] . Tobacco mosaic virus is very versatile since not only does it infect and multiply in tobacco leaf protoplasts but also in yeast protoplasts [37, 38] . The recently discovered plasmolytic uptake of living bacteria by isolated protoplasts [39, 40] may mean that new endosymbiotic relationships can be established between micro-organisms and higher plants thus permitting an even greater diversity of development of protoplasts and plant modification than ever previously envisaged.

References

1. Cocking, E. C. (1972). *A. Rev. Pl. Physiol.*, 23, 29.
2. Cocking, E. C. (1973). *In:* Dynamic Aspects of Plant Ultrastructure (Robards, A. W., ed.). McGraw Hill, New York. In press.
3. Evans, P. K. and Cocking, E. C. (1973). *In:* Plant Tissue and Cell Culture (Street, H. E., ed.). Blackwell's Botanical Monographs. In press.
4. Cocking, E. C. (1973). *In:* Protoplastes et Fusion de Cellules Somatiques Végétales, *Coll. intern. C.N.R.S.* **212**, 327.
5. Cocking, E. C. (1960). *Nature, Lond.*, **187**, 927.
6. Takebe, I., Otsuki, Y. and Aoki, S. (1968). *Pl. Cell Physiol.*, **9**, 115.
7. Power, J. B. and Cocking, E. C. (1970). *J. Exp. Bot.*, **21**, 64.
8. Evans, P. K., Keates, A. G. and Cocking, E. C. (1972). *Planta (Berl.)*, **104**, 178.
9. Jensen, R. G., Francki, R. I. B. and Zaitlin, M. (1971). *Pl. Physiol.*, **48**, 9.
10. Potrykus, I. and Durand, J. (1972). *Nature, New Biol.*, **237**, 286.
11. Schilde-Rentschler, L. (1972). *Z. Naturforsch.*, **27b**, 208.
12. Withers, L. A. and Cocking, E. C. (1972). *J. Cell Sci.*, **11**, 59.
13. Miller, R. A., Gamborg, O. L., Keller, W. A. and Kao, K. N. (1971). *Can. J. Genet. Cytol.*, **13**, 347.
14. Power, J. B., Frearson, E. M. and Cocking, E. C. (1971). *Biochem. J.*, **123**, 29P.
15. Willison, J. H. M. and Cocking, E. C. (1972). *Protoplasma*, **75**, 397.
16. Bhojwani, S. S. and Cocking, E. C. (1972). *Nature New Biol.*, **239**, 29.
17. Villanueva, J. R. and Garcia Acha, L. (1971). *In:* Methods in Microbiology, Vol. 4 (Booth, C., ed.). Academic Press, London and New York.
18. Bajaj, Y. P. S. (1972). 3rd International Symposium on Yeast Protoplasts, Salamanca, Spain 1972.
19. Atkinson, A. W. and Gunning, B. E. S. (1972). *Planta*, **107**, 1.
20. Power, J. B., *In:* Evans, P. K. and Cocking, E. C. (1973). *In:* Plant Tissue and Cell Culture (Street, H. E., ed.). Blackwell's Botanical Monographs. In press.
21. Power, J. B., Cummins, S. E. and Cocking, E. C. (1970). *Nature, Lond.*, **225**, 1016.
22. Willison, J. H. M., Grout, B. W. W. and Cocking, E. C. (1971). *Bioenergetics*, **2**, 371.
23. Grout, B. W. W., Willison, J. H. M. and Cocking, E. C. (1973). *Bioenergetics*, **4**, 585.
24. Johnson, R. T. and Rao, P. N. (1971). *Biol. Rev.*, **46**, 97.
25. Power, J. B. and Cocking, E. C. (1971). *Sci. Prog., Oxf.*, **59**, 181.
26. Steward, F. C. (1971). *A. Rev. Pl. Physiol.*, **22**, 1.
27. Takebe, I., Labib, G. and Melchers, G. (1971). *Naturwissenschaften*, **58**, 318.
28. Nagata, T. and Takebe, I. (1971). *Planta (Berl.)*, **99**, 12.
29. Nitsch, J. P. and Ohyama, K. (1971). *C.R. Acad. Sci. Paris, Serie D*, **273**, 801.
30. Grambow, H. J., Kao, K. N., Miller, R. A. and Gamborg, O. L. (1972). *Planta*, **103**, 348.
31. Kameya, T. and Uchimiya, H. (1972). *Planta*, **103**, 356.
32. Carlson, P. S., Smith, H. H. and Dearing, R. D. (1972). *Proc. natn. Acad. Sci. U.S.A.*, **69**, 2292.

33. Knox, R. B., Willing, R. R. and Ashford, A. E. (1972). *Nature, Lond.*, 237, 381.
34. Cocking, E. C. (1970). *Int. Rev. Cytol.*, **28**, 89.
35. Coutts, R. H. A., Cocking, E. C. and Kassanis B. (1972). *J. Gen. Virology*, 17, 289.
36. Coutts, R. H. A. (1973). *In:* Protoplastes et Fusion de Cellules Somatiques Végétales, *Coll intern. C.N.R.S.* 212, 353
37. Coutts R. H. A., Cocking, E. C. and Kassanis, B. (1972). *Nature Lond.*, **240**, 466.
38. Coutts, R. H. A. (1972). 3rd International Symposium on Yeast Protoplasts, Salamanca, Spain, 1972.
39. Davey, M. R. and Cocking, E. C. (1972). *Nature, Lond.* 239, 455.
40. Davey, M. R. (1972). 3rd International Symposium on Yeast Protoplasts, Salamanca, Spain, 1972.

Isolation, Fusion and Culture of Protoplasts of Petunia

I. POTRYKUS

Institut für Botanische Entwicklungsphysiologie der Universität Hohenheim Stuttgart, Germany

Although the isolation of protoplasts from higher plants was reported as early as 1892 [1] and fusion of isolated protoplasts was observed in 1937 [2], the current interest in this subject is high because of the recent progress in manipulating protoplasts since the pioneer work of Cocking [3], who introduced the method of enzymatic isolation into the work with higher plants. Protoplasts have been isolated enzymatically from a wide range of different tissues: meristematic tissue, such as root tips [4], differentiated tissue such as solanum fruit [5], mesophyll of leaves [6], avena coleoptiles [7], cotyledones [8], storage roots [9], ovules [10] and flower petals [11], and from callus cultures [12] and suspension cultures [13]. Protoplasts regenerate a new cell wall [14], are capable of nuclear [15] and cellular [16] division, can be taken into culture and can be grown up to flowering plants [17, 18]; they can take up macromolecules, viruses and small particles [19, 20], and they can be induced to fuse [4, 11]. Isolated protoplasts of higher plants thus are of increasing interest for plant physiologists, and pure and applied geneticists. They can be used as systems for experiments on problems of biochemical genetics, gene regulation, cell differentiation, somatic hybridization (intra- and interspecific) and organelle transplantation. Isolated protoplasts have, in comparison with isolated single cells, two main advantages: first, after cell wall removal protoplasts can easily take up macromolecules, small particles and viruses which cannot readily overcome the barrier of the cell wall. Protoplasts thus allow influence of the cellular metabolism in a very direct way by metabolites or antimetabolites, or by transfer of genetic material, under conditions which can be standardized. In addition it is possible to transplant cell organelles like nuclei [37] and chloroplasts [38]. Second, protoplasts can be induced to fuse, thus enabling the combination of genotypes of even very far related species, and, as they can be isolated out of distinct and out of distinctly far differentiated cells of the same plant, they enable the combination of genotypes of different phases of differentiation. Thus, studies using protoplasts of higher plants may open up new possibilities in plant breeding and in the experimental biology of higher plants [11]. Much information can be gained from protoplasts directly, but to obtain maximum information from such studies it is necessary to grow the protoplasts to flowering plants. This has been successfully done up till now with Tabaccum [17, 18], Daucus [21] and Petunia [22].

I. Isolation

Vacuolated protoplasts can be isolated mechanically or enzymatically, meristematic protoplasts can be isolated only by enzymatic methods. Until 1960 protoplasts of higher plants have been isolated only by mechanical methods: the cells of the differentiated tissues were plasmolysed and cut into thin slices [1]. Thus protoplasts could be liberated directly or by slight deplasmolysis from cells whose cell wall had been opened by the blade without damaging the protoplast within it. This method yields only a few protoplasts. The introduction of the enzymatic method by Cocking stimulated in a high degree the work with protoplasts of higher plants, because this method enables large scale isolation of viable protoplasts. This method, by which in an appropriate osmotic environment the middle lamella and the cell wall are digested with the aid of hydrolytic enzymes, has many advantages: (a) a larger number of protoplasts may be obtained; (b) less osmotic shrinking of the cytoplasm is required; (c) no cells are broken as a result of cutting; (d) multinucleate protoplasts can be isolated which have arisen as a result of spontaneous fusion during the actual treatment of the tissue with the wall degrading enzymes and (e) protoplasts from a more meristematic type of cell in which plasmolysis does not readily occur, can be obtained [23]. It is possible to combine mechanical and enzymatic methods: isolating single cells from appropriate tissue by grinding the tissue in a potter and treating the isolated cells with cellulase. But as we did not succeed in isolating single cells from Petunia in this way, we used the pure enzymatic method.

For the degradation of the cell wall substances we chose, after having tested several different pectinases and cellulases, the cellulase "Onozuka SS" (All Japan Bioch. Co., Nishinomiya, Japan) derived from the fungus *Trichoderma viride*. The preparation contains cellulase C_1 (attacking native and crystalline cellulose), cellulase C_x (attacking amorphous cellulose), cellobiase, xylanase, glucanase, pectinase, lipase, phospholipase, nucleases, and several other enzymes as well as potent mycolytic enzyme activity consisting of β-1,3-glucanase and chitinase [23]. For the hydrolysis of the pectic substances of the middle lamella we used "pectinase" (31660 from Serva Feinbiochemica, 69 Heidelberg, Germany) or "Macerozyme", a crude pectinase preparation from All Japan Bioch. Co.

The success of isolating vital protoplasts from higher plants depends on and has to be adapted to many parameters—type of organ and physiological condition of the plant, method of sterilization, nature and concentration of plasmolyticum, preparation and lot of enzyme, concentration of enzyme, use of enzymes simultaneously or successively, pH, time of incubation, temperature, technique for overcoming the cuticula of the epidermis (peeling off the lower epidermis [17], or cutting of the organs into small slices [11], or digesting away the cuticula with appropriate enzymes [24]), shaking or not shaking during incubation, preincubation in plasmolyticum or not, incubation in enzyme plus plasmolyticum or plus culture medium. Each tissue must be investigated systematically in relation to the optimum conditions for protoplast release [23], and these conditions have to be adapted throughout the year to the changing season and the varying physiological conditions of the plant.

For the isolation of protoplasts from mesophyll cells of leaves of *Petunia hybrida* var. *Cyatyp* [25], grown in a greenhouse under 16 h light, the present optimum conditions are the following: (a) physiological conditions—young, vigorously growing plants, some days before developing flowers; just expand leaves under good turgor; (b) season—April till July; (c) sterilization—depends to a high degree on the season and the weather. During March it was sufficient to wash the leaves in 6 changes of sterilized water. Standard method: 10 min in 10% v/v of a saturated solution of sodium hypochloride followed by 30 min in 6 changes of sterilized water. (d) Isolation technique: the leaves are cut into thin cross sections and are washed in 3 changes of water. Thereafter the sections are incubated for a short time under vacuum in a solution of 2% pectinase (Serva 31660) in 0.40 M mannitol pH 5.4 (1 g tissue per 20 ml enzyme solution) and are then incubated under normal pressure for 15 min at 35° C. They are then transferred into a solution of 2% cellulase "Onozuka SS" in 0.40 M mannitol, pH 5.8 and incubated for 25 min at the same temperature. At the beginning of the cellulase incubation and 10 min later the tissue is gently shaken to liberate the isolated cells from the macerated tissue. The remaining 15 min are sufficient to transform all liberated single cells into protoplasts. The complete removal of the cell wall is tested by staining with "Calcofluor White" [26] which causes bright blue fluorescence of cellulose. The protoplasts are collected by filtering the suspension through a wire gauze, centrifuging it at about 70 xg for 2 min, and washing the protoplasts 2 times in 0.40 M mannitol and once in culture medium. This method yields about 10^7 protoplasts per gramme fresh weight of tissue. If the whole procedure is not carried out under aseptic conditions, 80-90% of the protoplasts are vital, whereas under aseptic conditions only 50-60% are intact. Since the usefulness of protoplasts lies in their further manipulation and culture, it is most important to expose the protoplasts for a minimum period to the action of the crude cell wall degrading enzymes [23]. The incubation time of 40 min, necessary for liberation of appropriate amounts of protoplasts from leaves of Petunia, is the shortest time ever reported for any plant.

With similar methods we isolated protoplasts from ovules [10] and petals [27] of *Petunia hybrida,* from leaves and petals of *Torenia fournieri* and *Torenia baillonii* [11], and from petals of *Nemesia strumosa* and *Mimulus hybrida* [27], (Figs 1 and 2).

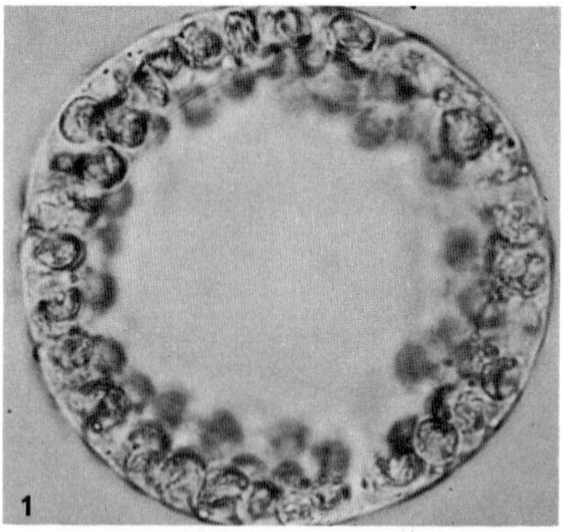

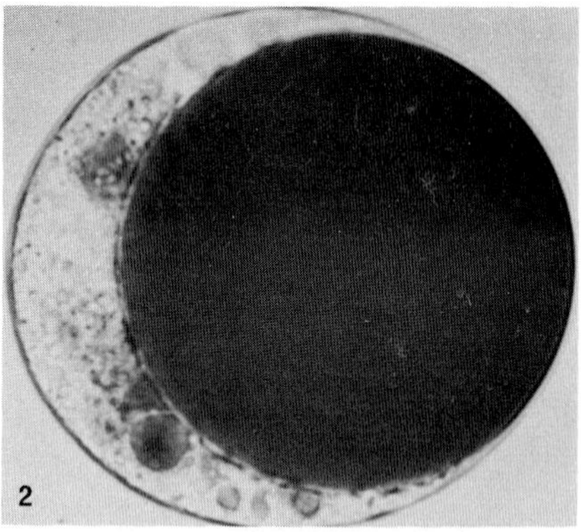

Fig. 1. Chloroplast-containing protoplast from mesophyll of leaves from *Petunia hybrida* var. *ck359A5* [29], x1225.

Fig. 2. Anthocyanin-containing protoplast from epidermis of petals from *Nemesia strumosa* var. *Firebird,* x1225.

II. Fusion

The possibility of combining somatic cells by fusing naked protoplasts is a promising area for experiment. The fusion of protoplasts from differentiated cells could become a helpful technique for studying problems of differentiation, and the fusion of protoplasts from distantly related plants is hoped to lead to great advances in pure and applied genetics.

Fusion of isolated protoplasts from higher plants was reported in 1937 [2], but the real beginning of widespread efforts on this problem arose from the work of Power *et al.* [4] who succeeded in fusing meristematic protoplasts of different cereals in 1970. Since that time a number of reports (see ref. [23]) have been published on spontaneous intra-specific fusion (fusion of protoplasts from neighbouring cells of the same tissue by enlargement of the plasmodesmata during the process of enzymatic isolation) and on induced intra- and interspecific fusion (fusion of already isolated protoplasts), of which the latter is of the most interest according to the goals mentioned above.

The method used for fusion experiments is in most cases that of Power *et al.*, i.e. fusion induced by sodium nitrate. With protoplasts isolated from meristematic tissue or from cells in suspension culture or callus culture, sodium nitrate may lead to fusion rates which are satisfactory but with protoplasts from differentiated tissues with a large central vacuole, especially with the chloroplast containing protoplasts from leaves, interspecific fusion is irregular and very low. It is these protoplasts which are the most interesting because they can be cultured and grown to plants. Carlson *et al.* [28] succeeded in growing hybrid plants from hybrid protoplasts derived by fusion, basing on a selective system of culture media on which only hybrids could grow. Thus, developing selective systems is one possibility for overcoming the disadvantage of low fusion rates. However, as such selective systems are not available for any desired plant, and as the fusion technique with sodium nitrate is restricted to protoplasts with nearly equal isoosmotic pressures, I thought it worth while to look for other fusion-inducing agents.

This agent should have the following five characteristics: first, its action should be independent of the osmotic pressure of the cell since the pressure varies between different differentiated cells of the same plant and equal differentiated cells of plants from different species. The fusing of protoplasts with different osmotic pressures by sodium nitrate is very difficult, if not impossible, because the main condition for this technique is a slight hypotonic medium. Second, the agent should lead to a high rate of agglutination followed by fusion. Third, it should agglutinate and fuse differentiated protoplasts as well as undifferentiated ones. Fourth, it should agglutinate protoplasts from any species of higher plants with any other one.

Finally, it should maintain the viability of the protoplasts.

To test the fusion efficiency of substances, protoplasts were isolated by modifying the above described standard method, which are easily to distinguish and to identify in mixtures with others by their coloured vacuoles or plastids. The following plants and tissues were used: (a) Petals from *Petunia hybrida* var. *Cyatyp* [23] (salmon red anthocyanin), *Petunia hybrida* var. *x275P* [29] (blue violet anthocyanin), *Petunia hybrida* var. *34d10* [22] (colourless), *Torenia fournieri* [11] (intensive violet anthocyanin), *Torenia baillonii* [11] (orange chromoplasts), *Nemesia strumosa* var. *Firebird* (fire red anthocyanin, blue anthocyanin), *Nemesia strumosa* var. *Orange prince* (orange crocetin), *Nemesia strumosa* var. *Blue bird* (light blue anthocyanin), *Tabaccum Sanderi* (carmine anthocyanin and chloroplasts), *Mimulus hybrida* (yellow chromoplasts). (b) Leaves from *Petunia hybrida* var. *Cyatyp, Petunia hybrida* var. *x275P, Torenia fournieri* (these protoplasts contain chloroplasts which differ in number, size, starch and colour, but which are not easily distinguished), *Petunia hybrida* var. *cK359A5* [29] (cells contain mutated white plastids). Protoplasts from (a) and (b) were isolated from highly differentiated cells with extremely large central vacuoles. Their osmotic pressure differed from about 0.22 M mannitol to 0.46 M mannitol. (c) Very young leaves and petals from *Petunia hybrida* var. *Cyatyp.* These protoplasts contain a high amount of cytoplasm, small vacuoles or none, and up to five large plastids.

The protoplasts were isolated separately and were adapted to 0.40 M mannitol. No agglutination or fusion occurred in any combination (red × green, red × yellow, red × blue, . . .) if they were mixed in mannitol. The same was true if they were washed and mixed in sodium nitrate [4], or mixed and centrifuged in sodium nitrate [11], or mixed (or mixed and centrifuged) in culture medium, or mixed and centrifuged in the salt combination of Erikson [30], or mixed in culture medium and kept in the refrigerator [31], or mixed and kept under high temperature [32]. That means that no, or no detectable, fusion occurred with the fusion methods so far used.

However, if lysozyme, set to the same pH of 5.8 as the other media, was added to a pure suspension of protoplasts, or to a suspension of any mixture of protoplasts, in a mannitol medium, then the protoplasts agglutinated and finally fused (Figs 3-7). No agglutination took place in lysozyme plus culture medium or salt solutions. The time necessary for fusion and the rate of agglutination depends on the concentration of the lysozyme and the titre of the

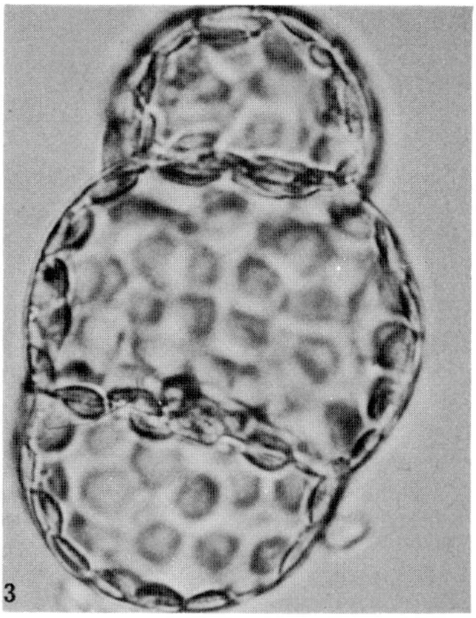

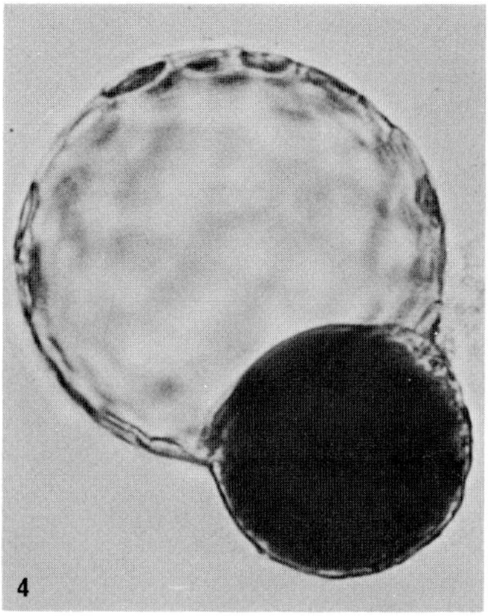

Figs 3 and 4. Fig. 3: Lysozyme induced intraspecific fusion of protoplasts from leaves of *Petunia hybrida* var. *Cyatyp*, x800. Fig. 4: Lysozyme induced interspecific fusion: *Petunia hybrida* var. *Cyatyp* (leaf, chloroplasts) x *Nemesia strumosa* var. *Fireking* (petal, anthocyanin), x800.

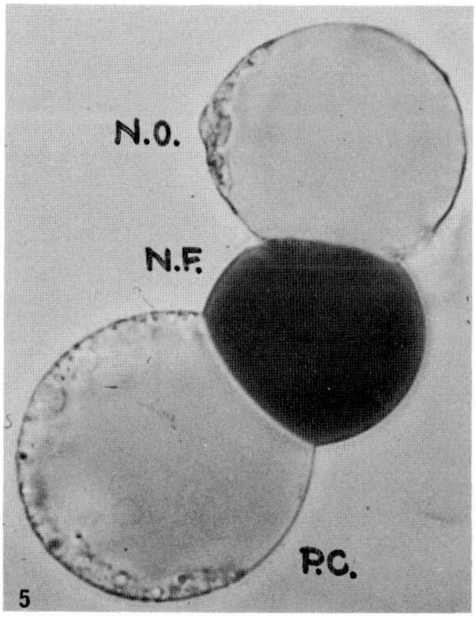

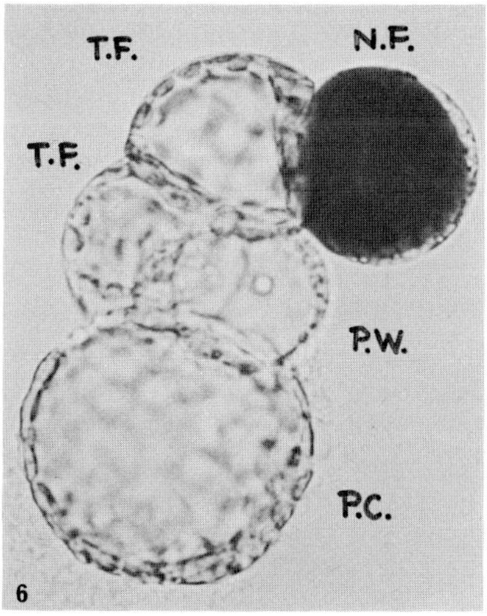

Figs 5 and 6. Fig. 5: Lysozyme induced interspecific agglutination: *Nemesia strumosa* var. *Orangeprince* (orange: N.O.) x *Nemesia strumosa* var. *Fireking* (fire red: N.F.) x *Petunia hybrida* var. *Cyatyp* (violet: P.C.), x800. Fig. 6: Lysozyme induced interspecific agglutination and fusion: *Petunia hybrida* var. *Cyatyp* (chloroplasts with starch: P.C.) x *Petunia hybrida* var. *cK359A5* (white plastids: P.W.) x *Torenia fournieri* (chloroplasts: T.F.) x *Nemesia strumosa* var. *Fireking* (red anthocyanin: N.F.), x 800.

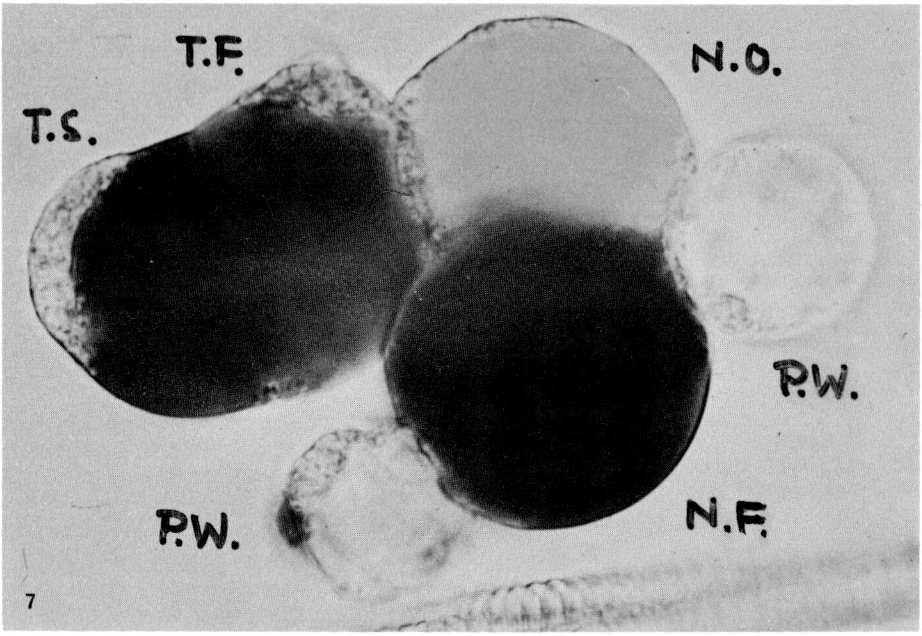

Fig. 7. Lysozyme induced interspecific agglutination and fusion: *Tabaccum Sanderi* (red anthocyanin + chloroplasts: T.S.) x *Torenia fournieri* (violet anthocyanin + plastids: T.F.) x *Nemesia strumosa* var. *Orangeprince* (orange crocetin: N.O.) x *Nemesia strumosa* var. *Fireking* (fire red anthocyanin: N.F.) x *Petunia hybrida* var. *34d10* (colourless: P.W.), x1120.

protoplasts. There is no difficulty in agglutinating all protoplasts of a suspension with high concentrations (0.5%) of lysozyme, but a high concentration of lysozyme leads to a destruction of the protoplasts. We have obtained the best results till now with 0.02% of lysozyme and a protoplast titre of about 10^5 ml^{-1}. Best results means that agglutination took place mainly between two protoplasts and that the protoplasts survived for several hours. The problem now is to stop the action of lysozyme as soon as the fusion has been induced. Experiments on solving this are in progress.* The action of lysozyme seems to be, at least partially, that of a polyanion, which attracts the negatively charged protoplasts, thus causing their tight agglutination. Therefore it should be worth while to test other polyanions for their fusion efficiency.

* I am indebted to the participants of the EMBO-workshop on "Plant cell protoplasts and their molecular and biological implications" (held October 1972 in Tübingen), especially to Prof. Takebe and Prof. Watts for discussions on the problem of the action of lysozyme.

Lysozyme seems to be (as far as I can say after some initial experiments) a nearly ideal agent for interspecific fusion experiments if it is possible to stop its toxic action—its effect is independent of the osmotic pressure of the protoplast; it leads to a very high agglutination rate; it agglutinates and fuses protoplasts at different phases of differentiation; it agglutinates and fuses protoplasts from very distantly related plants.

III. Culture

Protoplasts are very interesting objects by themselves and much information on cellular biology can be gained directly from them. In the field of applied genetics, and surely not only in this field, the culture of protoplasts to flowering plants is a basic condition for any profound work.

Since publication of the works by Steward *et al.* [33] and Vasil *et al.* [34], who had cultured isolated single cells of carrot and tobacco to complete plants, we could hope that it should be possible to get plants from isolated protoplasts too. Since the culture conditions for single cells of carrot and tobacco had been very well established since that time, it could be expected that the goal would be reached first with protoplasts isolated from these two species. And so it proved. In 1971, Takebe *et al.* [17] reported on the successful culture of diploid tobacco from isolated protoplasts, followed by Nitsch *et al.* [18] with haploid tobacco and by Grambow *et al.* [21] with carrot. Takebe and Nitsch used protoplasts isolated from leaves, Grambow cultured protoplasts isolated from cells in suspension culture. In our laboratory we have now succeeded, together with J. Durand, G. Donn, and D. Hess, to grow isolated protoplasts from leaves of *Petunia hybrida* to flowering plants. Thus Petunia became the first "normal" plant, from which protoplasts have been grown to plants, where "normal" means that the culture conditions for this species had to be developed completely from the beginning.

To develop a culture medium for a new species systematically means a lot of work, time and some fortune, because so many parameters, such as combinations and concentrations of about 8 "major elements", 8 "minor elements", vitamins, phytohormones, osmotica, pH, and culture conditions, have to be varied. Therefore we preferred first to test established culture media. The first protoplast division and callus formation took place in a Nitsch medium [36], which had been modified by the addition of "Reibsaft" [35]. Finally we succeeded in developing a completely

TABLE I
Composition of the culture medium for protoplasts from *Petunia hybrida*

Major elements	$(mg.1^{-1})$	Minor elements	$(mg.1^{-1})$
A. *Mineral salts*			
NH_4NO_3	270	H_3BO_3	2
KNO_3	1480	$MnSO_4.H_2O$	5
$MgSO_4.7H_2O$	340	$ZnSO_4.7H_2O$	1.5
$CaCl_2.2H_2O$	570	KJ	0.25
KH_2PO_4	80	$Na_2MoO_4.2H_2O$	0.10
$FeSO_4.7H_2O$	27.8	$CuSO_4.5H_2O$	0.015
Na_2EDTA	37.3	$CoCl_2.6H_2O$	0.010
B. *Organic constituents*			
Meso-inositol	100 $mg.1^{-1}$	Sucrose	0.05 M
Folic acid	0.4 $mg.1^{-1}$	Mannitol	0.3 M
Glycine	1.4 $mg.1^{-1}$		
Nicotinic acid	4.0 $mg.1^{-1}$	2,4-D	1.4 $mg.1^{-1}$
Thiamine-HCl	4.0 $mg.1^{-1}$	6-BA	0.4 $mg.1^{-1}$
Pyridoxine-HCl	0.7 $mg.1^{-1}$		
Biotine	0.04 $mg.1^{-1}$	pH 5.6	

defined medium by combining and modifying three of the established culture media [22].

For culture the isolated protoplasts (Fig. 8) were washed and suspended in liquid or in soft agar (0.6%) culture medium [17] to a final protoplast titre of 2-8. 10^4 ml^{-1}. They were cultured in plastic petri dishes (5 mm in diameter) in portions of 4 ml or in microchambers at 28° C with 16 h light of about 700 lux. Nearly all protoplasts developed a new cell wall within 3 days, increased in volume and changed to some egg shape (Fig. 9). The first cellular divisions appeared after 5 days of culture (Fig. 10), and after 2 weeks small calli have developed at a rate of about 2-4% of the initially taken protoplasts (Fig. 11). That is not a very high rate and work is in progress to find better conditions for the isolation or for the culture. However, the rate is high enough to get between 8,000 to 32,000 calli per petri dish (Fig. 12). At this phase of development the suspension is diluted with 5 ml of the same culture medium but without mannitol and with 2% of sucrose. Two weeks later the calli are big enough to be transferred on to a solidified medium (0.6% agar) which contains the mineral salts and vitamins of Nagata-Takebe [26], 2% sucrose and variable concentrations of phytohormones. Best growth of calli was obtained with 2 $mg.1^{-1}$ NAA and

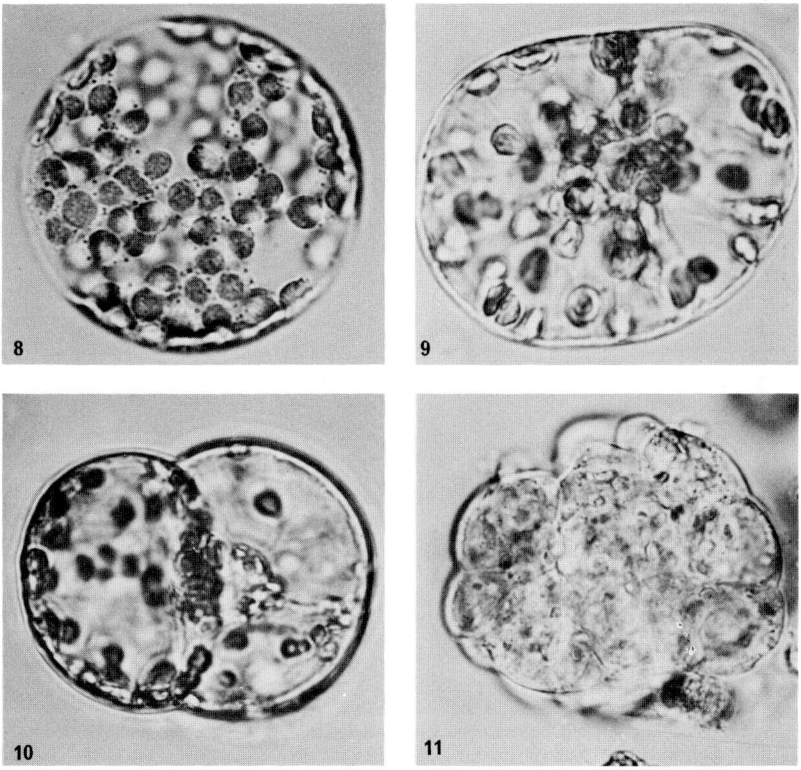

Figs. 8-11. Fig. 8: Mesophyll protoplast from leaf of *Petunia hybrida* var. *Cyatyp*, just after isolation, x800. Fig. 9. Protoplasts after 3 days of culture, x700. Fig. 10. First cellular division after 5 days of culture, x650. Fig. 11. Small callus of about 30 cells, derived from one protoplast after 2 weeks of culture, x500.

0.1 $mg.l^{-1}$ 6-BA (6-benzylaminopurine) (Fig. 13). On the same medium differentiation starts as soon as the calli have developed to about 5 mm in diameter. Though the calli of *Petunia hybrida* are not dependent in their differentiation on a very precise combination and concentration of phytohormones, the best differentiation to shoots was obtained with the following combinations (NAA/6-BA in $mg.l^{-1}$): 2/0.5; 1/0.2; 2/0.2; 0.5/0.5 (Fig. 14). It is also possible to direct the differentiation to root formation by using a medium containing 0.1 $mg.l^{-1}$ NAA and no 6-BA (Fig. 15). If young shoots are placed on this medium they very quickly differentiate (within

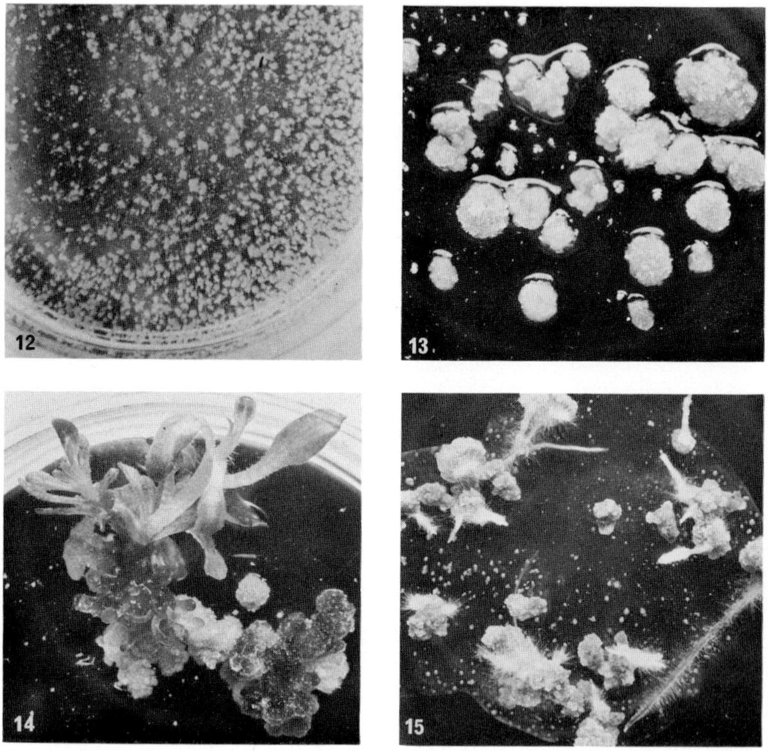

Figs 12-15. Fig. 12: Suspension of young calli in 5 cm petri dish after 4 weeks of culture, about 2 times natural size. Fig. 13: Calli on agar medium, 3 months after the isolation of the protoplast, about ×2. Fig. 14: Induction of shoot differentiation on agar medium with 1.0 mg.1^{-1} NAA and 0.2 mg.1^{-1} 6-BA, about ×2. Fig. 15: Induction of root differentiation on agar medium with 0.1 mg.1^{-1} NAA and without 6-BA, about ×2.

two weeks) a number of roots (Fig. 16). Transferred to soil the young plants grow to flowering plants, which have the same characters as the plants from which the protoplasts have been isolated (Fig. 17).

Acknowledgements

This work was supported by a grant from "Bundesminister für Bildung und Wissenschaft" to Professor D. Hess, to whom I am indebted for the plant material and encouraging discussions.

Figs 16 and 17. Fig. 16: Induction of root formation by small shoots which had differentiated on callus (see Fig. 14) on agar medium with 0.1 $mg.1^{-1}$ NAA and without 6-BA, about natural size. Fig. 17: Flowering plant which had been cultured from isolated protoplast from cells out of the mesophyll from leaves of *Petunia hybrida* var. *Cyatyp.* The plant resembles in all characters the plant from which the protoplasts had been isolated.

References

1. Klerker, J., Oefvers (1892). *Vet.-Akad. Forhdlg.*, **9**, 463.
2. Michel, W. (1937). *Arch. exp. Zellforschung*, **20**, 230.
3. Cocking, E. C. (1960). *Nature, Lond.*, **187**, 962.
4. Power, J. B., Cummins, S. E. and Cocking, E. C. (1970). *Nature, Lond.*, **225**, 1016.
5. Gregory, D. W. and Cocking, E. C. (1965). *J. Cell. Biol.*, **24**, 143.
6. Takebe, I., Otsuki, Y. and Aoki, S. (1968). *Pl. Cell Physiol.*, **9**, 115.
7. Ruesink, A. W. and Thimann, K. V. (1966). *Science, N.Y.*, **154**, 280.
8. Cocking, E. C. (1961). *Biochem. J.*, **82**, 12.
9. Pinto da Silva, P. G. (1969). *Naturwissenschaften*, **56**, 41.

10. Potrykus, I. (1971). *Naturwissenschaften*, **58**, 328.
11. Potrykus, I. (1971). *Nature, New Biol.*, **231**, 57.
12. Keller, A. W., Harvey, B., Gamborg, O. L., Miller, R. A. and Eveleigh, D. E. (1970). *Nature, Lond.*, **226**, 280.
13. Schenk, R. U. and Hildebrandt, A. C. (1969). *Crop Sci.*, **9**, 629.
14. Pojnar, E., Willison, J. H. M. and Cocking, E. C. (1967). *Protoplasma*, **64**, 460.
15. Erikson, T. and Jonasson, K. (1969). *Planta*, **89**, 85.
16. Nagata, T. and Takebe, I. (1970). *Planta*, **92**, 301.
17. Takebe, I., Labib, G. and Melchers, G. (1971). *Naturwissenschaften*, **58**, 318.
18. Nitsch, J. P. and Ohyama, K. (1971). *C. R. Acad. Sci. Paris*, **273**, 801.
19. Takebe, I. and Otsuki, Y. (1969). *Proc. natn. Acad. Sci. U.S.A.*, **64**, 843.
20. Cocking, E. C. (1970). *Int. Rev. Cytol.*, **28**, 89.
21. Grambow, H. J., Kao, K. N., Miller, R. A. and Gamborg, O. L. (1972). *Planta*, **103**, 348.
22. Durand, J., Potrykus, I. and Donn, G. (1973). *Z. Pflanzenphysiol.* **69**, 26.
23. Cocking, E. C. (1972). *A. Rev. Pl. Physiol.*, **23**, 29.
24. Schilde-Rentschler, L. (1972). *Z. Naturforsch.*, **27**, 208.
25. Hess, D. (1963). *Planta*, **59**, 567.
26. Nagata, T. and Takebe, I. (1970). *Planta*, **99**, 12.
27. Potrykus, I. (1972). *Phytomorphology*. **22**, 91.
28. Carlson, P. S., Smith, H. H. and Dearing, R. D. (1972). *Proc. natn. Acad. Sci. U.S.A.*, **69**, 2292.
29. Potrykus, I. (1970). *Z. Pflanzenzüchtung*, **63**, 24.
30. Erikson, T. (1971). *Coll. Int. C.N.R.S.*, **193**, 297.
31. Fodil, Y., Esnault, R. and Trapy, G. (1971). *C. R. Acad. Sci. Paris*, **273**, 727.
32. Power, J. B. and Cocking, E. C. (1971). *Sci. Progr.*, **59**, 181.
33. Steward, F. C., Mapes, M. O., Kent, A. E. and Holsten, R. D. (1964). *Science, N.Y.*, **143**, 20.
34. Vasil, V. and Hildebrandt, A. C. (1965). *Science, N.Y.*, **150**, 889.
35. Potrykus, I. and Durand, J. (1972). *Nature New Biol.*, **237**, 286.
36. Hess, D. and Potrykus, I. (1972). *Naturwissenschaften*, **59**, 273.
37. Potrykus, I. and Hoffmann, F. (1973). *Z. Pflanzenphysiol.* **69**, 287.
38. Potrykus, I. (1973). *Z. Pflanzenphysiol.* (In press.)

The Use of Protoplasts in Plant Biology

G. MOREL, J. P. BOURGIN and Y. CHUPEAU

Physiologie Végétale, Centre National de Recherches Agronomiques, Versailles, France

You are all aware, I am sure, of the fantastic advances brought to animal biology by the discovery of somatic fusion by Barski *et al.* [1] at the Institut Gustave Roussy, of Villejuif, in 1960. It has made it possible to apply the genetic approach, which has proved to be so fruitful in biological science, to the higher organisms.

The genetic, first limited to the study of the transmission of hereditary characters, has brought a revolution in very little related sciences, such as biochemistry and has been the basis of new ones, like molecular biology.

The genetic investigations, based on crosses and analysis of the recombinations, are very difficult to apply to many higher organisms which do not have more than one generation each year. It is limited by the complexity of sexual processes and by the incompatibility between litle related species. For obvious reasons, one cannot apply it to humans. This is why the human species is still very little known from a genetical point of view.

Somatic fusion in the hands of scientists like Ephrussi and Weiss [2], and Harris [3] has made possible the genetical study of man and mammals. It is now possible for many enzymes to assign to a given chromosome a given locus. Somatic fusion has proved to be a very useful tool in tackling the fundamental problem of cell differentiation which, up to now, seemed to escape all analysis. It has become possible to fuse two differentiated cells: cells having the same genetic information but engaged irreversibly in a different pathway with different morphology and different functions. Finally, the fusion of cancer and normal cells in the hands of Harris [3] and Klein and Wiener [4] has brought some very exciting results.

With higher plants one could expect still more spectacular results since in many instances it is possible to regenerate a whole plant from a single cell. Unfortunately many pitfalls and technical problems arose and progress has been very slow. The First International Colloquium on Plant Protoplasts and Cell Fusion, organized in Versailles in September, 1972, was designed to discuss the significant advances and the basic difficulties in handling isolated protoplasts for cell transformation and cell fusion.

In this report, I will discuss briefly some of our experiments on cell fusion and try to point out the main advances in the field of plant protoplasts.

At this point, it might be pertinent to report the question asked by one of the eminent animal biologists at the meeting: "What do you need plant protoplasts for?". The answer given by the lectures during the Colloquium was that we want to get rid of the cell wall to do what you have been doing with the animal cells, namely:

(I) Cell fusion, not mainly to study the formal genetics of plants, but mostly to expand the gene pool and give plant breeders access to genes not obtainable by conventional plant breeding processes.

(II) To modify the plant cell by bringing into it new materials, not only informative molecules like DNA and viruses, but also large units, such as organelles, chloroplasts or even bacteria.

(III) To study the wall formation.

We shall now examine these three topics.

I. Protoplasts Fusion

Plant protoplasts have proved to be much more difficult to fuse than animal cells. Two kinds of fusion have been described:

(a) Spontaneous fusion between adjacent cells occurring during the process of enzymatic digestion of the wall through unbroken plasmodesmata as shown by Withers [5] on a very elegant electron microscope preparation.

(b) Induced fusion. Sodium nitrate, as shown by Power *et al.* [6], has proved very useful for initiating protoplast fusion, possibly by modification of the electric charge on the plasmalemma. On the other hand, Kao *et al.* [7] found that an optimum concentration in Ca ions is also essential.

Heterocaryons have thus been obtained by heterospecific fusion between species far apart, like crab-grass and soybean [8]. In some cases, the percentage of protoplasts engaged in infusions might be as high as 25% [9].

Can spontaneous fusion take place between entirely isolated cells? This has been questioned. Let us now discuss some experiments that we have done to clear up this point [10].

The protoplasts were prepared from tobacco leaves c.v. Wisconsin 38 or Samsun. The age and conditions in which the plants are cultivated are very critical. The best results are obtained in winter or spring with young plants having 10 to 12 developed leaves, watered with a standard nutrient solution. The general experience shows that in summer time it becomes exceedingly difficult to obtain protoplasts from the leaves, even when the temperature is carefully controlled. It may be that the light intensity makes the wall more resistant to the enzymatic digestion; the calcium content of the leaf might also interfere. According to Coic and Lesaint [11], this content may vary from 33 meq. up to more than 150, when one goes from young to adult leaves.

The best results are obtained with fully expanded leaves, taken in the middle of the stem. Standard conditions, already described in detail elsewhere, have been used for the isolation. The leaf epiderm is stripped off and the leaf fragments are plasmolysed and incubated in a 0.7 M solution of mannitol or sorbitol containing 0.5% of macerozyme and 4% of Onozuka cellulase, plus 0.3% of potassium dextran sulphate at pH 5.5.

After 2 or 3 h of incubation at 35° C, the plant tissues are gently shaken, and the protoplast suspension centrifuged at 100 g for 3 min. The pellet containing the protoplasts is resuspended in the same medium, washed three times, after which it is formed almost exclusively of protoplasts (Fig. 1). The protoplasts are then plated on

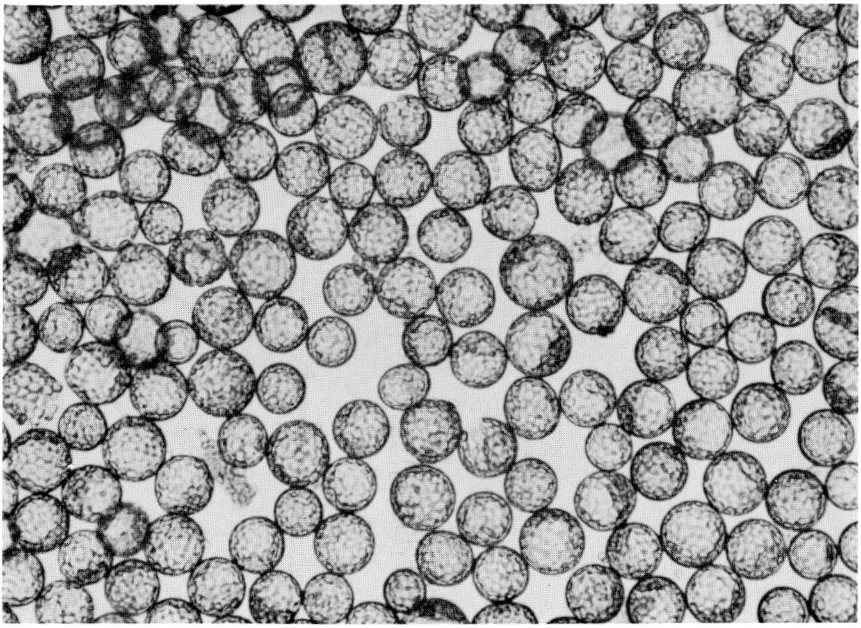

Fig. 1. Protoplast suspension after preparation and washing.

a slightly modified Murashige-Skoog [12] agar medium, containing 2% sucrose, 13% mannitol, 10^{-6} g/l of 2,4-dichlorophenoxyacetic acid and benzylaminopurine.

In these conditions, cytological examination shows that from 5 to 20% of the protoplasts have fused and are multinucelated; the number of nuclei varying from two a dozen (Figs 2, 3, 4).

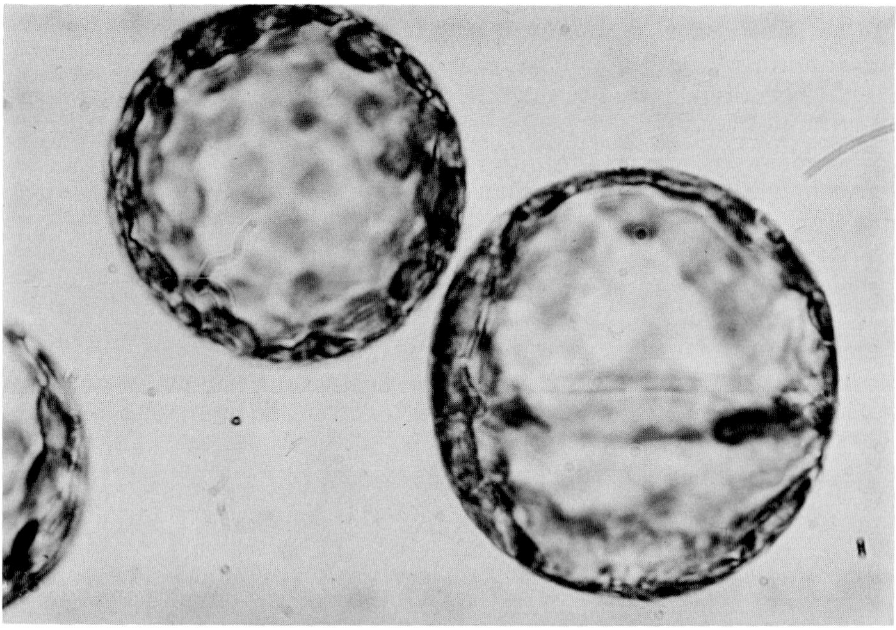

Fig. 2. Isolated "normal" protoplast and protoplast produced by spontaneous fusion of two protoplasts.

The evolution of these protoplasts can easily be followed by direct examination under microscope. After a few days, the smaller, presumably uninucleated or binucleated, divide and from 10 to 40% make calluses. The incompletely fused aggregates die rapidly. The largest units, containing at least five nuclei, have been followed for about one month. A strong increase of volume is observed and the cell wall regeneration is easily detected by plasmolysis but no cell division has been seen in these giant protoplasts.

In other experiments, the cells have been first isolated by macerozyme then centrifuged and incubated on cellulase in the same conditions as the leaf fragments. In that case, one observes also a

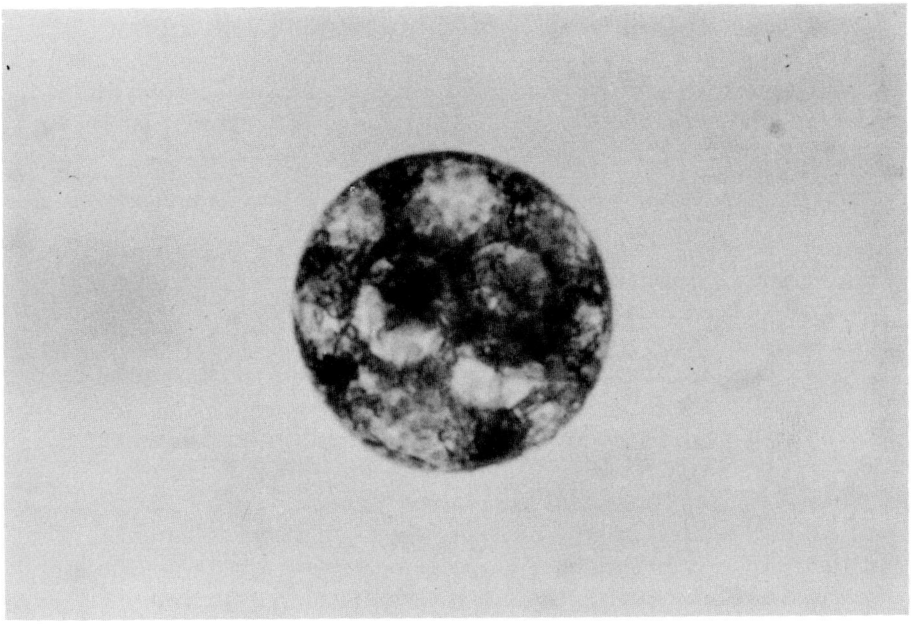

Fig. 3. "Giant" protoplast produced by spontaneous fusion of *n* protoplasts.

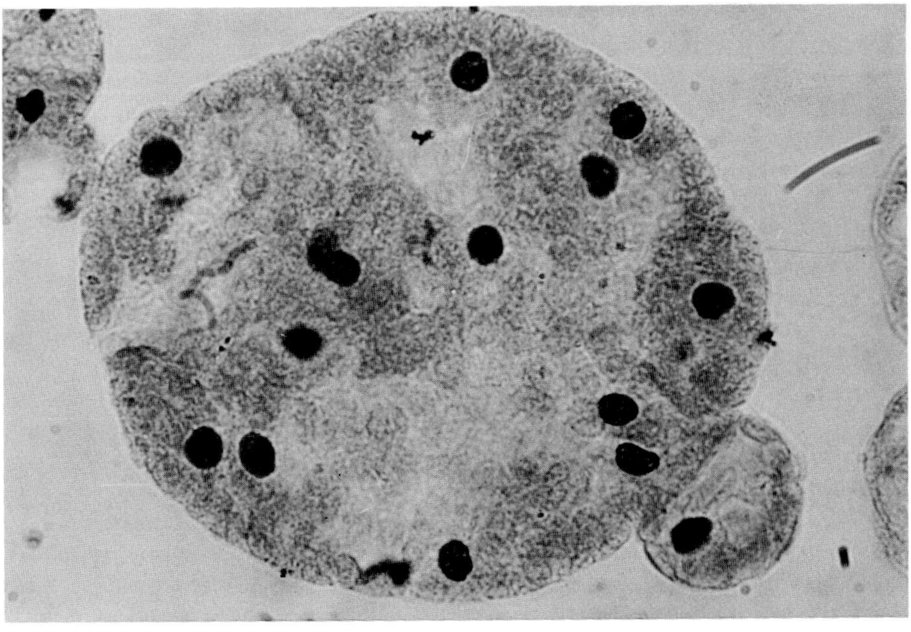

Fig. 4. "Giant" protoplast after staining.

certain amount of fusion although the plasmodesma have been broken. This seems to indicate that fusion can occur spontaneously between entirely isolated protoplasts. We plan now to study this in more detail, using the chlorophyll-less cells of variegated tobacco as marker.

II. Hybrid Cells

In order to obtain somatic hybrids, not only the cytoplasm but also the nuclei have to fuse. So far, nuclear fusion has never been observed directly. The incidence must be very low.

In an experiment to attempt to obtain somatic hybrids by mixing the protoplasts of the leaves of two tobaccos easy to distinguish by the colours of the leaves (Yellow green and Yellow Crittenden) Melchers [13] plated more than 5.10^7 protoplasts but did not get any clear-cut evidence of hybrid. He concluded that if there are fusions at all, the frequency is less than 0.068%.

One must bear in mind that even with animal cells, where the incidence of fusion is high, it is necessary to have a way of selecting the hybrid from the parents. This was first achieved because the hybrid cells had a selective advantage in the mixed culture but as this case is rather uncommon, a selective medium had to be found in which only the hybrid cell would grow and not the parents. So, it becomes possible to recover a hybrid cell with an incidence of less than one in 10^6. This is obtained by the use of mutants. In 1964, Littlefield [14] grew, in the presence of aminopterin, two kinds of mutants, each deficient in one enzyme necessary for growth in the presence of aminopterin. When the hybrid's cells were formed they were able to grow by mutual complementation, each parent supplying the enzyme lacking in the other. Littlefield found that one hybrid cell was formed per 2.10^5.

Unfortunately, with plants, besides two or three *Arabidopsis,* we do not have metabolic mutants at the present time, so we must find something else. The first parasexual hybrid was announced by Carlson [15]. Carlson took advantage of the fact that the cells of some hybrid tobacco, such as *N. gluca* × *N. langsdorfii,* are able to grow without any growth regulators while the parents need both an auxin and a cytokinin.

The protoplasts of the two parents were fused and plated on a medium deprived of growth regulators; about 25% of the protoplasts were involved in a fusion event, but only 33 hybrid cells were recovered from a mixture of more than 10^7 protoplasts of *N. glauca* and 10^7 of *N. langsdorfii*; so in this case, the rate of nuclear fusion is about 3 for 2.10^7. This first experiment shows clearly that no further progress will be made in this field, unless effective screening techniques are devised.

III. Modification of the Plant Cell

In the past few years, it has been shown that plant protoplasts can uptake foreign material by endocytosis. Mayo and Cocking [16] have studied the uptake of inert material, like latex sphere by tomato protoplasts. Takebe [17] has shown that virus like TMV particles can also be taken into tobacco leaves' protoplasts. This has opened a new way of modifying the plant cell by the introduction of cell replicating particles or informative molecules.

It has been postulated that green plants originated in the Pre-cambrian times through symbiotic unions between free autotrophic algae and various heterotrophic eukaryotes [18]. Indeed much less evoluted symbionts are known such as the protozoan, *Paramecium bursaria,* known as the green Paramecium because numerous photosynthetic algae inhabit the single cell, Muscatine *et al.* [19] or the *Glaucosystis nostochinearum* studies by Lefort [20], where a *Chlorococcus* has, instead of chloroplast, a blue-green alga in the cytoplasm, which has lost its cell wall. So it is reasonable to think that some organelles can be introduced into a foreign cytoplasm and can develop. From a practical point of view, it would be very desirable to introduce the highly efficient photosynthetic apparatus, such as the one of tropical grasses or of the soybean mutant studied by Kek *et al.* [21], into the cell of other plant crops. One could question the possibilities of such transfer. However, this has now been achieved by Carlson [22], who obtained a perfectly normal green plant by introducing a tobacco chloroplast into a protoplast from an albino mutant.

In the same line Giles [23], has been studying the maize Iojap. As the result of a mutation, the leaves have three kinds of chloroplasts: the normal green mutant, white, and a mixture of both. Fusion of green Iojap protoplast with white induces the formation of small light green plastid in the white-deficient cytoplasm as a result of complementation due to the normal chromosomes donated by the Iojap nucleus.

The introduction of a nitrogen fixating organism in non-legume would also be highly desirable, but the problem of species specificity in *Rhizobium* has so far limited this application even in different legumes. Davey and Cocking [24] have shown that whole bacterium can be taken up into protoplasts. This gives an experimental approach to unravelling the problem of species specificity and of the nitrogen fixation.

As far as direct transformation is concerned with higher plants, many attempts have been made but few positive results have been so far obtained. Even in bacteria, where the phenomenon has been discovered and extensively studied, transformation is a rare event and few instances are known.

In higher plants, the most studied case is the so-called crown-gall tumour transformation. Crown-gall is a plant cancer induced by a specific bacterium, *Agrobacterium tumefaciens.* It has been indirectly shown that during the process of tumour formation permanent information is transferred from the bacteria to the plant cell [25]. After the transformation one finds in the plant cell new enzymatic activities present in the bacteria but never found in plant tissues. Various hybridization experiments between normal plant tumour and *Agrobacterium* DNA have shown conclusively that part of the bacterial genome is integrated in the plant cell [26]. At the same time, when the bacterial strain is lysogenetic, the phage is also incorporated and can be recovered after years of aseptic culture of the tumour tissue [27, 28], so clearly a foreign DNA molecule can be incorporated in a plant genome. But despite many attempts, no conclusive results have been obtained by supplying the plant cell with purified *Agrobacterium* DNA.

Can the use of protoplasts help us to understand this phenomenon? Several attempts have already been made. The whole bacterium, like *Rhizobium*, can be taken up by a protoplast but usually dies. This is not surprising since normally the *Agrobacterium,* unlike the *Rhizobium,* stays in the intercellular spaces and never penetrates into the cell cytoplasm.

Heyn and Schilperoort [29] have studied the fate of $^{32}P^3H$ labelled *A. tumefaciens* DNA after incubation with normal tobacco cells in suspension culture. In that case, the protoplasts were then isolated and used to measure DNA penetration to avoid the error due to the adsorption of DNA on the cell walls. Only 0.145% DNA taken up was found to penetrate into the cell. This technique may also allow us to study the distinction between the DNA taken up in the cytoplasm and in the nucleus.

In a preliminary experiment, Schilde Rentschler [30] was unable to obtain transformation of tobacco protoplasts either by the entire *Agrobacterium* or by bacterial DNA.

So, although the crown-gall is the only clear case of transformation in higher plants, it has not been possible so far to duplicate the normal transformation process brought about by the *Agrobacterium* by the use of DNA either with plant cell or protoplasts.

In these conditions, what should we think of the other cases of transformation? In bacteria, according to Hotchkiss [31] the donor DNA molecules can transfer individual or closely linked genetic determinants into suitable competent recipient cells making an intermediate which in bacteria at least is a molecularly hybrid DNA.

In plants, nothing similar has been established. Hess [32, 33] claims that he has been able to transform white flowering *Petunia* into red-flowering by DNA of the red flowering strain, but the results are not reproducible, due to the difficulty of penetration of the macromolecule through the cell wall. To clarify the situation, he studied the conditions of uptake of DNA by *Petunia* protoplasts and has found that the foreign DNA was aparently taken up by the protoplasts without any extensive degradation.

On the other hand, Ledoux [34] has shown conclusively that highly polymerized DNA from various micro-organisms, such as *E. coli, Micrococcus lysodeiktycus,* or *Streptomyces coelicolor,* can be easily taken up by the roots of barley seedling or the seeds of *Arabidopsis thaliana.* An heteroduplex DNA of intermediate density is formed and duplicated during the whole life of the plant and found in the progeny.

In the case of *Arabidopsis,* the foreign DNA stored in the cotyledons is translocated at the flowering time in the flowers and the seeds.

The extraction and denaturation of this heteroduplex shows that the foreign DNA seems to be bound at the end of a host DNA chain, since by sonication two picks of different density were obtained after ultracentrifugation whereas only one pick was observed after alkaline denaturation (Fig. 5).

If one compares these extensive studies of Ledoux with the bacterial transformation, one sees clearly that we are dealing with two different phenomena and that, with plant cells molecularly hybrid, DNA has never been observed.

One can question the necessity of going through protoplast isolation and culture if free DNA is so easily taken up by the whole plant. The limited experience on this subject shows that DNA uptake by entire cells is not always so high since Heyn and Schilperoort [29] recovered only 0.125% of the *Agrobacterium* DNA taken up by the tobacco cell, almost all being adsorbed on the cell wall.

In other cases, such as Price [35], the DNAse activity on the cell wall is such that almost all the DNA is broken up before it can get

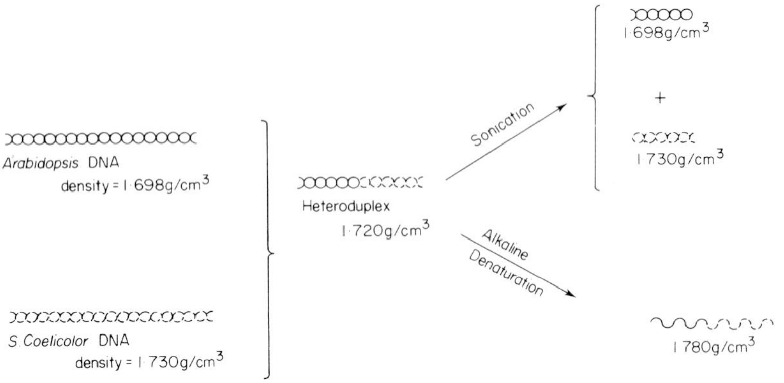

Fig. 5. Drawing after Ledoux *et al.* [34]. Model of molecule containing *Arabidopsis* and exogenous DNA.

into the cell. In these cases, the use of protoplasts could be most helpful, preliminary experiments (for instance those of Ohyama [36]) showing that as much as 1% can be taken up by protoplasts.

IV. Cell wall formation

Naked cells should constitute an ideal tool for the study of a new wall formation. It has been known since the first observations of Pojnar [37] that the isolated protoplasts form a new wall in a few hours, visible in U.V. microscope with chalcofluor.

This wall formation has been studied by several workers, especially Roland and Prat [38], and Willison [39], and I would like now to briefly summarize their work. A few hours after isolation, the activity of the Golgi apparatus seems to increase. The dyctyosomes emit amorphous vesicles of polysaccharidic nature. Large bodies arise in the vesicles in the perinuclear region. They may be exocytosed from the protoplast to be incorporated in the new wall, but most of the precursors diffuse into the external medium.

The formation of the new wall begins by the appearance of small isolated scales outside the plasmalemma which are at first interrupted and do not cover the whole protoplast surface. A multilamellar wall component is formed which precedes the usual cellulosic wall. Cellulose is layed down later between the plasmalemma and this multilamellar envelope and the electron microscope observation clearly indicates that on naked cell most of the wall material not stopped by a barrier diffuses into the external medium and is lost.

These observations show that, in the normal cell, the wall constitutes a matrix on which the fibrils are disposed. Several enzymes, namely phosphatases, are in the same way secreted in the periplasm. They are normally stopped by the wall and contribute to its formation. On protoplast, many of them also diffuse in the medium; this may explain why, in some cases, the new wall formation is such a slow process.

References

1. Barski, G., Sorieul, S. and Cornefert, F. (1960). *C.R. Acad. Sci. Paris,* **251**, 1825.
2. Ephrussi, B. and Weiss, M. C. (1969). *Sci. Am.,* **220**, 26.
3. Harris, H. (1970). Cell Fusion. Clarendon Press, Oxford.
4. Wiener, F., Klein, G. and Harris, H. (1972). *In:* Protoplasts and Fusion of Somatic Plant Cells, p. 325. Colloques Internationaux du CNRS No. 212. Paris, 1973.
5. Withers, L. (1972). *In:* Protoplasts and Fusion of Somatic Plant Cells, p. 517. Colloques Internationaux du CNRS No. 212. Paris, 1973.
6. Power, J. B., Cummins, S. E. and Cocking, E. C. (1970). *Nature, Lond.,* **225**, 1016.
7. Kao, K. N., Gamborg, O. L., Michaylak, M. and Miller, R. A. (1972). *In:* Protoplasts and Fusion of Somatic Plant Cells, p. 207. Colloques Internationaux du CNRS No. 212. Paris, 1973.
8. Giles, K. L. (1972). *Pl. Cell Physiol.,* **13**, 207.
9. Carlson, P. S., Smith, H. H. and Dearing, R. D. (1972). *Proc. natn. Acad. Sci. U.S.A.,* **69**, 2292.
10. Bourgin, J. P., Chupeau, Y., Missonier, C. and Morel, G. (1972). *In:* Protoplasts and Fusion of Somatic Plant Cells, p. 429. Colloques Internationaux du CNRS No. 212. Paris, 1973.
11. Coic, Y., Lesaint, C., Piollat, M. T. and Lelandais, M. (1969). *Ann. Physiol. vég.,* **11**, 5.
12. Murashige, T. and Skoog, F. (1962). *Physiol. Pl.,* **15**, 473.
13. Melchers, G. (1972). *In:* Protoplasts and Fusion of Somatic Plant Cells, p. 367. Colloques Internationaux du CNRS No. 212. Paris, 1973.
14. Littlefield, J. W. (1964). *Science, N.Y.,* **145**, 709.
15. Carlson, P. (1972). *In:* Protoplasts and Fusion of Somatic Plant Cells, p. 496. Colloques Internationaux du CNRS No. 212. Paris, 1973.
16. Mayo, M. A. and Cocking, E. C. (1969). *Protoplasma,* **68**, 223.
17. Takebe, I. and Otsuki, Y. (1969). *Proc. natn. Acad. Sci. U.S.A.,* **64**, 843.
18. Margulis, L. (1971). *Sci. Am.,* **225**, 2.
19. Muscatine, L., Karakashian, S. J. and Karakashian, N. W. (1967). *Comp. Biochem. Physiol.,* **20**, 1-12.
20. Lefort, M. (1965). *C.R. Acad. Sci. Paris,* **261**, 233.
21. Kek, B. W., Dilley, R. A., Freeman Allen, C. and Biggs, S. (1970). *Pl. Physiol.,* **46**, 692.
22. Carlson, P. S. (1972). *In:* Protoplasts and Fusion of Somatic Plant Cells, p. 495. Colloques Internationaux du CNRS No. 212. Paris, 1973.
23. Giles, K. L. (1972). *In:* Protoplasts and Fusion of Somatic Plant Cells, p. 485. Colloques Internationaux du CNRS No. 212. Paris, 1973.

24. Davey, M. R. and Cocking, E. C. (1972). *In:* Protoplasts and Fusion of Somatic Plant Cells, p. 327. Colloques Internationaux du CNRS No. 212. Paris, 1973.
25. Goldmann-Ménagé, A. (1970). *Ann. Sci. nat. Bot.*, **12**, 223.
26. Schilperoort, R. A., Veldstra, H., Warnaar, S. O., Mulder, G. and Cohen, J. S. (1967). *Biochim. biophys. Acta,* **145**, 523.
27. Parsons, L. C. and Beardsley, R. E. (1968). *J. Virol.*, 2, 651.
28. Tourneur, J. and Morel, G. (1970). *C.R. Acad. Sci. Paris,* **D270**, 2810.
29. Heyn, R. F. and Schilperoort, R. A. (1972). *In:* Protoplasts and Fusion of Somatic Plant Cells, p. 385. Colloques Internationaux du CNRS No. 212. Paris, 1973.
30. Schilde-Rentschler, L. (1972). *In:* Protoplasts and Fusion of Somatic Plant Cells, p. 479. Colloques Internationaux du CNRS No. 212. Paris, 1973.
31. Hotchkiss, R. D. (1971). *In:* Informative Molecules in Biological Systems, p. 21. North-Holland Publishing Co., Amsterdam.
32. Hess, D. (1969). *Pflanzenphysiol.*, **60**, 348.
33. Hess, D. (1972). *In:* Protoplasts and Fusion of Somatic Plant Cells, p. 343. Colloques Internationaux du CNRS No. 212. Paris, 1973.
34. Ledoux, L., Huart, R. and Jacobs, M. (1971). *In:* Informative Molecules in Biological Systems, p. 159 (Ledoux, L., ed.). North-Holland Publishing Co., Amsterdam.
35. Ledoux, L., personal communication.
36. Ohyama, K., Miller, R. and Gamborg, O. L. (1972). *In:* Protoplasts and Fusion of Somatic Plant Cells, p. 423. Colloques Internationaux du CNRS No. 212. Paris, 1973.
37. Pojnar, E., Willison, J. H. M. and Cocking, E. C. (1967). *Protoplasma,* **64**, 460.
38. Rolland, J. C. and Prat, R. (1972). *In:* Protoplasts and Fusion of Somatic Plant Cells, p. 243. Colloques Internationaux du CNRS No. 212. Paris, 1973.
39. Willison, J. H. M. (1972). *In:* Protoplasts and Fusion of Somatic Plant Cells, p. 215. Colloques Internationaux du CNRS No. 212. Paris, 1973.

The Role of Lipids in Mitochondrial Membrane Function

ANTHONY W. LINNANE, K. A. WARD, I. T. FORRESTER, J. M. HASLAM and D. T. PLUMMER*†

Biochemistry Department, Monash University, Clayton, Victoria, Australia

I. Introduction

The study of biological membranes has proved difficult largely due to the multiplicity of components making up a given membrane, and more particularly because there are a number of degrees of freedom for the organization of components within a membrane. The mitochondrial membranes of *Saccharomyces cerevisiae* are a good illustration of the dynamism of membranes. For example depending upon the growth conditions of the cells the pre-existing mitochondrial membranes may contain none of the membrane bound cytochromes of the terminal electron transport chain, through to very high (or intermediate) levels of these components. The mitochondria of yeast are ideal models for membrane study as they can be readily manipulated in composition and function by both physiological and genetic means. *S. cerevisiae* a facultative anaerobe, does not require oxidative metabolism to grow on fermentable substrates, and hence situations which result in defective organelles can be experimentally exploited. The inability of *S. cerevisiae* to synthesize either unsaturated fatty acids (UFA) or sterols in the absence of the molecular oxygen [1, 2] allows the incorporation into its membranes of a wide range of these compounds directly from the growth medium [3, 4]. Hence the effects of changes in the lipid composition of anaerobically grown cells can be studied. In addition, a number of faty acid desaturase mutants have been found which are unable to synthesize UFA even in the presence of oxygen [5]. These auxotrophs can be maintained by the inclusion of an unsaturated fatty acid (UFA) supplement in the growth medium and

* On leave from Biochemistry Department, Chelsea College, University of London, United Kingdom.

† D. T. Plummer thanks the Wellcome Foundation for a Travel Grant in Aid.

therefore the lipid composition of the mitochondrial membranes of these organisms can be varied under aerobic conditions.

In this brief review we wish to consider a number of instances as outlined below in which the influence is explored of a reorganization of the mitochondrial membrane on systems such as oxidative phosphorylation and the mitochondrial translation process.

II. The Influence of the Unsaturated Fatty Acid Composition of Mitochondrial Membranes on Oxidative Phosphorylation

The formation of respiratory competent mitochondria by *S. cerevisiae* requires the presence of molecular oxygen, and hence the experimental manipulation of the cellular unsaturated fatty acid composition by anaerobic growth is not possible in the study of oxidative phosphorylation. This difficulty may be overcome by using the fatty acid desaturase mutant of *S. cerevisiae,* KD 115, whose UFA composition may be varied by growing the organism in media containing defined supplements of fatty acids [6]. In this way the mitochondrial UFA content of this organism can be varied from about 75% UFA in cells grown in a medium supplemented with excess amounts of UFA (ca. 3 mg/ml) down to about 8% UFA in cells grown in a medium containing small UFA supplements (ca. 50 μg/ml).

Initial experiments indicated that when strain KD 115 was grown on low UFA supplements, the organism was unable to utilize non-fermentable substrates for growth, despite the finding that the cells contained a full complement of the mitochondrial cytochromes and could respire at rates comparable with those of normal cells [6]. It was subsequently shown that UFA-depleted mitochondria retain respiratory activity but lose the capacity for oxidative phosphorylation *in vivo* and *in vitro* (Table I) [7, 8]. With either pyruvate-malate or succinate as substrate, the P/O ratios obtained for mitochondria containing about 70% UFA are normal and approximate 1.5, whereas with mitochondria containing about 20% UFA phosphorylation was undetectable These studies were extended [9, 10] to further characterize the phosphorylation lesion by examination of three aspects of mitochondrial ion transport: (a) the energy-dependent accumulation of potassium ions in the presence of valinomycin; (b) the active ejection of protons energized by respiration and (c) the passive uptake of protons by non-respiring mitochondria.

(a) Mitochondria (75% UFA) isolated from the mutant strain KD 115 grown in the presence of an excess of unsaturated fatty acids,

TABLE I
The influence of the unsaturated fatty acid content of mitochondrial membranes on respiration and P/O ratios

	Mitochondria	
	High UFA content	Low UFA content
Total fatty adic, percentage unsaturated	68	21
Respiration	145	135
P/O ratio:		
Substrate, succinate	1.4	<0.1
Substrate, pyruvate-malate	1.6	<0.1

The mutant yeast strain KD 115 (deficient in fatty acid desaturase) was grown to the stationary phase on a glucose-yeast extract medium containing either 3 mg/ml or 0.05 mg/ml of Tween 80 as a source of UFA. The mitochondria were isolated as described elsewhere [8]. The unsaturated fatty acids consist almost exclusively of oleic and palmitoleic, a complete analysis of the lipids has been published elsewhere [7]. All values included in the Table are the means of three experiments, respiration values are expressed as ng atoms oxygen/min/mg mitochondrial protein.

when incubated in the presence of potassium ions and an energy source (provided either by succinate oxidation or ATP hydrolysis) contain a resting level of 50-80 ng ions of potassium per mg of mitochondrial protein (Fig. 1). When the permeability of these mitochondrial membrane to potassium ions is increased by adding valinomycin there is a rapid accumulation of about a further 45 ng ions of potassium per mg protein. The valinomycin-stimulated uptake of potassium ions is energy-dependent; it is abolished by dinitrophenol with either succinate or ATP as the energy source (Fig. 1a), or by the ATPase inhibitor oligomycin when ATP is the energy source (Fig. 1b). The accumulation of potassium ions by these organelles is thus a typical mitochondrial energy-linked process. However mitochondria isolated from the UFA-depleted cells of strain KD 115 are not able to accumulate potassium ions in the presence of either ATP or succinate and valinomycin (Fig. 2). The mitochondria prior to the addition of valinomycin contain a relatively low initial level of potassium (about 40 ng ions per mg protein), and when valinomycin is added to the incubation medium, the mitochondria actually lose potassium ions. The addition of dinitrophenol or oligomycin is without effect on the potassium content of the mitochondria.

(b) We have also studied the rapid proton ejection induced by the addition of oxygen to mitochondria incubated anaerobically in the

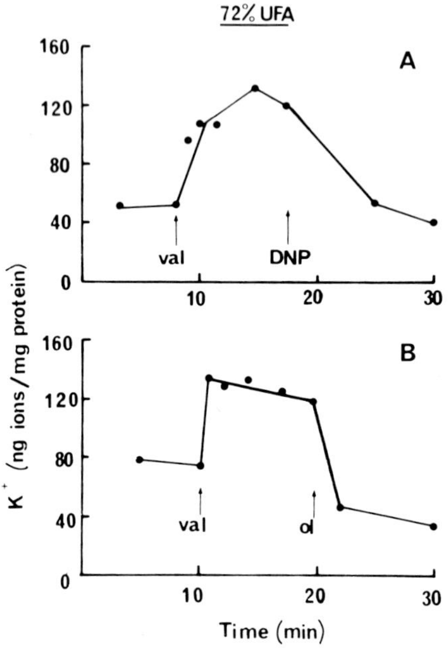

Fig. 1. The uptake of potassium ions by mitochondria isolated from strain KD 115 supplemented with excess unsaturated fatty acids. A fatty acid desaturase mutant of *S. cerevisiae*, strain KD 115, was grown on 2% glucose-1% yeast extract-salts medium containing as a UFA supplement, Tween 80 (3 mg/ml) [8]. Mitochondria (UFA = 72%) were isolated as described in [8], and incubated in a small flask in 10 ml of medium consisting of sorbitol (0.5 M), Tris-maleate (10 mM), KCL (1 mM), EDTA (0.5 mM), bovine serum albumin (2 mg/ml) and Tris-acetate (3 mM). Other additions, where indicated, were valinomycin (0.02 μg), Tris-ATP (10 mM), Tris-succinate (3 mM), 2,4-dinitrophenol (0.1 mM), and oligomycin (100 μg/ml). The pH of all reagents was adjusted to 6.5. The flask was shaken at 30° C samples were removed at intervals and the potassium content of the mitochondria measured by atomic absorption spectroscopy as described in [10]. (a) The energy source is succinate or ATP. (b) The energy source is ATP. Each point represents the average of two samples.

presence of an oxidizable substrate as described for animal mitochondrial systems by Mitchell and Moyle [11]. In these studies we have shown that mitochondria of normal fatty acid composition eject protons in an energy dependent manner [10]. However, mitochondria of low UFA content (ca. 20%) have lost the ability to actively eject protons, and behave essentially as uncoupled mitochondria.

(c) The inner membrane of normal coupled mitochondria from

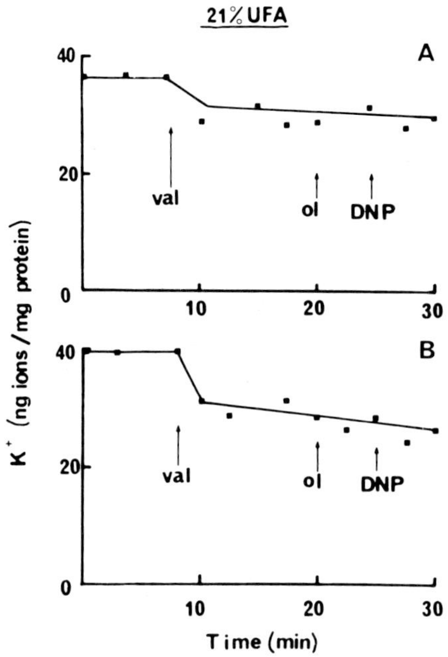

Fig. 2. The uptake of potassium ions by mitochondria isolated from strain KD 115 supplemented with limiting unsaturated fatty acids. *S. cerevisiae,* strain KD 115 was grown as described in the legend of Fig. 1, except that the UFA supplement was 50 μg/ml Tween 80. Mitochondria (UFA = 21%) were isolated and potassium content measured as described in Fig. 1. (a) The energy source is succinate. (b) The energy source is ATP.

yeast is highly impermeable to protons, as shown in Fig. 3(a), and this property is essential for coupled phosphorylation [11, 12]. When a small amount of acid is added to a lightly buffered suspension of mitochondria of normal UFA content, the pH of the solution drops abruptly and then slowly rises due to the removal of protons from the free solution as they passively enter the mitochondrial matrix. If an uncoupler of oxidative phosphorylation such as CCP (Carbonyl cyanide m-chlorophenyl-hydrazone) is added after the acid it greatly stimulates the entry of protons into the mitochondria. If the uncoupler is added before the acid the overshoot in acidity is greatly diminished as it allows a very rapid equilibration of protons across the inner mitochondrial membrane to take place. The UFA depleted mitochondria behave very similarly to uncoupled mitochondria (Fig. 3b), there is a very rapid equilibration between the

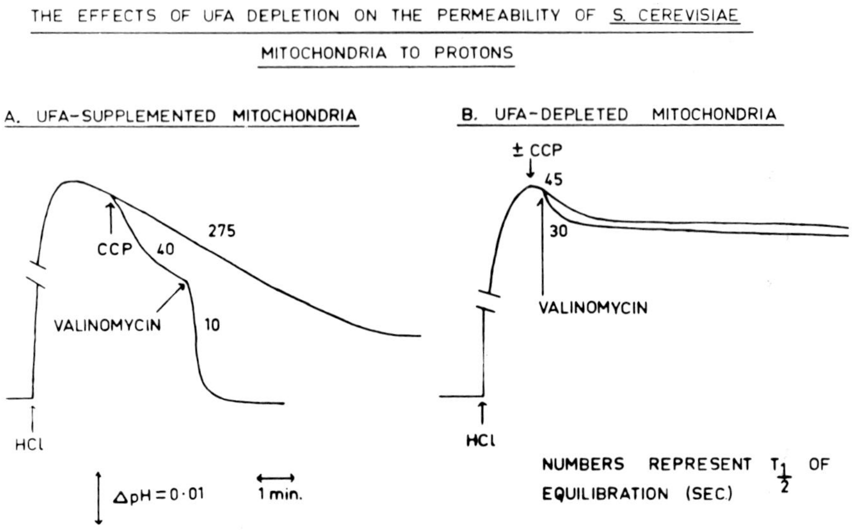

Fig. 3. The effects of unsaturated fatty acid depletion on the passive permeability of yeast mitochondria to protons. *S. cerevisiae,* strain KD 115 was grown as described in Fig. 1 on medium supplemented with 3 mg/ml or 0.05 mg/ml of Tween 80. Mitochondria were isolated from both batches of cells and were incubated (10 mg protein) aerobically at 30° C in 3.0 ml of medium containing KCl (125 mM), glycylglycine (2 mM, pH 7.0), EDTA (0.5 mM, pH 7.0) bovine serum albumin (2 mg/ml) and antimycin A (5 μg). Other additions where indicated are valinomycin (0.2 μg), carbonyl cyanide m-chlorophenylhydrazone (10 μM) and HCl (0.5 μmole). Proton absorption was measured as described in [10]. The numbers above the curves represent the half-times for the passive equilibration of the protons.

mitochondrial matrix and the external solution. The UFA depleted organelles have little permeability barrier to proton entry.

Cation transport is closely linked to energy conservation in mitochondria and may be the primary process in oxidative phosphorylation [12]. The ability to maintain a proton gradient across the yeast mitochondrial membrane is a necessary condition for ATP synthesis, and large changes in the permeability of the membrane to protons would collapse this gradient and drain off energy which would otherwise be used to synthesize ATP. The uncoupling of oxidative phosphorylation in the UFA-depleted mitochondria is therefore readily explained by membrane permeability changes. We consider the change in the membrane permeability of mitochondria of low UFA content to be the exclusive consequence of a change in lipid composition of the membranes [8]. All the proteins necessary for oxidative phosphorylation are apparently present in the UFA-depleted mitochondria. Oxidative phosphorylation can be reconstituted in non-growing whole cells by UFA-supplementation of the growth medium, even in the presence of the antibiotics cycloheximide and chloramphenicol, which together inhibit both cytoplasmic and mitochondrial protein synthesis. Thus protein synthesis, either mitochondrial or cytoplasmic, is unnecessary for the lipid-induced recoupling of oxidative phosphorylation [8].

III. Mitochondrial Ribosomes and Membrane Association

The isolation of mitochondrial ribosomes, especially from *S. cerevisiae* has proved a difficult task. It has emerged from studies on the isolation of the ribosome that there is a close physical and functional relationship between the mitochondrial ribosome and the mitochondrial membrane. The ribosome is firmly attached to the membrane and simple detergent treatment under conditions sufficient to release ribosomes from the rough endoplasmic reticulum results in only a small yield of mitochondrial ribosomes; it is necessary to use a combination of detergent (e.g. 1% Triton X-100) and high salt concentrations (e.g. 0.5 M NH_4Cl) to achieve even some dissociation of the ribosome from the membrane. This association between the ribosome and the inner mitochondrial membrane is in keeping with the apparent function of the mitochondrial protein synthesis system, which is responsible for the production of a limited number of inner membrane proteins, and it can be clearly recognized in electron micrographs of intact yeast cells.

In cells fixed with a glutaraldehyde-acrolein mixture and stained with uranyl acetate and lead citrate, the ribosomes are readily visualized as densely stained granules while the membranes of such cells are outlined in negative contrast. Figure 4a is a thin section of a whole yeast cell showing the occurrence of ribosomes within the mitochondria and cytoplasmic ribosomes distributed throughout the cytoplasm. Examination of the mitochondria at higher magnifications (Fig. 4b and c) reveals that both classes of ribosomes are closely associated with the mitochondrial membranes. The cytoplasmic ribosomes are in direct contact with, and presumably attached to, large areas of the outer mitochondrial membrane. Such an arrangement suggests that the mitochondrial proteins synthesized by the cytoplasmic system may be formed specifically by these attached ribosomes. The new proteins can then be envisaged as being inserted into the matrix space and membranes by a process analogous to that

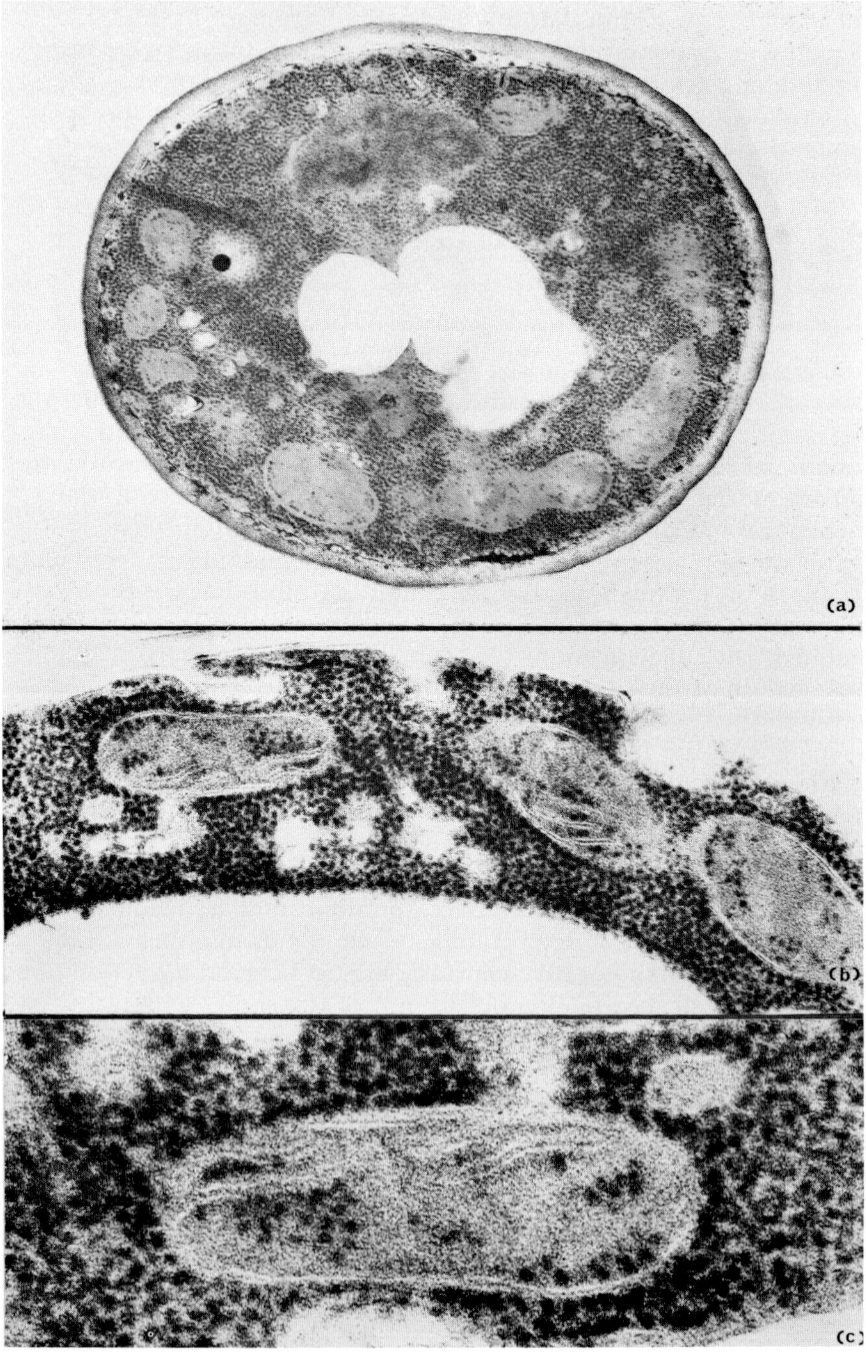
(a)
(b)
(c)

known to occur in the rough endoplasmic reticulum.

The mitochondrial ribosomes can be easily distinguished from the associated cytoplasmic ribosomes, as the former are associated only with the inner face of inner membrane or lie in the matrix (Fig. 4c).

Consideration of the difficulty of releasing the mitochondrial ribosomes from the organelles lead us to conclude that the matrix ribosomes are also probably membrane bound. An artist's drawing of Fig. 4c is shown in Fig. 5a; Fig. 5b is a possible representation of this mitochondrial organelle in three dimensions showing the inner membrane of the mitochondrion (not visualized in Fig. 4c) lying at right angles to the electron beam.

Another approach which we have been utilizing in examining the status of the association of the mitochondrial membrane with the ribosome has been to examine the effect consequences of mitochondrial membrane phase changes on protein synthesis.

It has recently been shown that the Arrhenius plots of many membrane enzyme reactions (log. enzyme activity/reciprocal of the absolute temperature) show a discrete discontinuity at specific temperatures corresponding to lipid phase changes indicating a change in Arrhenius activation energies [13]. The temperature at which this phase change occurs is a function mainly of the lipid composition of the membrane; the bound enzyme systems are affected if there is an interaction between the membrane lipids and the particular system. Adopting this approach we first studied the interdependence between the membranes of the rough endoplasmic reticulum and their ribosomes. We selected this system to serve as a control for the interpretation of results obtained with mitochondria. The studies show that ribosomes bound to the membranes of the endoplasmic reticulum are affected by the membrane phase changes while the Arrhenius activation energy of reaction of free ribosomes

Fig. 4. (a) Thin section of a whole yeast cell. The mitochondrial membranes are visualized in negative outline and the ribosomes appear as dense darkly staining granules. The yeast cells were fixed for 4 h in 5% glutaraldehyde-2% acrolein in 0.1 M cacodylate buffer pH 7.2. They were then dehydrated with acetone and embedded in an Epon Araldite mixture. Sections were cut and stained with uranyl acetate followed by lead citrate. (b) Thin section of part of a yeast cell. The association between the mitochondrial ribosomes and the inner membrane is clearly shown as is that of the outer mitochondrial membrane and cytoplasmic ribosomes. (c) A high magnification of a single mitochondrion and surrounding cytoplasmic ribosomes. The organelle shown is an enlargement of that shown in Fig. 4b. A number of the mitochondrial ribosomes appear to be localized in the matrix as well as attached to the inner membrane.

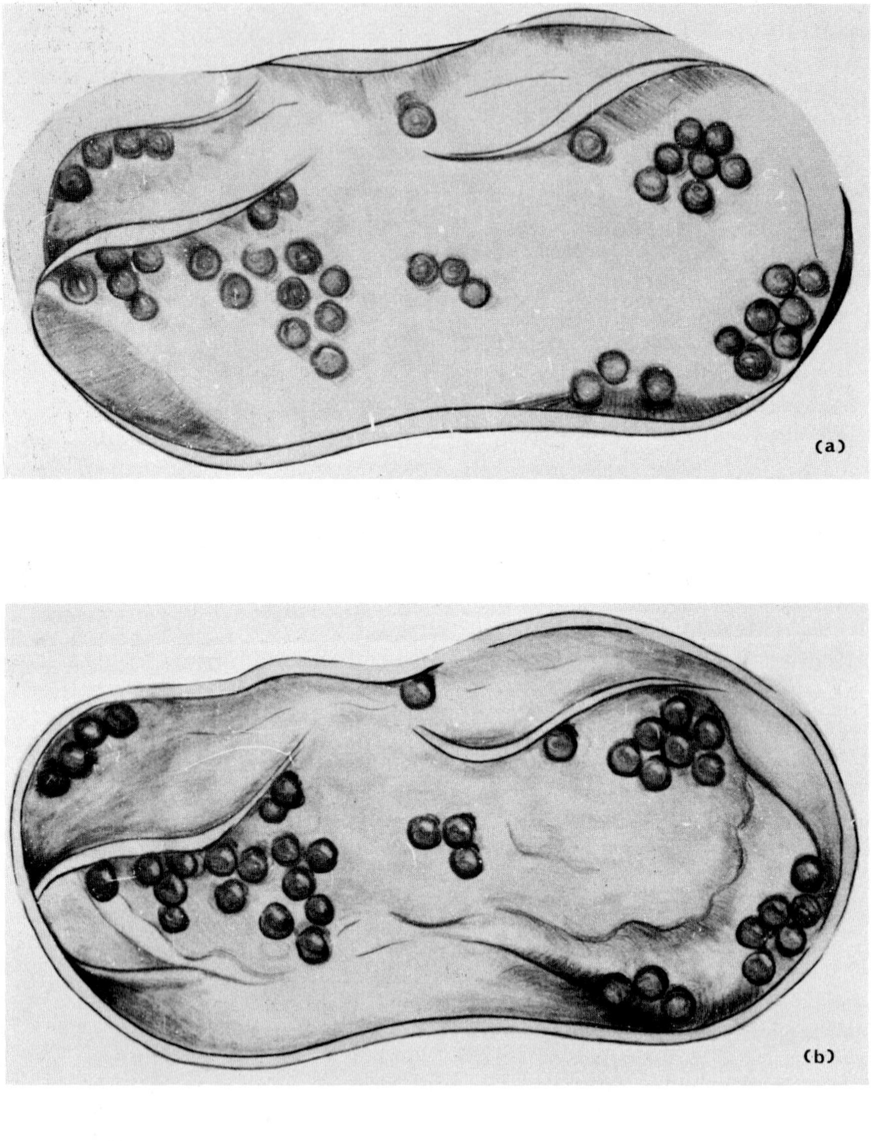

Fig. 5. (a) Drawing on the mitochondrion shown in Fig. 4c. (b) An interpretation in three dimensions of the mitochondrion shown in Fig. 4c. The ribosomes which appear free in the matrix space in Fig. 5a are shown here in the drawing as attached to the inner mitochondrial membrane which lies at right angles to the electron beam and hence is not visualized in the electron micrograph Fig. 4c.

separated from the membranes is not affected by temperature changes [14]. The work was then extended to include a consideration of the state of the mitochondrial ribosome. Such an investigation shows that mitochondrial protein synthesis by organelles isolated from tissue sources as diverse as yeast, rat liver, sweet potato, corn roots, all show a discontinuity in the Arrhenius plot [15] indicating that the mitochondrial ribosome is bound to the mitochondrial membrane. Further the studies indicate that temperature induced phase changes in the lipids of the membrane affect the functioning of the ribosomal system and that such a relationship may be common to all mitochondria.

IV. Mitochondrial Lipid Composition and Translation

The results obtained from studies of Arrhenius plots of mitochondrial protein synthesis indicate that direct alteration of the lipid components of the mitochondria may have a considerable influence on translation. In that regard the findings appear to be consistent with our earlier results in which we reported that protein synthesis could not be detected in mitochondrial fractions isolated from anaerobically grown lipid depleted cells (sterol and UFA contents about one-seventh of normal) although this activity was normal in organelles isolated from lipid supplemented anaerobically grown cells [16]. Further it appeared that the inability of the lipid depleted mitochondria to incorporate amino acids into protein was most probably due to a deficiency of mitochondrial ribosomes, since mitochondrial rRNA could not be readily detected in the organelles [17].

While aware that our fractions consisted of damaged mitochondria it was considered that this was unlikely to be a crucial point in the interpretation of the data on the occurrence of ribosomes in the fractions as disruption of normal mitochondria did not release the mitochondrial ribosomes. However it has become necessary to revise that conclusion in the light of the development of a milder method for the preparation of mitochondria from lipid depleted cells. This procedure enables us to isolate the organelles from the lipid depleted cells in a largely intact form. Essentially the new method for the preparation of mitochondria is similar to that described previously [8] except the yeast protoplasts prior to rupture are suspended in a solution of high sorbitol concentration (1.2 M) and then ruptured in this medium in a French pressure cell at low pressure; the organelles are subsequently isolated in a sorbitol medium adjusted to 0.8 M.

Figure 6a and b compare mitochondrial fractions isolated from lipid depleted cells by our previously described procedure and our most recent modification of it.

The material isolated by our earlier procedures consists largely of empty vesicles and membrane fragments (Fig. 6a). However the mitochondrial fraction prepared in 0.8 M sorbitol solution contains many intact organelles with dense matrix as well as limited amounts of swollen mitochondria and some membrane contamination (Fig. 6b).

Mitochondrial rRNA isolated from normal aerobically grown cells has been shown to have a guanine plus cytosine (G + C) base content of about 27% and the % G + C of whole mitochondrial RNA is very similar at about 28% [17]. Cytoplasmic ribosomes of yeast have a G + C content of 47%. It is therefore a simple matter to identify, and distinguish between, these two sources of RNA. The RNA fractions extracted from the mitochondria prepared from aerobically grown cells and cells grown anaerobically on lipid supplemented medium, have G + C contents of about 30% indicating that the RNA is largely mitochondrial in nature (Table II). Examination of the two fractions prepared from lipid depleted cells by the previous and more recent procedures shows that they have % G + C contents of about 47% and 28% respectively (Table II). It is therefore apparent that the more intact structures contain large amounts of mtDNA transcripts whereas the damaged fraction does not, but rather it is grossly contaminated with cytoplasmic ribosomal material. Consideration of the amounts of RNA associated with mitochondrial fractions prepared from the anaerobically grown lipid supplemented and depleted cells as well as aerobically grown cells is of some interest; it shows that the lipid supplemented structures contain low levels of mitochondrial RNA compared with the organelles prepared from aerobically grown cells and lipid depleted cells (Table II). The lipid depleted organelles apparently synthesize increased amounts of mitochondrial RNA.

Fig. 6. Mitochondrial fractions isolated from lipid depleted *S. cerevisiae* cells grown anaerobically on a galactose-yeast extract medium. (a) Mitochondrial fraction isolated by a procedure as previously described [8]. (b) Mitochondrial fraction prepared as described in the text. The fractions were fixed in 5% glutaraldehyde-0.8 M sorbitol-0.1 M phosphate pH 7.2 for 60 min, then transferred to 1% OsO_4 - phosphate pH 7.2 for 60 min, subsequently they were dehydrated and embedded in araldite.

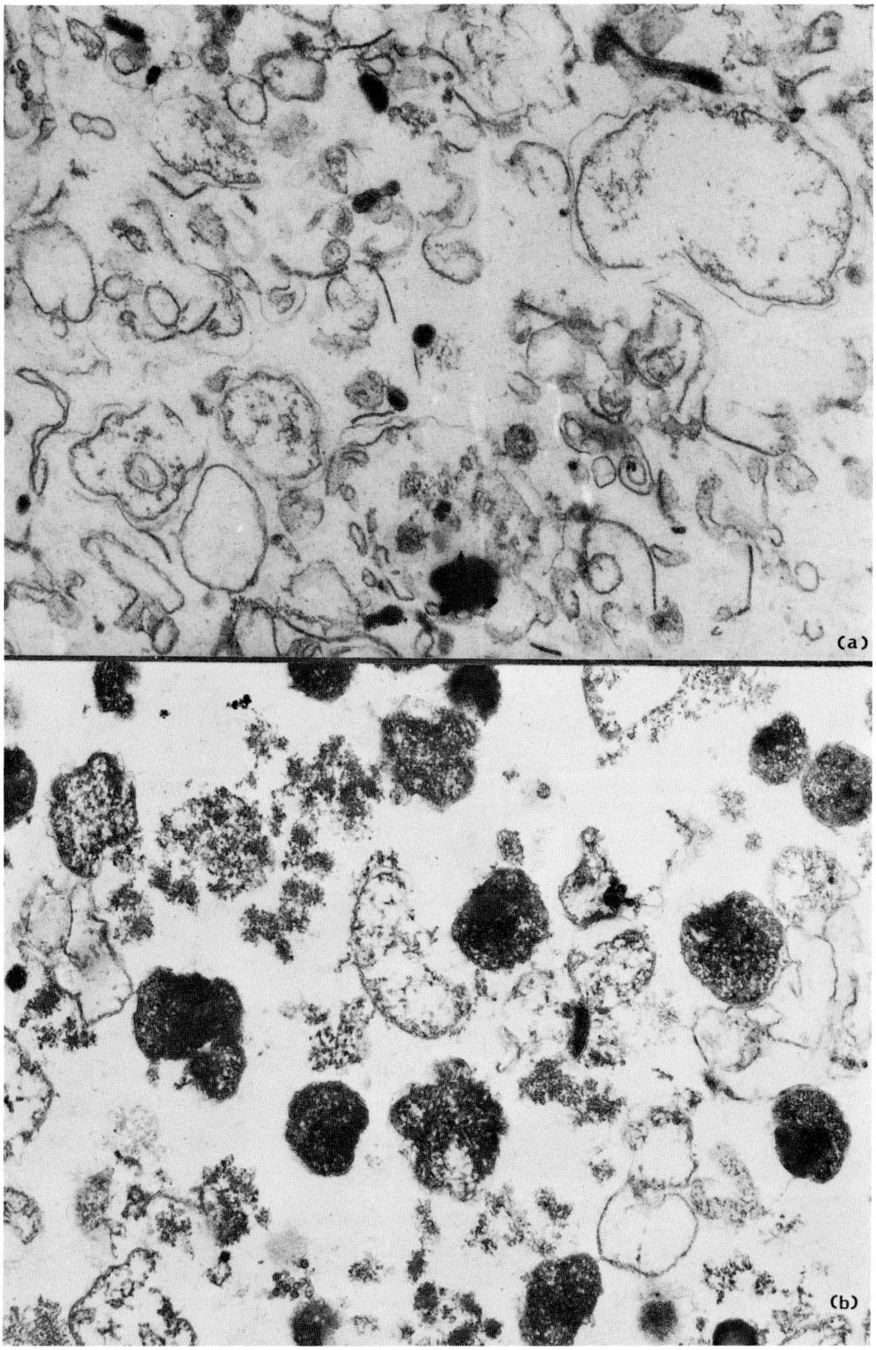
(a)
(b)

TABLE II
The effect of membrane lipid composition on mitochondrial protein synthesis and mitochondrial RNA

Growth conditions	Mitochondrial lipids UFA	Sterol	Mitochondrial protein synthesis	Mitochondrial RNA G + C (%)	Yield (μg/mg protein)
AER-GAL	72	55	80	27.1	46
AN-GAL + T + E	68	46	42	30.5	25
AN-GAL	13	18	undetectable	45.0	21
AN-GAL*	11	10	undetectable	27.8	43

Mitochondria were isolated from anaerobically grown lipids supplemented (An-Gal + T + E) and lipid-depleted (An-Gal) cells and from aerobically grown cells (Aer-Gal) as described previously [8]. * The mitochondria were isolated from these lipid depleted cells as described in the text. The major carbon source in the growth media was galactose. Mitochondrial protein synthesis is expressed as p moles ^{14}C-leucine incorporated/20 min incubation/mg mitochondrial protein. The determination of the amount of RNA in the various mitochondrial preparations, and the RNA nucleotide analyses were carried out as described previously [17].

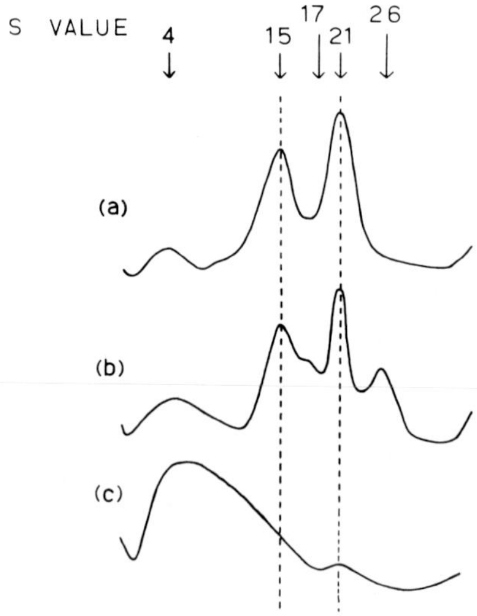

Fig. 7. Sucrose density gradient centrifugation profiles of mitochondrial RNA prepared from mitochondrial fractions of: (a) cells grown aerobically on galactose-yeast extract medium; (b) cells grown anaerobically on galactose-yeast extract-Tween 80-ergosterole medium; (c) cells grown anaerobically on galactose-yeast extract medium (lipid depleted). The total mitochondrial RNA extracted from the different preparations were analysed on 15-30% linear sucrose gradients. The direction of sedimentation in the figure is from left to right.

We would appear to have earlier been in error in concluding that *in vivo* the lipid depleted mitochondria do not contain ribosomes. However, one dilemma still remains, namely that *in vitro* assay of the intact lipid depleted structures still fails to detect any capacity of the organelles to incorporate amino acids into protein (Table II). Perhaps these *in vitro* results can be explained by the data presented in Fig. 7 which shows the nature of the RNA extracted from the different mitochondrial preparations. The mitochondria of the aerobically grown cells and lipid supplemented anaerobically grown cells is much as expected, they contain normal mitochondrial 23 *S* and 15 *S* rRNA together with some cytoplasmic rRNA (26 *S* and 17 *S*) and 4-5 *S* RNA. However the RNA extracted from the lipid depleted organelles is extensively degraded, little 23 *S* or 15 *S* rRNA is obtained, rather a broad RNA band corresponding to about 4-12 *S* in size is observed. These organelles apparently possess an active RNase which degrades the rRNA during the process of isolation of the mitochondrial fraction. Another, but perhaps more unlikely explanation would be that even *in vivo* the protein synthesizing system is non-functional, the mtDNA transcripts being degraded by an activated RNase. This RNase would normally be inactive *in vivo* in mitochondria of normal lipid composition with well formed membranes. Certainly it has been our experience that there is a latent RNase in the mitochondria of normal cells and damage to the mitochondrial membranes leads to its activation and subsequent rRNA degradation. We are currently continuing our investigations of the lipid depleted systems with a view to providing a more detailed explanation of our findings.

References

1. Andreasen, A. A. and Stiers, T. J. B. (1953). *J. Cell. Comp. Physiol.*, 41, 23.
2. Bloomfield, D. K., Bloch, K. (1960). *J. Biol. Chem.*, 235, 337.
3. Bloch, K., Baronowsky, B., Goldfine, H., Lennarz, W. J., Light, R., Norris, A. T. and Scheuerberant, G. (1961). *Fed. Proc.*, 20, 921.
4. Proudlock, J. W., Wheeldon, L. W., Jollow, D. J. and Linnane, A. W. (1968). *Biochim. biophys. Acta*, 152, 434.
5. Resnick, M. A. and Mortimer, R. K. (1966). *J. Bacteriol.*, 92, 597.
6. Proudlock, J. W., Haslam, J. M., Linnane, A. W. (1969). *Biochem. Biophys. Res. Commun.*, 37, 847.
7. Proudlock, J. W., Haslam, J. M. and Linnane, A. W. (1971). *J. Bioenergetics*, 2, 327.
8. Haslam, J. M., Proudlock, J. W. and Linnane, A. W. (1971). *J. Bioenergetics*, 2, 351.

9. Linnane, A. W., Haslam, J. M. and Forrester, I. T. (1972). *In:* Biochemistry and Biophysics of Mitochondrial Membranes, p. 523 (Azzone, G. F., Carafoli, E., Lehninger, A. L., Quagliariello, E. and Siliprandi, N., eds). Academic Press, New York.
10. Haslam, J. M., Spithill, T. W., Linnane, A. W., Chappell, S. B. (1973). *Biochem. J.* In press.
11. Mitchell, P. and Moyle, J. (1967). *In:* Biochemistry and Mitochondria, p. 53 (Slater, E. C., Kaninga, Z. and Woytczak, L., eds). Academic Press, New York and Polish Scientific Publishers, Warsaw.
12. Mitchell, P. (1966). *Biol. Rev. Cambridge*, **41**, 455.
13. Raison, J. K. (1973). *J. Bioenergetics*, **4**, 285.
14. Towers, N., Raison, J. K., Kellerman, G. and Linnane, A. W. (1973). *Biochim. biophys. Acta*, **287**, 301.
15. Towers, N., Kellerman, G. M., Raison, J. K. and Linnane, A. W. (1973). *Biochim. biophys. Acta*, **299**, 153.
16. Watson, K., Haslam, J. M., Veitch, B. and Linnane, A. W. (1971). *In:* Autonomy and Biogenesis of Mitochondria and Chloroplasts, p. 163 (Boardman, N. K., Linnane, A. W. and Smillie, R. M., eds). North Holland Press, Amsterdam.
17. Forrester, I. T., Watson, K. and Linnane, A. W. (1971). *Biochem. Biophys. Res. Commun.*, **37**, 996.

AUTHOR INDEX

Numbers in brackets are reference numbers and are included to assist in locating references in the text where authors' names are not mentioned. Numbers with an asterisk indicate the pages on which references are listed in full

C

G

H

I

J

M

N

W

Y

Z

SUBJECT INDEX